Air Pollution's Toll on Forests and Crops

A WORLD RESOURCES INSTITUTE BOOK

Air Pollution's Toll on Forests and Crops

Edited by
James J. MacKenzie and
Mohamed T. El-Ashry

Yale University Press   New Haven and London

Designed by James J. Johnson and set in Melior Roman types by The Composing Room of Michigan. Printed in the United States of America by Vail-Ballou Press, Binghamton, New York.

Library of Congress Cataloging-in-Publication Data

MacKenzie, James J. (James John), 1939–
    Air pollution's toll on forests and crops / James J. MacKenzie and Mohamed T. El-Ashry.
        p.   cm.
    Bibliography: p.
    Includes index.
    ISBN 0–300–04569–7 (cloth)
        0–300–05232–4 (pbk.)
    1. Trees—United States—Effect of air pollution on.   2. Trees—Europe—Effect of air pollution on.   3. Crops—United States—Effect of air pollution on.   4. Crops—Europe—Effect of air pollution on.   5. Trees—Wounds and injuries.   6. Crops—Wounds and injuries.   7. Forest declines.   8. Trees—Losses.   9. Crop losses.   10. Air quality management.   I. El-Ashry, Mohamed T.   II. Title. SB745.M33   1990
632'.19—dc20   89–9018

The paper in this book meets the guidelines for permanence and durability of the Committee on Production Guidelines for Book Longevity of the Council on Library Resources.

10   9   8   7   6   5   4   3   2

Contents

Preface

The scientific evidence is growing that air pollution—primarily ozone and acid deposition—is causing extensive damage to vegetation in both Europe and the United States. In the United States, extensive injury to both trees and crops has been documented. Unless measures are taken to reduce pollutant levels, the nation's air pollution problems are going to steadily worsen in the coming decades, leading to further forest damage and crop losses, not to mention ill health effects, water and soil acidification, damage to materials, and visibility degradation.

In this book, scientific evidence is reviewed that air pollution is contributing to the damage of forest trees and crops. The individually authored chapters presenting this evidence were written by some of the most respected researchers in their fields. The chapters were commissioned by the World Resources Institute as part of an extensive two-year policy review of air pollution's effects on trees and crops.*

In chapter 1 we provide an introduction to the issues. Damage either proved or strongly suspected to be related to air pollution is summarized along with the scientific evidence linking this damage to air pollution.

In chapter 2, Reinhard F. Huettl, a researcher at Albert-Ludwigs University, Freiburg, West Germany, presents a chronology and description of the recent European forest decline problem—similar to, though more advanced than, the U.S. problem. Early reported forest damage is described along with various symptoms appearing in Europe in the late 1970s. Current scientific thinking about the causes of the problem in Europe is reviewed.

In chapter 3, Paul R. Miller, a researcher with the Department of Agricul-

*James J. MacKenzie and Mohamed T. El-Ashry, *Ill Winds: Airborne Pollution's Toll on Trees and Crops* (Washington, D.C.: World Resources Institute, 1988).

ture's Forest Service, offers an extensive review of the data documenting ozone's injury to pines in California. Symptoms and trends of the declines are described along with the levels, sources, and trends of ambient air pollutants.

In chapter 4, Robert Ian Bruck of North Carolina State University summarizes forest decline symptoms in the southeastern United States. He reviews the evidence—much of it based on his own research—that air pollution is contributing to the dramatic decline of red spruce on Mount Mitchell, North Carolina. The status of the yellow pines in the Southeast is also covered.

Chapter 5, by Arthur H. Johnson of the University of Pennsylvania and Thomas G. Siccama of Yale University, reviews the forest decline problem in the Northeast. The authors review the scientific evidence on the relative roles of anthropogenic (for example, air pollution) and natural factors (such as weather extremes, drought, and insects) in contributing to the observed high-elevation decline of red spruce.

In chapter 6, Walter Heck of North Carolina State University, chairman of the National Crop Loss Assessment Network's Research Management Committee, presents an encyclopedic review of the damages to crops from air pollution, primarily ozone. The experimental techniques for determining crops losses are described along with the results for many crops. Estimates of national losses are presented.

In chapter 7, Lauraine G. Chestnut and Robert D. Rowe, of RCG/Hagler, Bailly, summarize the results of economic analyses of the benefits from reducing air pollution. They describe how estimates of damages from air pollution are made and summarize the results of major studies on the costs of pollution related to health injury and visibility degradation.

In the concluding chapter we suggest policy recommendations for reducing air pollution levels. The sources of the pollutants—ozone and acid deposition—are reviewed along with expected trends in the absence of further controls. A comparison of the geographical sources of pollution with the states showing vegetative damage is presented.

Our policy analysis has two special features. First, discussion and analysis are not confined to a single pollutant. Examined instead is the evidence that multiple pollutants—especially such secondary pollutants as acidic compounds and ozone—are contributing to the problem, both separately and together. Second, policy recommendations are made within a broad framework of both short- and long-term national energy planning, not in terms of narrower efforts to meet ambient air standards by regulating individual sources of pollution. This latter approach simply fails to take into account growth in the number of pollution sources, as well as the long-range transport and interactions among pollutants. The recommendations made here do take into account such other critical issues as national energy security, failure to attain clean air standards in our major cities, and climate protection.

Acknowledgments

This book represents the culmination of a two-year World Resources Institute research effort on the effects of air pollution on forests and crops. This research program, including support for the commissioned chapters that comprise most of this book, was supported by grants from the Joyce Foundation, the Geraldine R. Dodge Foundation, and the John D. and Catherine T. MacArthur Foundation, for which we are grateful.

During this study we have consulted with many scientists from Europe, Canada, and the United States. We are especially grateful to members of our Advisory Panel (listed in the back of the book), who provided helpful guidance as well as constructive comments on drafts of the chapters.

Our thanks to Jessica Mathews, Robert Repetto, and Bill Moomaw for their comments on the policy sections of the book, to Kathleen Courrier for editorial advice, to Linda Starke for her patience and skill in editing the book, to Hyacinth Billings and Jane Andrle for their assistance in preparing the figures, and to Beverlyann Austin, Laura Lee Dooley, Donna Pirlot, and Maggie Powell for their assistance with the manuscript. Thanks also to Michael P. Walsh and Susan Garbini for their reports on transportation and clean coal technologies, respectively, and to Chris Bernabo for many helpful discussions on forest decline.

Finally, our gratitude goes to Gus Speth for his overall advice and guidance.

Air Pollution's Toll on Forests and Crops

1

Tree and Crop Injury:
A Summary of the Evidence

JAMES J. MACKENZIE AND
MOHAMED T. EL-ASHRY

Widespread damage to trees and crops is not a late–1980s phenomenon. Such natural causes as drought, unseasonal heat or cold, high winds, diseases, and insects have long been known to injure and kill both.

Scientists have also recognized for many decades that air pollution can injure vegetation (Bormann 1985, Heck chapter 6, McLaughlin 1985, Smith 1981). Historically, smelters, power plants, and other large "point sources" of pollution have acutely damaged vegetation downwind; usually, high concentrations of sulfur dioxide or fluoride (generally as hydrogen fluoride) were at fault. Within the past 30 years, such photochemical oxidants as ozone have also been shown to cause vegetation damage.

For 10 to 25 years now, an unusually large number of dead and damaged trees have been observed in central Europe and the United States, though at first no obvious causes were identified. Visible foliar symptoms first appeared in Germany in the late 1970s and by 1980 were observed in many other European countries. In Europe, the symptoms developed more rapidly than in the United States. Now, some trees in all important species at all elevations show some visible symptoms of decline (GEMS 1987, Huettl chapter 2, Nilsson and Duinker 1987, WRI/IIED 1986).

In the United States, tree injury and crop damage are occurring across the entire nation, from California to Maine. Tree mortality has been most extensive in California and high in the Appalachian Mountains from North Carolina up through New England; costly crop injury is occurring through much of the U.S. breadbasket.

California's losses, mostly pines, began in the late 1940s; Appalachia's, mostly spruce, began showing symptoms of decline as early as 1960. Growth reductions among low-elevation red spruce in the East, among commercial yellow pines in the Southeast, and in New Jersey's Pine Barrens have also been

1

observed. Growth reductions and damages have been documented for sensitive eastern white pines too. And sugar maples in the Northeast and Canada are losing their crowns and dying prematurely.

The causes of these damages can seem baffling. Forests are complex ecosystems, acted upon by nature and man alike. Most declines probably reflect multiple stresses acting together rather than a comparatively easy-to-discover single cause. Many contributing causes may lead to decline *only* collectively.

In Europe, scientists generally agree that air pollutants number among the primary causes of forest decline (Huettl chapter 2, Prinz 1987, Schuett and Cowling 1985). No other factor can explain the near-synchronous decline of so many different species over so vast an area. As a result, considerable research has been devoted to understanding the possible role of multiple air pollutants (especially acid deposition and ozone).

In the United States, a somewhat similar research program (the National Acid Precipitation Assessment Program), under way since 1982, was launched to identify the causes and effects of acid deposition and related pollutants. In some U.S. declines, the principal causes are already well known. Losses of ponderosa and Jeffrey pines in southern California and of eastern white pines are two cases in point. In both these declines, ozone has been shown to contribute to tree damage and to increase susceptibility to mortality from such other natural factors as drought, insects, and weather extremes. In contrast, there is still scientific uncertainty over the declines of red spruce and Fraser fir at high elevations in the eastern United States and of the sugar maple in the Northeast, though evidence summarized in this report indicates that—at least in the case of the fir-spruce decline—air pollution is an important contributing factor.

How air pollution affects agricultural crops was the question asked in a major EPA-sponsored research program begun in 1980, the National Crop Loss Assessment Network (NCLAN). This recently completed research demonstrated that ozone is the principal air pollutant reducing crop productivity. At current concentrations, ozone is causing annual losses in excess of $3 billion for major crops alone.

In this chapter we summarize the evidence that multiple air pollutants are injuring trees and crops in the United States. Much of the evidence that we cite is drawn from the following six chapters of this book.

Damage to Forests and Crops

Many influences shape the overall health and growth of trees and crops. Some of these influences are natural: competition among species, changes in precipitation, temperature fluctuations, insects, and disease. Others result from air pollution, use of pesticides and herbicides, logging, land-use practices, and other human activities. With so many possible stresses at play, determining precisely which are to blame when trees die in large numbers or crop yields fall is difficult indeed (Smith 1985). But the research summarized in this chapter and in a related report strongly implicates air pollution as an important contributing factor.

Crop failures are usually easier to diagnose than are widespread tree declines. By nature, agricultural systems tend to be highly managed and ecologically simpler than forests. Also, much larger resources have been devoted to developing and understanding agricultural systems than to understanding natural forests. Relative to agricultural science, forest ecology is still in its infancy, and many years of painstaking field research may be required to fully grasp the etiology of specific forest declines.

Historically, observed injuries to forest trees and crops have prompted scientific research to uncover the causes. In some cases, such as pine damage and mortality in southern California, the principal initiating agent was eventually identified: ozone air pollution. In others—including the ongoing decline of red spruce and Fraser fir in the eastern United States—intense research is still under way. (Even with the spruce-fir decline, though, evidence indicates that air pollution is an important predisposing factor.)

Forest Damage

Numerous forest declines have hit both Europe and North America over the past 100 to 200 years. At least 5 regional declines have occurred in Europe, and at least 13 others have taken place in North America in this century alone (Cowling 1985). The symptoms in these declines have varied widely, and some appear in just one tree species (Cowling 1985). Airborne chemicals are considered important (or potentially so) in 6 of these episodes: the decline of pines in the San Bernardino Mountains of California; white pine mortality in the eastern United States; Europe's *Waldschaeden* (described below); the red spruce and Fraser fir declines in the eastern United States; the decline in growth of yellow pines in the southeastern United States; and the recent symptomatically distinct decline of sugar maples in the northeastern United States (Cowling 1985, Woodman and Cowling 1987). Natural factors—disease, insects, competition, weather extremes, and so forth—were the principal causes of the remaining declines and, indeed, may well have played a role in the others.

Historically, high concentrations of air pollution have contributed to the death of trees near large power plants and other industrial facilities sources (NAPAP IV 1987). In the early twentieth century, a copper smelter in Trail, British Columbia emitted more than 10,000 tons of sulfur dioxide per month. In a region south of the plant the pollution killed or severely damaged more than 60 percent of the trees for 33 miles; 30 percent of the trees were dead or in morbid condition for 52 miles (Miller chapter 3). In the early 1950s in Spokane, Washington, an aluminum ore reduction plant's fluoride emissions killed all the trees within a 3-square-mile area near the plant (Miller chapter 3). Significant foliar damage was observed in a 50-square mile area.

Over the past three decades, a number of forest declines observed in the western United States (especially California), in central Europe, and along the

Appalachian Mountains from North Carolina to New Hampshire and Vermont appear to differ significantly from those of the past (Cowling 1985, Johnson and McLaughlin 1986, NAPAP IV 1987). These declines and their curious symptoms stimulated scientists' current concern over forest health.

Pines in San Bernardino. The San Bernardino forest, some 75 miles east of Los Angeles, is a predominantly mixed conifer forest of ponderosa pine, Jeffrey pine, sugar pine, white fir, and incense cedar (Miller chapter 3, NAPAP IV 1987). Damage to ponderosa pine trees in the San Bernardino National Forest was first noted in the early 1950s. By 1962, an estimated 25,000 acres of the mixed-conifer forest had been injured. Subsequent surveys (1969) showed that ponderosa and Jeffrey pines on 100,000 to 160,000 acres of forest were moderately to severely damaged (USEPA 1986).

The symptoms observed in the San Bernardino forest were described as chlorosis (yellowing) of older needles, leading to premature senescence. The problem was dubbed "x-disease" because its cause eluded scientists at the time. Although 1946 and 1953 were the driest years then on record, the symptoms did not match those caused by drought, insects, or disease. Moreover, ostensibly healthy pines and other more drought-sensitive species were growing alongside diseased trees. The damaged pines showed decreased radial growth and reduced tolerance to the western pine beetle and other stresses. Later surveys revealed similar damage in the Laguna Mountains east of San Diego, and in the Sierra Nevada and San Gabriel Mountains (Miller chapter 3). In 1983, about one-fourth of the trees on established plots in the Sequoia National Forest showed a yellow mottle on new needles, compared with 14 percent in 1975. More recent surveys have shown that ponderosa and Jeffrey pines are suffering increasing damage over time. Whereas 48 percent of all surveyed trees showed some injury in 1980– 82, 87 percent did by 1985 (Miller chapter 3). Later research established that ozone is the principal cause of x-disease and that ponderosa and Jeffrey pine are particularly sensitive to this pollutant (Woodman and Cowling 1987).

White Pines in the Eastern United States. White pine trees grow widely throughout southeastern Canada and the eastern United States. In New England, white pines grow from sea level up to 500 meters in elevation; in the southern Appalachians, they can be found up to 1200 meters (NAPAP IV 1987).

Within the past few decades researchers have found that sensitive white pines are being damaged by ozone over much of the eastern United States (NAPAP IV 1987). Most of the injury occurs during June and July, the high-ozone season. Visible foliar injury begins with needle flecking and proceeds until necrotic bands develop and the needle tip dies. Sensitive white pines in high-ozone regions grow less in height, diameter, and needle length than more resistant genotypes (NAPAP IV 1987). Over the long term, some researchers believe, ozone exposures will eliminate sensitive genotypes of the species (NAPAP IV 1987).

Forest Declines in Central Europe. The multispecies forest decline that central Europe is now witnessing was first recognized on low mountains in the late 1970s (WRI/IIED 1986). Silver fir (*Abies alba*) in the Federal Republic of Germany's Black Forest began to display unusual symptoms of disease—loss of needles from the inside of the branches outward and from the bottom upward to the crown—and to die in large numbers (U.S. Congress 1984). Soon afterward, Norway spruce (*Picea abies*) began to show similar symptoms: chlorosis of the needles (clearly linked with nutritional deficiencies, especially of magnesium, calcium, zinc, potassium, and manganese) (Prinz 1987), and defoliation of the older trees. Pine was next to show symptoms (thinned crowns), followed in 1982 by the hardwoods—mainly beech and oak. The crown leaves of beech trees were yellowing and dropping off in early summer, and leaves and branches were forming abnormally. Discoloration and premature leaf fall have now been reported for essentially all major forest tree species in West Germany (Ebasco 1986).

Although damage in West Germany first appeared at elevations of 800 meters and higher, it has now spread to lower lying areas. Damage is worst on west-facing slopes, which face prevailing winds. Although trees of all ages are affected, older trees show the most symptoms.

Responding to the increasing damage observed, West Germany began forest surveys in 1982. By 1984, the surveys had become standardized and applied uniformly in all eleven German states. The 1983 survey showed that 34 percent of West Germany's trees were affected. In 1984, 1985, and 1986, the percentages climbed to 50, 52, and 54, respectively, although only about one in five of these damaged trees lost more than 25 percent of its leaves (Huettl chapter 2). (See table 1.1.) Fir trees showed the heaviest damage: 83 percent exhibited some symptoms and over 60 percent showed moderate to severe damage. Injury to oak, beech, and spruce has also been substantial.

Symptoms of *neuartige Waldschaeden*—literally, new type of forest damage—have now been detected over most of central Europe on all types of soils and at all elevations. (*Waldschaeden* has generally replaced the older term *Waldsterben*—forest death—as a description of the European problem.) By 1983, injury was detected in Switzerland, with about 14 percent of all Swiss forests showing some symptoms of decline (Huettl chapter 2); by 1984, 34 percent were damaged. A 1985 forest survey in Austria found that 22 percent of forests were damaged, 4 percent moderately to heavily. In France, according to a 1984 survey, some trees on 22 percent of the coniferous forest area and 4 percent of the deciduous forest area displayed pronounced damage. In Holland, a 1986 survey showed that 29 percent of forest areas were marked by moderate to heavy symptoms of decline.

Table 1.1. *Forest damage in 1986 in West Germany by species and damage class (in percent).*

Damage stage	Spruce	Pine	Fir	Beech	Oak	Others	Total
Undamaged	45.9	46.0	17.1	39.9	39.3	65.8	46.3
Slightly damaged	32.4	39.5	22.5	41.2	41.2	24.5	34.8
Moderately damaged	20.1	13.1	49.1	17.5	18.7	8.5	17.3
Seriously damaged or dead	1.6	1.4	11.4	1.4	0.8	1.2	1.6
All damage classes	54.1	54.0	82.9	60.1	60.7	34.2	53.7

Source: Waldschadenserhebung 1986, Bundesministerium fuer Ernaehrung, Landwirtschaft and Forsten, table 5.

Red Spruce and Fraser Fir in the Eastern United States. Red spruce can be found from North Carolina through Maine. In the Southeast, this tree grows only at higher elevations (1500–2000 meters); in the Northeast, its natural habitat ranges from sea level up to 1500 meters (NAPAP IV 1987).

All along the eastern mountain chain, the radial growth rate of red spruce has declined sharply. In the Northeast, these reductions began in the early 1960s (Hertel 1988). Continuing into the mid-1980s, they have affected spruce stands of widely differing ages and disturbance histories, and so are unlikely to represent the natural decrease in growth expected of an even-aged stand of second-growth spruce (Johnson and Siccama chapter 5).

Many distinct declines of red spruce in the Northeast have been documented over the past century. In the Northeast, the most recent decline has been under way for 25 years at elevations over 800 meters, and the number of living trees has decreased by 50 percent or more (Hertel 1988). On some mountains, live basal area (the total horizontal area of the tree trunks, measured at breast height) has dropped by 60 to 70 percent.

Less severe damage to spruce is also occurring at lower elevations. Foliar symptoms, growth reductions, and some red spruce mortality have been documented throughout the eastern spruce-fir range (NAPAP IV 1987). On some islands off Maine, spruce exhibit chlorosis, needle loss, "stork's nest" crowns, and extensive crown thinning (Johnson and Siccama chapter 5).

Major damage to conifers has also been observed in the southeastern Appalachian Mountains. Surveys begun during 1983 on Mount Mitchell in North Carolina and five other southern Appalachian peaks show that growth of red spruce and, to a lesser extent, Fraser fir fell markedly at elevations over 1920 meters beginning in the early 1960s (Bruck chapter 4). Damage and ring-width suppression are worst on west-facing slopes. By 1987, almost half the red spruce and Fraser fir on Mount Mitchell's west-facing slopes were dead (Bruck chapter 4). No significant pathogens have infested the crowns, trunks, or roots of red spruce trees in the southern Appalachians, though scientists attribute the immediate cause of death of many of the fir trees to the balsam wooly adelgid (Bruck chapter 4). High mortality has also been observed in the spruce-fir forests of Tennessee, Virginia, and North Carolina.

Yellow Pines in the Southeast. Yellow pines, including loblolly and slash pine, grow over much of the South. Some 42 million hectares (160,000 square miles) of these trees grow in Florida, Georgia, South and North Carolina, and Virginia. Data from the most recent Forest Inventory Analysis Assessment completed by the Forest Service in the Southeast revealed that the average annual radial growth rates of most naturally regenerated yellow pines under 16 inches in diameter have declined by 30 to 50 percent over the past 30 years (Sheffield et al. 1985). Also, pine mortality has increased sharply, from 9 percent in 1975 to 15 percent in 1985 (Sheffield et al. 1985). Overall, annual mortality of pine-growing stock has increased by 77 percent. Occurring without any other visible symptoms, abnormal reductions in growth in natural stands were termed "worrisome" by the Forest Service (Sheffield and Cost 1987).

Although changes in the pine forests can be detected using FIA data, this information is of limited help in determining what causes growth reductions (Sheffield and Cost 1987). Most likely, atmospheric deposition, increased stand density and age, competition with hardwoods, drought, reductions in the water table, changes in soil conditions, and disease are among the causal factors (Sheffield et al. 1985).

Sugar Maple in the Northeast. Sugar maple, a major hardwood species, grows on upland sites in the Great Lake states, New England, Canada's maritime provinces, and Ontario and Quebec. In the late 1970s, sugar maple in southeastern Canada began experiencing crown dieback and elevated mortality rates (NAPAP IV 1987). Leaves yellowed, autumn colors and leaf drop came prematurely, branchlets began dying from the top of the crown downward, bark peeled on main branches, and trees eventually died. Crown dieback symptoms have also been reported for sugar maple in Vermont, Pennsylvania, New York, and Massachusetts. (Slumps in maple syrup production in 1986 and 1987 are, however, generally attributed to unfavorable weather.) In some regions, yellow birch, American beech, white ash, white spruce, and balsam fir are also showing symptoms of decline.

Sugar maples may be hardest hit in Quebec, where an aerial survey in 1985 showed that 52 percent of all maple stands in the province showed symptoms of decline (Mitchell 1987), compared with 28 percent the year before. Earlier maple declines (in 1934, 1944, 1952, and 1956–58) were brief and ended in recovery. In some parts of Quebec, decline is occurring where no insect attacks have been recorded and no recovery is in sight (NAPAP IV 1987).

The shallow-rooted sugar maple grows best on deep, fertile, moist, well-drained soils, and it can tolerate only a narrow range of moisture conditions and very little site disturbance (Mitchell 1987). Damage is severest in highly humid environments or on mountaintops where soils are thin. Affected soils are rather acidic, with low levels of calcium and magnesium and high levels of iron and aluminum (Mitchell 1987). Maple dieback is less pronounced on deep soils without serious nutrient deficiencies.

Crop Damage

The value of U.S. crops, grown on approximately 331 million acres in 1985, totaled $76 billion (USDOC 1986). In terms of both value and acreage harvested, corn, soybeans, wheat, and hay are the four most important American crops (ITFAP 1986).

As with forests, the health and productivity of these crops are subject to a wide variety of natural and man-made factors, including insects, disease, and air pollution. Heavy use of pesticides notwithstanding, weeds, diseases, and pests still destroy an estimated 37 percent of preharvest crops (NAPAP IV 1987). Drought, excessive moisture, and frost can reduce a farm's maximum yield by as much as 69 percent in a given year (NAPAP IV 1987).

Air pollution's effects on crops were first noticed late in the nineteenth century near such major point sources as smelters that emitted large amounts of sulfur dioxide. Only in the last 30 years has more widely spread damage been associated with lower concentrations of air pollution, primarily ozone.

Natural factors—pests, disease, drought—usually induce visible symptoms that are generally well-understood. Ozone's effects, however, can be visible or invisible, and visible injury may be either acute or chronic (Heck chapter 6). Acute symptoms usually appear within one or two days after high exposures of several hours. Chronic symptoms result from longer term low-concentration exposures. Visible effects on leaves include changes in shape, discoloration, and necrosis. Typical ozone injury appears as flecks (small, bleached necrotic areas) or stipples (small pigmented areas) (Heck chapter 6). Subtler effects include growth reductions or such physiological changes as those in chlorophyll content.

How much the yields of four important U.S. crops—soybeans, corn, wheat, and peanuts—would increase if current ozone concentrations were reduced has been estimated by agricultural expert Walter W. Heck and his colleagues (Heck et al. 1982). Using validated dose-response models for each crop, researchers estimated that bringing ozone concentrations down to 0.025 parts per million (ppm) would boost wheat production by 8 percent, soybean production by 17 percent, corn production by 3 percent, and peanut production by 30 percent. These four crops account for about 64 percent of the total cash value of U.S. agricultural output, and the economic benefit from increased production was estimated at about 9 percent of the value of the crops—$3.1 billion in $1978, $5.4 billion in $1987.

The forest and crop damages summarized here affect a major portion of the country (see figure 1.1). (Only those states with annual agricultural losses of more than $100 million [$1987] are shown.) As the figure makes clear, much of

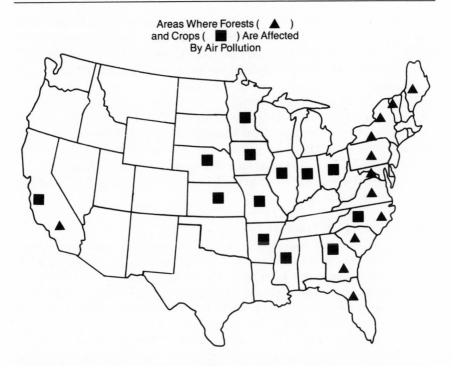

Figure 1.1. Areas where air pollution affects forest trees and agricultural crops. Source: Authors.

the nation is already experiencing vegetative injury of one form or another. Affected forests are primarily on the west and east coasts while crop injury is occurring through much of the Midwest. In short, the problem is national in scope and will require a national response if it is to be solved.

Links between Air Pollution and Damage to Forests

Air pollution can injure trees either acutely, as it historically has done around smelters and other large point sources, or more gradually at lower concentrations. Today's understanding of how air pollution can damage trees has evolved over the past 40 years through research prompted by observed damage at specific sites. Applying this research to recent declines, scientists have identified many mechanisms by which air pollution can contribute to tree injury and death.

In the high-elevation Appalachian Mountain environments where red spruce and Fraser fir are declining, air pollution concentrations are high, substantially greater than at nearby lower elevations. At these elevations, the 24-hour average ozone concentrations are typically twice those at lower neighboring elevations (NAPAP IV 1987), primarily because ozone concentrations do not fall at night as they do at lower elevations. Acid deposition on these mountains from rain, clouds, fog, and dry deposition is also high. Typical annual sulfate deposition rates at lower elevations in the eastern United States are 20–30 kilograms per hectare (NAPAP III 1987). In 1986, sulfate deposition on Whitetop Mountain, Virginia, for April through December was 200–400 kilograms per hectare, about ten times the annual rate at lower elevations (Hertel 1988). Similarly, annual hydrogen ion (H^+) deposition—a direct measure of acidity—at low elevations in the eastern United States ranges from 0.3 to 0.6 kilograms per hectare (NAPAP III 1987). On Mount Mitchell, the annual H^+ deposition rate is between 2.0 and 4.6 kilograms per hectare, roughly ten times that at lower elevations (Saxena et al. 1989). At high elevations, where the damage is worst, trees may be covered in highly acidic, high-ozone fogs and clouds for up to 3000 hours per year.

Effects of Ozone on Trees

In the early 1950s, when needle damage to ponderosa pine was first observed in southern California's San Bernardino Mountains, healthy and yellowed diseased trees grew side by side (Miller chapter 3). Drought was ruled out because the damage progressed from the bottom to the top and from the inside out (the opposite of drought's normal effects) and trees more drought sensitive than ponderosa pine showed no symptoms.

Ozone from the nearby Los Angeles basin—known to damage plants—became the primary suspect in the pine decline. To test their suspicion, scientists enclosed various branches of affected trees in chambers and treated them, respectively, with ambient air, carbon-filtered air, and filtered air with measured amounts of ozone added. Branches in the filtered air improved, while the other two showed new or increased injury. These and later fumigation experiments confirmed that ozone caused the damage.

Fogs and mists like those found in high-elevation sites in the eastern United States make it easy for ozone to enter the stomata of leaves and needles. By damaging wall membranes of mesophyll cells containing chlorophyll, ozone reduces photosynthesis (NAPAP IV 1987). In studies of eastern white pine trees, researchers found that ozone injured needles and decreased tree growth (USEPA III 1986). For sensitive white pines, chronic ozone exposures reduced annual growth up to 70 percent.

For several hardwood and conifer species, the greater the ozone concentration and dose, the greater the reduction in photosynthesis (Reich and Amundson 1985). Indeed, seedling experiments suggest that ambient concentrations of ozone are stunting the growth of most, maybe all, conifer and hardwood forests in the eastern United States (Reich and Amundson 1985). In seedlings, photosynthetic activity can decline before any observable symptoms appear, so a lack of visible injury does not necessarily indicate a lack of ozone effects (NAPAP IV 1987). Preliminary results from the Boyce Thompson Institute in New York also suggest that ambient ozone concentrations may interfere with the hardening of red spruce, making them more susceptible to winter kill (Weinstein 1988).

Direct Effects of Acid Deposition on Trees

In most experiments on the direct impacts of acid deposition, young conifer and deciduous seedlings have been exposed for periods ranging from weeks to as long as 30 months to simulated acid precipitation of various types and amounts. In experiments with red spruce, scientists from the Boyce Thompson Institute exposed seedlings separately to mists of nitric acid, sulfuric acid, and a mixture of the two over a pH range of 2.5 to 4.5 (Jacobson and Lassoie 1989). Significant foliar damage (20 percent) was observed only from the sulfuric acid mist and only when the pH was at or below 2.6. As the acid droplets dried on needle surfaces, the pH decreased substantially (in one case to a value of 1.6), leading researchers to conclude that acid deposition may be important where evaporation follows wet deposition. In other studies, acid precipitation with pH values ranging from 2.0 to 4.7 was applied to jack pine, and root weight decreased as acidity increased (NAPAP IV 1987). In a study of acid mist's effects on the leaf cell structure of tulip poplar seedlings, a significant collapse of cells was observed at a pH 2.6. Still other studies showed that at pH levels of 2.6, conifers and hardwoods lost above-ground biomass (NAPAP IV 1987).

On Mount Mitchell, experiments with red spruce have demonstrated that acid deposition at ambient levels can damage the stomatal wax plugs of the needles (Bruck chapter 4). Since these plugs are believed to minimize water loss and facilitate gas exchange, such injury can retard growth. Other experiments on Mount Mitchell have shown that acidic rainfall leaches nutrients from needles (Bruck chapter 4). In these studies, the chemical composition of rainfall was compared with that of the rain dripping from the needles (throughfall) and of water flowing down the trunks (stemflow). Concentrations of magnesium, calcium, potassium, and sodium in the throughfall and stemflow proved higher than in the rainwater—a finding attributed to leaching from the needles. Elevated levels of sulfate and nitrate were also found in the throughfall and stemflow. Researchers found that essentially all the hydrogen ions (H^+) reaching the forest floor came directly from the rainfall. Thus, the acids reaching the needles are being neutralized: the hydrogen from the rain is exchanging with positive cations (magnesium, calcium, potassium, and sodium) in the needles.

Ozone and Acid Deposition Acting Together

In other experiments, conifer needles were exposed to both ozone and simulated acid precipitation. When 5-year-old Norway spruce trees in Germany were fogged twice weekly at a level of pH 3.5 while under continuous exposure to ozone levels of 100 or 300 parts per billion (ppb), the rate of magnesium loss from the needles was 20 percent greater with the 100 ppb ozone exposures than with the controls, and greater still at the 300 ppb level (Krause et al. 1983). Similar or greater leaching rates for potassium, calcium, nitrate, and sulfate were observed, leading researchers to conclude that high levels of ozone intensify nutrient loss from needles where acid fog is present.

Together, these studies indicate that at precipitation pH levels of 3.0 and above, ozone and acid deposition have few significant direct short-term effects on tree seedlings. Below pH 3.0, however, foliar injury appears and growth decreases, particularly in seedling roots. Notably, during 1986 the pH of cloud-water affecting the above-cloud forests in the eastern United States reached values of 2.2 (Mount Mitchell), 2.6 (Whitetop Mountain), and 2.8 (Whiteface Mountain) (Mohnen 1987).

Acid Deposition's Effects on Soils

Even more important than direct foliar damage are the changes that acid deposition can bring about in soils. In some soils, acid precipitation can deplete nutrients by leaching calcium, magnesium, and potassium. The replacement of these vital cations by hydrogen ions (H^+) and the mobilization of aluminum can accelerate soil acidification. Where the input of hydrogen ions is too great to be neutralized by the natural weathering of minerals in the soils that replenishes important nutrients, acidification would be expected. Also, unless enough elements are released in weathering to replace those lost by leaching, nutrient imbalances in trees will eventually occur.

In soils treated with strong acids, nutrient leaching and aluminum mobilization accelerate while litter decomposition slows (which slows nutrient recycling) (NAPAP IV 1987). "Exchangeable aluminum" (aluminum ions, Al^{3+}, that are not bound within rocks) can damage the fine roots of trees in soils where the ratios of aluminum to calcium or aluminum to magnesium are high (Huettl chapter 2). Elevated aluminum concentrations can block the root uptake of calcium and magnesium, leading to nutrient deficiencies (Huettl chapter 2). Excess aluminum can also impair water transport within the tree, increasing sensitivity to drought.

If acid deposition is leading to nutrient leaching from soils and otherwise upsetting the nutrient balance of trees, observed damage represents a serious, long-term threat since reversing the symptoms will be difficult at best. Researchers at Oak Ridge National Laboratory in 1986 evaluated acidic deposition's potential to alter forest soils and thus harm trees (Turner et al. 1986). Field experiments indicated that acid solutions will leach cations from soils, especially those near the surface, and magnesium and calcium have typically been found to be more heavily leached than potassium (Turner et al. 1986). Still, the results of many acid-irrigation experiments are indeterminate—perhaps, the authors surmise, because they last a comparatively short time. Acid deposition may temporarily increase growth as nitrogen fertilizes trees, but long-term deposition could result in nutrient leaching and eventual decline (Turner et al. 1986). So far, none of the experiments has been carried out long enough to test this hypothesis (Turner et al. 1986).

According to the Oak Ridge researchers, the soils in approximately 41 percent of eastern U.S. forests are susceptible to acidification. Actual nutrient leaching rates, stress the researchers, would depend on acid deposition rates, inputs of calcium, magnesium, potassium, and other base cations in deposition, the weathering rates of soils, the mobility of sulfate and nitrate ions, and various biological factors. As for soils within this highly susceptible category where substantial weathering has already occurred, the Oak Ridge team estimated that 18 percent of eastern U.S. soils fit. Obviously, nutrient loss in the eastern United States is potentially significant even though the true magnitude of the problem remains to be determined.

Actual nutrient deficiency in soils in the southeastern United States has been observed in transplant experiments using soils from Mount Mitchell and Mount Gibbs (Bruck chapter 4). In a study begun in 1986 by North Carolina State University researchers, soils from 2000-meter Mount Gibbs that were exposed to relatively higher deposition levels were transposed with soils from a 1740-meter site on Mount Mitchell. Red spruce seedlings were then planted at both elevations in both the transposed soils and control soils from the same sites. Needle analysis disclosed that the seedlings growing on the high-elevation soils had much lower nutrient levels than those growing on the low-elevation soils. The average uptake of potassium, magnesium, and calcium by the deficient trees was only 35, 47, and 49 percent, respectively, of that of the healthy seedlings. Soil analysis revealed that the levels of these nutrients were much lower in the high-elevation soils than in those from lower elevations.

Evidence also suggests that soil acidification is occurring in the northeastern United States and that trees there may be experiencing nutrient deficiency. In the Adirondacks, soil changes in the Huntington Forest, where spruce dieback is occurring, have been studied. In a thoroughgoing data review, George Tomlinson, a tree chemist with Domtar, Incorporated, Research Centre in Quebec, found very large net calcium losses from the soil as a result of acid leaching (Tomlinson 1983). The ratio of aluminum to calcium in the near-surface soils was so high that trees' fine roots were being damaged. In related research, Walter Shortle and Kevin Smith of the U.S. Forest Service examined aluminum's role in blocking calcium uptake by red spruce in the Northeast (Shortle and Smith 1988). These researchers concluded that aluminum in the soils limits the calcium supply of spruce trees, thereby contributing to their decline, and that continued acid deposition to these soils will make matters worse.

Excess Nitrogen Deposition

Nitrogen, absolutely vital to tree growth, is frequently the limiting nutrient in U.S. forests. Yet some evidence indicates that nitrogen deposition is excessive on the mountains where red spruce are declining. By one estimate, total wet deposition of nitrogen at a high-elevation New Hampshire site is seven times as great as at low elevations (Friedland, Hawley, and Gregory 1985).

Excessive nitrogen deposition can harm forests (Bruck chapter 4). Vigorous growth stimulated by nitrogen fertilization may lead to nutrient deficiency if other nutrients are not sufficient. Nitrogen compounds can adversely alter physiological and anatomical development (for example, winter hardening). Excessive quantities are believed to increase trees' susceptibility to freezing or desiccation in winter—a suspected contributor to red spruce decline (Friedland, Hawley, and Gregory 1985). Cold temperatures, whether from an early frost before winter hardening is completed, an exceptionally cold spell in winter, or a cold period following an early thaw, may harm red spruce (Friedland, Hawley, and Gregory 1985).

According to the evidence summarized here, air pollution—primarily ozone and acid deposition—can lead over time to direct foliar damage, to the leaching of nutrients from both trees and soils, and to physiological changes that make trees more vulnerable to normal stresses. Which nutrients will be deficient depends on soil conditions (Prinz 1987). As the trees weaken, climatic extremes or other natural stresses can cause further weakening and damage, nutrient deficiency, and tree death.

Role of Macronutrients in Tree Decline

Several macronutrients are vital to a tree's health and growth, and well-known symptoms arise predictably in their absence. Three key nutrients whose deficiency in soils has been linked with air pollution are magnesium, potassium, and calcium (Tomlinson 1986).

> *Magnesium (Mg).* Magnesium is a constituent of chlorophyll, which converts carbon dioxide into organic matter. Magnesium is mobile in trees. In conifers, magnesium from older needles moves to the newer outer needles if there is a deficiency in the soils. The older needles then turn yellow (a condition called chlorosis) and eventually die. In short, magnesium deficiency results in needle loss from the trunk outward and from the base upward.
>
> *Potassium (K).* Potassium is essential to tree growth. Without it, roots could not push their way through the soil; nor could the tree's bark expand as the tree grows radially outward. Potassium, like magnesium, is highly mobile, and without sufficient supply foliage begins to yellow, much as it does when magnesium is deficient.
>
> *Calcium (Ca).* Calcium is essential to the formation of cell walls and to the tree's radial and vertical growth. Calcium pectate forms the active cell walls of the cortex of the fine roots through which inorganic nutrients and water enter the tree. It is not mobile and moves to new growth only when supplies in the soil are adequate. When there is inadequate calcium, root development is poor, growth is reduced, and foliage is lost from the top down and inward from the ends of the branches—opposite to the pattern observed in magnesium and potassium deficiencies.

Links between Air Pollution and Damage to Crops

Historically, sulfur dioxide and hydrogen fluoride were the first air pollutants known to damage vegetation (Heck chapter 6, NAPAP IV 1987). Typically, they killed most vegetation within a few miles of smelters, electric power plants, or other large point sources. More widespread damage to vegetation from photochemical air pollution was first recognized in 1944 in the Los Angeles area. Photochemical oxidants (primarily ozone) now injure and damage crops across most of the United States (Heck chapter 6).

The air pollutants of greatest national concern to agriculture today are ozone (O_3), sulfur dioxide (SO_2), nitrogen dioxide (NO_2), and sulfates and nitrates. Of these, ozone is by far the most worrisome; the potential role of acid deposition at ambient levels remains to be determined. At present deposition rates, most studies indicate, acid deposition does no identifiable harm to foliage (Heck chapter 6). But, at lower-than-ambient pH levels, various impacts include leaf spotting, acceleration of epicuticular wax weathering, and changes in foliar leaching rates. When applied simultaneously with ozone, acid deposition also has been shown to reduce a plant's dry weight.

Identifying Ozone's Impacts on Crops

A three-stage procedure is used to estimate the total impact of air pollutants on crops (Heck chapter 6). First, the response of various crops to specific pollutant concentrations over time must be empirically determined and analytically described. Second, a crop-inventory data base must be devised. And, third, the crop inventories and findings on responses must be combined with a suitable air pollution data base to estimate total crop losses.

Evidence on how crops respond to pollutants can be gathered using several techniques. For example, crops can be subjected to varying controlled concentrations of pollutants to determine a dose-response relation. Alternatively, protective chemicals could be applied to some plants to assess the pollutant's impact on unprotected plants. If a pollution-resistant cultivar of a crop is available, the impact of varying pollutant concentrations can be assessed by comparing its hardiness with that of nonresistant cultivars. Scientists can also use open-top chambers to expose crops to controlled amounts of pollutants. Researchers have also estimated the lowest ozone concentrations and duration times that can damage leaves or reduce plant growth and yield (USEPA III 1986).

Research has shown that ozone enters crops through the microscopic openings on the leaves and that ozone exposures can lead to either visible or subtler effects (Heck chapter 6). The plant will not suffer injury if ozone concentrations are so low that it can detoxify the gas and its metabolites or otherwise repair the damage. Peaks in ozone concentrations are more likely to harm plants. Cell damage that is not repaired or compensated for ultimately leads to visible effects (such as changes in morphology and color), tissue death, or secondary effects (such as reduced plant growth, decreased yield or crop quality, and alterations in susceptibility to stress) (USEPA III 1986).

Visible symptoms can arise from either acute or chronic exposures. Acute symptoms—chlorosis, flecking, and stippling—appear within a day or two of short exposures (measured in hours) to high concentrations. Chronic exposures can cause chlorosis or other color changes and, eventually, cell death. But the symptoms of chronic ozone exposure are not reliable in diagnosis since they are easily mistaken for injury from diseases, insects, and other natural stresses.

So far, acute injury is the only certain form of ozone damage, and researchers now know how it affects the growth and yield rates of many plants. Yet, as figure 1.2 shows, some low-ozone concentrations and exposure times (those below the broken line on the graph) appear essentially harmless (USEPA III 1986). Indeed, the lower limit for reducing plant performance is an ozone concentration of 0.05 ppm for several hours daily for more than 16 days. For exposures of 10 days, the ozone threshold increases to about 0.10 ppm and, for 6 days, to about 0.30 ppm (USEPA III 1986).

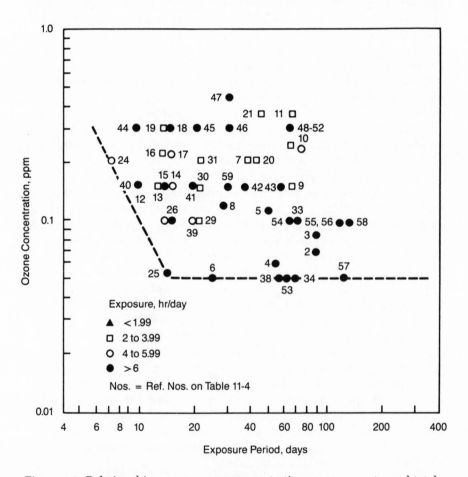

Figure 1.2. Relationship among ozone concentration, exposure rate, and total exposure period for observed reductions in plant growth.

Quantifying the Impacts

Exposures to ozone can have a number of measurable effects on plants generally and on crops specifically. Growth and yield fall off when ambient concentrations reach certain levels, and the quality of the usable product may also be impaired. These impacts were studied intensively between 1980 and 1987 as part of the government-sponsored National Crop Loss Assessment Network (NCLAN) review of ozone's effects on corn, cotton, peanuts, sorghum, soybean, wheat, alfalfa, barley, clover, tomato, and tobacco.

The goal of the NCLAN research was to estimate crop yield losses resulting from ambient ozone concentrations above naturally occurring levels. The dose-yield data collected for the 11 crops mentioned above were fitted to an empirical, nonlinear model. This model was then used to predict the crop yield losses as a function of the seasonal 7-hr/day mean O_3 concentration. (See table 1.2.) (In an ozone-free atmosphere, yield losses would be zero.) Figure 1.3 shows the yield losses for corn, wheat, cotton, soybeans, and peanuts as a function of the seasonal 7-hr/day ozone mean. Figure 1.4 indicates that 8 of 37 crops or cultivars are reduced by 10 percent at ozone concentrations between 0.045 and 0.049 ppm. More than half the crops are predicted to show a 10 percent loss at 7-hour seasonal mean concentrations below 0.05 ppm, the level prevailing in most agricultural regions (USEPA 1986). About 11 percent of the species or cultivars show a 10 percent loss below 0.035 ppm.

Estimating Productivity Losses

Data from various research programs clearly show that ambient concentrations of ozone are high enough in parts of the United States to impair plants' growth and yield (USEPA 1986). The economic losses from current ozone con-

Table 1.2. *Predicted yield losses (percent) at several seasonal 7-hr/day mean O_3 concentrations.*

| | Concentration (ppm) | | | |
Species	0.04	0.05	0.06	0.09
Barley	0.1	0.2	0.5	2.9
Bean, kidney	11.0	18.1	24.8	42.6
Corn	0.6	1.5	3.0	12.5
Cotton	4.0	6.9	10.0	20.0
Peanut	6.4	12.3	19.4	44.5
Sorghum	0.8	1.5	2.5	6.5
Soybean	7.3	12.1	17.0	30.7
Tomato	0.7	1.7	3.6	16.0
Winter wheat	3.5	6.9	11.1	27.4

Source: Heck chapter 6.

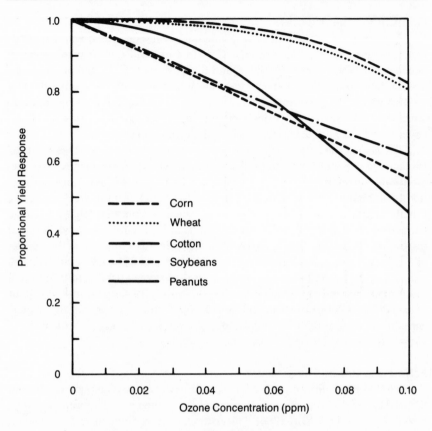

Figure 1.3. Predicted yield losses for corn, wheat, cotton, soybeans, and peanuts as a function of the 7-hour seasonal ozone concentration. Source: Heck chap. 6.

centrations can be estimated by combining crop inventory data, ambient ozone data, and the model linking various ozone concentrations to yield losses. An economic model is also required to convert yield losses into economic losses.

Ambient ozone levels in most agricultural regions are about 0.05 ppm (Heck et al. 1983). In one study, yield losses for soybeans, corn, wheat, and peanuts were calculated by comparing present ozone levels to a background level of 0.025 ppm. For these four crops, approximately $3 billion of productivity would be gained if current maximum ozone concentrations were reduced to 0.025 ppm (Heck chapter 6). Compared with a background ozone level of 0.025 ppm, present ozone levels probably lead to yield losses in U.S. crop production of 5–10 percent (NAPAP IV 1987).

In one economic assessment of crop losses, yield reductions for corn, soybeans, cotton, wheat, and peanuts were estimated (Kopp, Vaughan, and Hazilla

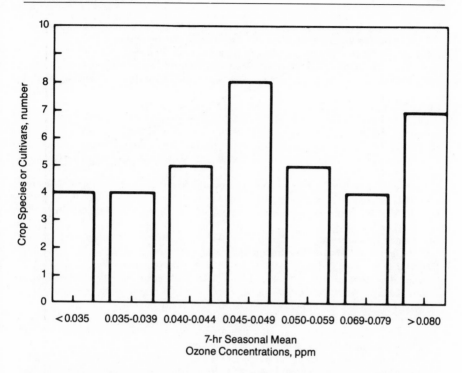

Figure 1.4. Number of crop species or cultivars (from a total of 37) showing a 10 percent yield loss as indicated 7-hour seasonal mean ozone concentrations. Source: U.S. EPA 1986.

1984). According to the researchers, reducing ambient ozone concentrations (seasonal 7-hour average) to 0.04 ppm would result in a $2.1 billion net benefit ($ 1987). If ozone concentrations increase to 0.08 ppm, net losses would total $5.2 billion ($ 1987). In another review, losses for six major crops that account for over 75 percent of U.S. crop acreage—corn, soybeans, wheat, cotton, grain sorghum, and barley—were examined (Adams, Hamilton, and McCarl 1984). Reducing 1980 ozone concentrations by 25 percent would yield benefits of $2.3 billion ($ 1987).

References

Adams, R. M., S. A. Hamilton, and B. A. McCarl. 1984. The economic effects of ozone on agriculture. PB85–168441/XAB. Springfield, Va.: National Technical Information Service.

Bormann, F. H. 1985. Air pollution and forests: An ecosystem perspective. *BioScience* 35 (July/August):434–41.

Cowling, E. B. 1985. Comparison of regional declines of forests in Europe and North America: The possible role of airborne chemicals. Paper presented at symposium, Air Pollutant Effects on Forest Ecosystems, May 8–9, at St. Paul, Minn. 217–234.

Ebasco Services Incorporated. 1986. Acid Deposition Studies. Prepared for the Business Roundtable Environment Task Force. New York: Ebasco Services.

Friedland, A. J., G. J. Hawley, and R. A. Gregory. 1985. Investigations of nitrogen as a possible contributor to red spruce (Picea rubens Sarg.) decline. In Symposium: Effects of Air Pollutants on Forest Ecosystems, May 8–9, 1985. Minneapolis: University of Minnesota Press.

Global Environment Monitoring System, International Co-operative Programme on Assessment and Monitoring of Air Pollution Effects on Forests. (GEMS) 1987. Forest damage and air pollution, report of the 1986 forest damage survey in Europe.

Heck, W. W., et al. 1982. Ozone impacts on the productivity of selected crops. In Effects of Air Pollution on Farm Commodities. Washington, D.C.: Izaak Walton League of America.

Heck, W. W., et al. 1983. A reassessment of crop loss from ozone. Environmental Science and Technology 17 (December): 573A–581A.

Hertel, G. D. 1988. Spruce-fir research cooperative technical report. Broomall, Pa.: U.S. Forest Service.

Interagency Task Force on Acid Precipitation (ITFAP). 1986. Report on the Crop Response Workshop, April 17–18. Washington, D.C.: NAPAP.

Jacobson, J. S. and J. P. Lassoie. 1989. Response of red spruce to sulfur- and nitrogen-containing contaminants in simulated acidic mist. In Proceedings of the Symposium on the Effects of Atmospheric Pollution on Spruce and Fir Forests in the Eastern United States and the Federal Republic of Germany. U.S. Forest Service.

Johnson, A. H. and S. B. McLaughlin. 1986. The nature and timing of the deterioration of red spruce in the northern Appalachian Mountains. In Acid Deposition Long-Term Trends. Washington, D.C.: National Academy Press.

Kopp, R. J., W. J. Vaughan, and M. Hazilla. 1984. Agricultural sector benefits analysis for ozone: Methods evaluation and demonstration. PB85–119477/XAB. Springfield, Va.: National Technical Information Service.

Krause, G. H. M., et al. 1983. Forest effects in West Germany. In Air Pollution and the Productivity of the Forest. Washington, D.C.: Izaak Walton League of America.

McLaughlin, S. B. 1985. Effects of air pollution on forests, a critical review. Journal of the Air Pollution Control Association 35 (5):512–534.

Mitchell, B. 1987. Air pollution and maple decline. Nexus 9 (Summer):1–13.

Mohnen, V. 1987. Exposure of forests to gaseous air pollutants and clouds. Research Triangle Park, N.C.: U.S. Environmental Protection Agency.

National Acid Precipitation Assessment Program (NAPAP IV). 1987. Interim Assessment, the Causes and Effects of Acid Deposition. Vol. 4, Effects of Acidic Deposition. Washington, D.C.: U.S. Government Printing Office.

National Acid Precipitation Assessment Program (NAPAP III). 1987. Interim Assessment, the Causes and Effects of Acid Deposition. Vol. 3, Atmospheric processes. Washington, D.C.: U.S. Government Printing Office.

Nilsson, S., and P. Duinker. 1987. The extent of forest decline in Europe: A synthesis of survey results. Environment 29 (November):4–31.

Prinz, B. 1987. Causes of forest damage in Europe: Major hypotheses and factors. Environment 29 (November):10–37.

Reich, P. B., and R. G. Amundson. 1985. Ambient levels of ozone reduce net photosynthesis in tree and crop species. Science 230:566–570.

Saxena, V. K., et al. 1989. Monitoring the chemical climate of the Mt. Mitchell State Park for evaluating its impact on forest decline. Tellus 41b (February):92–109.

Schuett, P., and E. B. Cowling. 1985. Waldsterben, a general decline of forests in central Europe: Symptoms, development, and possible causes. Plant Disease 69 (July):548–558.

Sheffield, R. M., and N. D. Cost. 1987. Behind the decline. *Journal of Forestry* 85 (January):29–33.

Sheffield, R. M., et al. 1985. *Pine Growth Reductions in the Southeast.* Resource Bulletin SE–83. Asheville, N.C.: U.S. Forest Service, Southeastern Forest Experiment Station. November.

Shortle, W. C., and K. T. Smith. 1988. Aluminum-induced calcium deficiency syndrome in declining red spruce. *Science* 240 (May 20):1017–1018.

Smith, W. H. 1981. *Air Pollution and Forests.* New York: Springer-Verlag.

Smith, W. H. 1985. Forest quality and air quality. *Journal of Forestry* (February):82–92.

Symposium on Air Pollutants Effects on Forest Ecosystems, St. Paul, Minn., May 8–9, 1985.

Tomlinson, G. H. 1983. Die-back of red spruce, acid deposition, and changes in soil nutrient status—a review. In *Effects of Accumulation of Air Pollutants in Forest Ecosystems.* Ed. B. Ulrich and J. Pankrath. Reidel.

Tomlinson, G. H. 1986. Nutrient deficiencies and forest decline. Canadian Pulp and Paper Association Annual Meeting, Montreal, Jan. 29, 1986.

Turner, R. S., R. J. Olson, and C. C. Brandt. 1986. Areas having soil characteristics that may indicate sensitivity to acidic deposition under alternative forest damage hypotheses. Oak Ridge, Tenn.: ORNL/TM–9917. Oak Ridge National Laboratory.

U.S Department of Commerce (USDOC). 1986. *Statistical Abstract of the United States, 1987.* Washington D.C.: U.S. Government Printing Office.

U.S. Congress. House Committee on Interior and Insular Affairs. Subcommittee on Mining, Forest Management, and Bonneville Power Administration. Statement of Baron Franz Riederer Von Paar. *Effects of Air Pollution and Acid Rain on Forest Decline.* Hearings, 7 June 1984.

U.S. Environmental Protection Agency (USEPA). 1986. *Air Quality Criteria for Ozone and Other Photochemical Oxidants.* Washington, D.C.: U.S. Government Printing Office.

Weinstein, L. H. Telephone conversation with James MacKenzie, February 1988.

Woodman, J. N., and E. B. Cowling. 1987. Airborne chemicals and forest health. *Environmental Science and Technology* 21 (2):120–126.

World Resources Institute and International Institute for Environment and Development (WRI/IIED). 1986. *World Resources 1986.* New York: Basic Books.

2

"New Types" of Forest Damages in Central Europe

REINHARD F. HUETTL

Recently increased environmental pollution has brought about a problem of enormous importance, concern, and complexity for humanity. The air, soil, and water have at least at places received the impact of a wide variety of toxic gaseous, liquid, and solid substances that individually and in combination adversely affect the environment (Kozlowski and Mudd 1975).

Although interest in the effects of pollutants on trees and forests has increased greatly in the last few years, apprehension about air pollution has been expressed for a long time. But concern previously focused on local forest damages directly related to short-distance emission sources. The main pollutants at that time were sulfur dioxide (SO_2) and particulates. With increasing industrialization, enhanced combustion of fossil fuels, and the recent "high-stack policy," the air pollution problem became far more serious and complex (Kozlowski and Constantinidou 1986).

Based on these essential considerations, this chapter will address recent forest damages commonly attributed to the negative impacts of air pollution in West Germany and neighboring countries.

Development of Forest Damages in West Germany and Neighboring Countries

Inventories have been carried out in central Europe to describe the degree and the development of forest damages.

The author wishes to acknowledge the profound guidance and support of Heinz Zoettl, director of the Institute of Soil Science and Forest Nutrition at the Albert-Ludwig-University in Freiburg, West Germany. I also appreciate the funding support supplied by the West German Ministry of Science and Technology at Bonn for the presented research projects. Furthermore, I acknowledge I. Mantel for her exceptional clerical work.

West Germany

Since the mid–1970s the so-called new types of forest damages have been observed in West Germany. The silver fir dieback (*Abies alba* Mill.) in the Black Forest and in the Bavarian Forest was soon followed by severe damages in Norway spruce (*Picea abies* Karst.) and Scots pine (*Pinus sylvestris* L.) stands. Since 1983 increased damages have also been found in deciduous trees such as European beech (*Fagus sylvatica* L.) and oak trees (*Quercus* spp.).

Annual terrestrial inventories have been carried out since 1983 to evaluate the damaged area and the development of the damages. But only since 1984 have the inventory parameters been uniformly applied for all eleven German states. Foliage losses are the main criteria to determine damage classes (table 2.1), and the results of the surveys are expressed as the percentage of forest areas containing trees with at least 11 percent (slight) loss of foliage.

Initially the inventory results suggested a fast increase in the percentage of areas showing at least slight loss of foliage. In 1983, 34 percent and in 1984 already 50 percent of the areas contained some trees showing slight to heavy damage symptoms. Thereafter the development stagnated. The damages increased in 1985 and in 1986 by 2.0 percent annually. Thus, in fall 1986, 54 percent of the total forest area in West Germany was believed to contain some trees with at least slight foliage loss. However, only 19 percent of the investigated forests was marked by distinct damages (foliage losses greater than 25 percent). The damages were and are commonly attributed to the adverse effects of acid rain and other air pollutants (Breloh and Dieterle 1986, Gussone 1986).

It seems worth noting that since this type of damage inventory has been carried out, a significant part of the scientific community involved in "new type"

Table 2.1. *Damage classes based on percentage of needle/leaf losses as applied in central Europe by International Cooperative Program for Assessment and Monitoring of Air Pollution Effects on Forests.*

Class	Needle/leaf loss (%)[a]	Damage
0	1–10	none
1	11–25	slight
2	26–60	moderate
3	60–99	severe
4	>99	dead

Source: Koehl 1987.

Note: In West Germany more than 200,000 trees are observed annually at > 300 plots based on a 4 × 4 km grid net (Koehl 1987).

[a]Pronounced foliar discolorations may elevate the damage class category.

forest decline research has stated that foliar losses can be caused by multiple factors and are thus of an unspecific nature.

The amount of healthy foliage of a particular tree species varies due to site conditions, genetic variability, and other factors, indicating that no "normal" leaf area can be assumed for a specific tree or stand. But a normal leaf area must be assumed in order to determine the amount of foliar losses because the former foliar status is in almost all cases unknown. It must therefore be concluded that the parameter foliage loss is not a reliable means for evaluating the decline of forest trees and stands when research is focused on determining specific causes (Huettl and Wisniewski 1987, Innes 1987).

Owing to the rapid development of the damages and the remarkable publicity that the *Waldsterben* (German for forest death) issue has received in West Germany, numerous comprehensive field and laboratory experiments have been established in all major forest decline areas since 1983. The results of these projects already allow a profound insight into the problem of new type forest damage, as will be illustrated later (cf. Forschungsbeirat Waldschaeden/ Luftverunreinigungen der Bundesregierung un der de Laender 1986). Similar research efforts have been initiated in all affected countries of central Europe.

Switzerland

No signs of new type forest damages were reported in Switzerland before 1982. But in 1983 loss of foliage was observed at various locations (Schmid-Haas 1985). A survey carried out that year in all Swiss forest districts indicated that about 14 percent of the forests showed some trees with at least slight loss of foliage.

As in West Germany, forest damage inventories based on a systematic sample system have been carried out in Switzerland since 1984. The intent is to obtain solid data of the vitality status of the forest and its changes on a comparable basis for the whole country. This program is called Sanasilva. The inventory procedure is similar to that used in West Germany (see table 2.1).

The results of this damage inventory indicate that in 1984, 34 percent of the investigated stands contained some damaged trees. However, 92 percent of the damaged stands contained trees with only slight symptoms (foliage losses less than 25 percent). Besides damages in stands of Scots pine, silver fir, and Norway spruce, oak and beech trees also showed symptoms.

As in West Germany, the development of visible symptoms in Switzerland was less dramatic in 1985 compared with the previous years. Also, the major increases in symptoms in Switzerland were in deciduous tree species. Of particular importance for the country is that symptoms occur mainly in high-elevation forests. Hence the important protection forests provide against avalanches, erosion, etc. is reduced considerably in the Alps (Bundesamt fuer Forstwesen und Landschaftsschutz 1986). In 1986, 52 percent of the investigated forests showed visible syumptoms. But only 16 percent of the stands contained trees with more than 25 percent foliage loss.

Austria

In 1984 for the first time forest damage inventories were undertaken in Austria. The inventory concept is based on systematically distributed permanent observation areas. There the development of the crown condition is observed in permanently marked trees. Owing to a lack of sufficient personnel, the 1984 inventory was carried out in only five of the nine Austrian states; it indicated that about 30 percent of the forest area revealed crown thinning, of which 6 percent was marked by distinct damages.

The nationwide inventory of 1985 indicated that 74 percent of the Austrian forests showed no symptoms, 22 percent of the investigated forests were marked by weak symptoms, and 4 percent by moderate to heavy damages. When the 1984 inventory data are compared with the results of 1985 (based on the five investigated states of 1984), an 8.0 percent reduction of damages can be seen.

In 1986 the total damages had increased by 5 percent. In certain regions the development was very pronounced—that is, only one year later, trees and stands that had formerly been slightly damaged showed moderate to heavy decline symptoms (Pollanschuetz 1987).

France

An inventory was first carried out in 1983 to evaluate the development of the forest damages in France. But this survey was confined to the area of the Vosges Mountains. The observation plots were laid from east to west, based on inventory axes. The identification of damages is similar to the procedure applied in West Germany (Scheifele 1985).

A first profound insight into the damage situation in the area of the Vosges Mountains was provided by the inventory results of 1984. The data indicated that about 22 percent of the coniferous trees as well as 4 percent of the deciduous trees were marked by pronounced damages.

Owing to these inventory data the observation area was enlarged in 1985. The data of this observation net showed that about 76 percent of the trees were healthy in 1985, 17 percent revealed damages of 10–20 percent needle loss (similar to the German damage class 1), and 7 percent were characterized by heavier damages (comparable to the German damage classes 2–4). The main damages were found in the Vosges Mountains as well as in the northern Alps.

The results of the 1986 survey revealed nearly the same level of damages as in 1985. The broad-leaved stands were less affected than the conifers. But the conifers were somewhat more damaged in 1986 than in 1985. However, the share of trees showing more than 25 percent foliage losses varied between 5 and 8 percent. The most damaged region is the Vosges Mountains, where 18 percent of the conifers show pronounced damage (Bonneau and Joliot 1987).

Benelux Countries

Only in the Netherlands and Luxembourg have forest damage surveys been carried out (Kroemer-Butz 1986). In Belgium, new type forest damages are reported only from some areas in the Ardennes Mountains (Weissen and van Praag 1984).

In 1986, 190,000 ha of the total 300,000 ha of the Dutch forest area were surveyed. Forty-one percent of the investigated forests did not reveal any decline symptoms. However, 29 percent were marked by moderate to heavy decline symptoms. Forest damage is severest in the southeast of the Netherlands, an area with an extremely high density of livestock.

Luxembourg's forests are still relatively healthy. In 1986, 80 percent of the forest area revealed foliar losses of less than 10 percent and were thus considered undamaged. Only 4 percent of all sample trees were characterized by more than 25 percent foliage loss.

Concluding Remarks

From the 1986 survey data, it seems obvious that within central Europe the Dutch and German forests show the greatest amount of new type forest damages. The inventory results of Switzerland, Austria, and France indicate similar development patterns and they vary within a range similar to that found in West Germany, even though the inventory methodology is modified to some extent in each country.

Visual and Nonvisual Symptoms

During recent years a wide variety of different symptoms has been described and included in the classification of new type forest damages. This section describes the most common damage symptoms for the three forest tree species most affected in central Europe.

Damages in Norway Spruce

As the Norway spruce is the most common and most important economic tree species in central Europe, it has been studied in many research projects prior to the recent declines. Recent changes in the vitality of this tree species can therefore be clearly described.

According to the German Forschungsbeirat für Waldschaeden/Luftverunreinigungen der Bundesregierung und der Laender (Research Council Forest Damages/Air Pollution of the Federal Government and the States 1986) the most important spruce damages can be differentiated into five types:

- yellowing in stands at higher elevations of the German central chains of mountains,
- crown thinning in stands at middle elevations of the German central chains of mountains,
- needle necroses in older stands in southern Germany,

- yellowing in stands on carbonate sites at higher elevations of the calcareous Alps, and
- crown thinning in coastal areas.

Yellowing in Stands at Higher Elevations of the German Central Chains of Mountains. This damage type is a truly new phenomenon. It is marked by magnesium (Mg) deficiency symptoms as found by Zech and Popp (1983) in the Fichtel Mountains, by Bosch et al. (1983) in the Bavarian Forest, by Zoettl and Mies (1983) as well as by Huettl (1985) in the Black Forest, and by Hauhs (1985) in the Harz Mountains.

The yellowing ("tip-yellowing") starts in the older needle-year classes of the lower and middle crown area. The yellowing occurs only on those needle and twig sides that are directly exposed to sunlight. After some time, which cannot be predicted, the needles often become necrotic and will finally fall off.

Generally the current shoots are not affected by the tip-yellowing. The discoloration is "transferred" to the next, still-green needle-year class when the new shoots are developed in spring. This is probably caused by translocation of the mobile Mg from older into younger needle tissue (Mies and Zoettl 1985). The height growth of moderately yellowed spruces is not reduced. But analyses of the volume growth of yellowed trees compared with healthy looking trees show decreased increment rates (Aldinger 1987, Roehle 1985; cf. Evers 1984). The heaviest needle losses are frequently found in the middle crown area, provoking the "sub-top-dying" symptom. Finally, the trees may die off completely. This happens generally in combination with frost, drought, or insect or needle cast fungi infestations.

In some cases when declining younger Norway spruce trees were observed for several years, the yellowing stagnated or was even reversed to some degree (Kandler et al. 1987). Such natural regeneration processes are probably related to more favorable climatic conditions, for example, higher and more evenly distributed precipitation during the vegetative period, allowing higher nutrient uptake, particularly on poor soils.

This damage type is found in younger as well as in older spruces. In mixed stands, the intermingled tree species such as silver fir, European beech, Scots pine, and Douglas fir can also be affected. For this as for all other decline types of Norway spruce associated with nutritional disorders, a pronounced correlation to the chemical soil conditions (the substrate-specific nutrient supply) can be found (Zoettl and Huettl 1986). Fink (1983), Parameswaran et al. (1985), and Huettl and Fink (1988) found changes in the needle tissues of tip-yellowed needles on a microscopic level, indicating damages within the vascular bundle (phloem collapses). This damage type occurs occasionally also in low-elevation Norway spruce stands on acid, base-poor soils (Horras 1986).

The same tip-yellowing phenomenon has been observed for comparable site conditions in Austria (OeDB 1986), France (Bonneau 1987), Belgium (Weissen and van Praag 1984), and the Netherlands (Boxman et al. 1987; van den Burg 1987; Roelofs et al. 1985).

Comparable symptoms have been reported by Bruck (1985) and Hoshizaki et al. (1987) for high-elevation red spruce forests and at low-elevation sites by Jagels (1986), as well as by Huettl and Wisniewski (1987) in low- and high-elevation forests in the eastern United States.

Crown Thinning at Middle Elevations of the German Central Chains of Mountains. This damage type is primarily characterized by a pronounced crown thinning. Yellowing symptoms may occur, particularly in the exposed crown area of older dominant spruces. It is found mainly on nutrient-poor sites at elevations of 400–600 meters above sea level. This phenomenon is also widespread and occurs in forest areas with relatively high SO_2 concentrations and relatively high wet deposition of hydrogen ions. During recent years an increase of this phenomenon has been observed.

Predominant and dominant trees in older Norway spruce stands are primarily affected by this damage type. Needle loss may occur in the lower crown area, in the middle crown area (sub-top-dying), as well as in the top (top-dieback). Undifferentiated combinations of these crown thinning types can be observed as well (Schroeter and Aldinger 1985). Frequently, however, the needle losses are rather evenly distributed in the crown area.

In contrast to the yellowing phenomenon in stands at higher elevations, needle loss is not necessarily preceded by yellowing symptoms. The process of needle loss generally begins in the older needle-year classes, which leads to a crown thinning from inside to outside. At the branches of the second order, needle loss starts at the twig base and proceeds to the tip, where generally one to four needle-year classes are left. Occasionally green needles may fall off.

Microscopic investigations into this phenomenon carried out by Ebel and Rosenkranz (1984, 1985) show that the crystalline epicuticular waxes of current needles were not obviously affected, but in the previous years' needles the waxes were eroded, particularly at the most exposed needle surfaces. Furthermore, the stomata were covered by rather compact wax material that in places showed cracks. Histological research indicates furthermore that changes of the chloroplast structures are present that are severer in older needles that show discoloration symptoms.

From a nutritional viewpoint, chemical needle analysis often indicates poor calcium (Ca) and insufficient Mg supply. Also phosphorus (P) and potassium (K) may be low at certain sites.

When investigating Norway spruces exhibiting this phenomenon, Feig and Huettermann (1985) found reduced contents of photosynthetically active pigments on a physiological and biochemical level. This might be related to insufficient K and Mg needle contents (Gerriets and Schulte-Bisping 1985). In 1941, Michael had already reported reduced foliar pigment contents in the case of Mg deficiency (Michael 1941).

Needle Necroses in Older Stands in Southern Germany. In late fall 1982 a very pronounced reddish needle discoloration phenomenon associated with crown thinning was observed in older Norway spruce stands in wide areas of southern Germany, especially in forests close to the Alps (Rehfuess and Rodenkirchen 1984). This damage type can easily be distinguished from other damage phenomena because it shows a unique development.

Beginning in mid-September various older needle-year classes, particularly in the lower and middle crown area, simultaneously start to become orange-yellow and then turn into reddish-brown after two more weeks. The short-term visible yellowing affects the whole needle and can thus be seen on all its sides. Younger needle-year classes are generally not affected except for the heavily shadowed crown areas in the lower tree part. After becoming necrotic and turning brownish, most of the older needles fall off beginning at the end of October.

This loss of needles can happen quickly but occasionally brown needles may be retained for some months. The youngest shoots in the outer crown area and in the crown top stay generally green and healthy. The crowns of older spruces become particularly thinned due to the massive shedding of older needles, proceeding from inside to outside and from the lower crown area to the top. The thinning phenomenon becomes especially evident when needle shedding occurs for consecutive years. This damage type occurs predominantly in stands more than 60 years old.

Even though heavy crown thinning may occur, the death of trees has been observed only when further stresses were added, such as insect or fungi infestations. In various areas Rehfuess (1983b) observed the development of adventitious shoots and described this as a natural regeneration process.

In southern Germany, where this damage type is frequently found, no long-term growth depression in spruce has been observed. So far no coincidence has been detected between amount of needle loss and growth development of single trees as long as needle losses are below 40–50 percent. This indicates clearly that sparse trees can produce the same growth increment as their neighbors with dense crowns. This is probably due to the fact that the shedding affects primarily older needles of the shadowed crown area, which contribute only little to the production yield (Schulze et al. 1977).

In contrast to the yellowing phenomenon in Norway spruce stands at higher elevations, the necroses phenomenon is found in trees that are sufficiently supplied with nutrients as well as in deficient stands. However, Zoettl and Huettl (1985) found that this damage type occurred at a significantly increased rate in stands with low or deficient K supply.

Microscopic investigations of necrotic needles proved that they were infected by various needle cast fungi (*Lophodermium piceae, Rhizosphaera kalkhoffii*). These needle cast fungi were analyzed by Butin and Wagner (1985), Koenig (1983), Kowalski and Lang (1984), and Schwenke et al. (1983). This observation is also consistent with former research findings (for example, Baule and Fricker 1967) indicating that trees marked by low or deficient K supply are less resistant to fungal infestations.

Much earlier, Hartig (1889) as well as Nobbe (1893) and Neger (1924) described this phenomenon, but never has this damage type been observed over such a wide area. Also surprising is that only older stands are affected, whereas adjacent younger stands show no such signs of damages.

Yellowing in Stands on Carbonate Sites at Higher Elevations of the Calcareous Alps. The condition of spruce forests at higher elevations of the calcareous Alps gives reason for serious concern. The forest damage inventory of 1986 found that crown thinning has increased considerably in this area. At some places, older stands have started to break down completely.

The damages in spruce stands that grow on shallow soils derived from carbonate parent material on steep slopes and ridges are marked by specific needle discolorations, heavy needle losses, and a characteristic constellation of site factors.

The condition of the spruce stands in this area is of concern because the natural regeneration of site-adapted trees as well as the establishment of mixed young stands under the canopy of the dying older stands is impeded significantly by overpopulations of various wild game species. Owing to this unnatural impact, grassy stand gaps void of trees are common. This condition may easily lead to enhanced erosion problems. Along with the flooding due to the heavy precipitation experienced recently in the Alps of Austria, Switzerland, northern Italy, and West Germany, snow movement must be considered in this context. Thus, erosion endangers the complete forest ecosystem in this region.

Spruce forests on southernly exposed steep slopes and on shallow rocky alkaline soils derived from lime and dolomite parent materials at elevations above 1000 meters above sea level were always characterized by slow growth. In addition, needle discolorations associated with nutrient deficiencies were and are common in these spruce forests. In particular, the older needles are frequently marked by a chlorotic discoloration that starts at the tip. But only those branches exposed to sunlight are affected. This yellowing is often followed by necroses that also start at the tip of the needle. As known from extensive research work, these symptoms are typical of a K defiency.

In addition to this deficiency, another nutrient disorder can be observed under these stand and site conditions. In this case the current shoots are affected and reveal a whitish to yellowish discoloration. The older needle-year classes stay generally green. Kreutzer (1970) and others attribute this phenomenon to a manganese (Mn) deficiency. In rare cases iron (Fe) deficiencies were also observed, which show a very similar symptomatology. The Mn deficiency symptoms are most evident during the nonvegetative period. As Mn and Fe are rather immobile nutrient elements in the phloem, it is not surprising that deficiency symptoms are first observed in current tissue. Often the needles do not die off but become green in the following year due to an accumulation of Mn in older needle-year classes. This damage type was also detected at lower elevations under comparable site conditions, for example, in the Swabian Mountains (Huettl 1985).

As indicated above, these discoloration pattens have been known for a long time in these areas. But starting in 1981 an unusual increase of the yellowing during summertime was observed. This decline is at places associated with distinctly reduced tree vitality as well as with premature losses of older needles. Even the rapid death of single trees or smaller tree groups may occur. As with the tip-yellowing in Norway spruce at higher elevations on acidic sites, this damage type also occurs in all age classes. Again, the observed nutrient deficiencies are clearly correlated to the chemical substrate and are thus site specific.

In addition to the described Mn (Fe) and K deficiencies (which only in few cases occur in combination), nitrogen (N) and P supply is frequently rather low or even deficient. Owing to the chemical substrate condition, Ca and Mg contents are generally optimal. Butin and Wagner (1985) detected various needle cast fungi in declining stands. In addition to this, root fungus infestations are common at these sites. Observations of this damage type have also been made in Switzerland by Flueckiger et al. (1984), in Austria by OeDB (1986), and in France by Bonneau (1986).

Crown Thinning in Coastal Areas. As reported by the Forschungsbeirat für Waldschaeden/Luftverunreinigungen der Bundesregierung und der Laender (1986) aerial damage inventories showed in 1983 that one-third of the older spruces (more than 60 years old) were marked by distinct damages (damage classes 2 to 4) in coastal forests.

Investigations of growth parameters in undamaged spruces of the Wingst area revealed that since the mid-1960s a pronounced growth decline has been evident. The warm, dry year of 1976 caused a further significant growth decrease. Since then, annual volume growth based on earlier reference data has been reduced by 40–60 percent.

Yearly root biomass was determined in slightly damaged and heavily damaged stands between 1983 and 1985. The roots were separated in vital and damaged fine roots. The values of dry biomass varied between 2800 and 6900 kilograms per hectare. In both stands, fine roots were evenly distributed in the mineral soil down to a depth of 40 centimeters (cm) in February 1983. But the percentage of damaged roots was much higher in the stand that showed heavy damage symptoms than in the less damaged trees. The samples taken in July 1984 and 1985 indicated a much less deep rooting of vital fine roots into the mineral soil, which was even more pronounced in the more heavily damaged stand. Simultaneously, the fine root biomass increased in the organic matter and in the upper mineral soil horizons.

Foliar nutrient element analyses generally indicate sufficient nutrient supply. In some cases N, P, and particularly Mg and K may be low or even deficient.

The macroscopic symptoms that characterize this damage type are very similar to those described in the section on crown thinning in stands at middle elevations of the German central chains of mountains. However, discoloration symptoms have been observed only in very few declining stands of the coastal area.

Similar crown thinning damages are observed in coastal areas of Belgium and the Netherlands.

Damages in Silver Fir

Owing to its site demands, silver fir is mainly found in southern Germany (for example, the Bavarian Forest and Black Forest), in parts of Austria, in Switzerland, and in France (for example, the Vosges Mountains). The observed symptoms are very similar in all these regions and are therefore discussed here only for southern Germany.

Silver fir dieback has been reported for more than 100 years (Wachter 1978). It is, however, unclear whether the former declines are identical to the recent damages. At least in some cases former declines were caused by bark beetle infestations or other biotic damages. However, to some degree the recent decline is also caused by heavy deer browsing.

In 1919, Neger described fir decline symptoms similar to those occurring today (Neger 1919). Silver fir diseases were previously observed in various regions periodically, but no reports indicate the extent of the damages. Brandl (1985) demonstrated that no extensive fir decline was observed in the Black Forest in former times.

As far as the recent fir dieback is concerned, the first pronounced symptoms were observed in the early 1970s in the Bavarian Forest (Seitschek 1981). Since then, damages increased more or less rapidly.

Whereas needle yellowing and necroses are the dominant factors considered in differentiating damage types in spruce, discolorations in fir are much less common and reveal a much less regular spatial and temporal distribution.

Needle Yellowing. As in spruce, older needle-year classes in declining firs may be marked by a yellowish discoloration. All age classes of trees are affected, but younger trees generally reveal a higher frequency of this phenomenon. Also in silver fir the tip-yellowing is caused by a Mg deficiency. The same site parameters are related to this decline type as described in the first section on Norway spruce.

Needle Necroses. These symptoms can be observed in firs of all age classes. Various patterns may occur. Necroses can be found in older needles, generally preceded by acute yellowing. Furthermore, all needle-year classes may turn necrotic. This symptom is occasionally observed in springtime. A more common symptom in older fir trees is a tip necrosis.

Crown Thinning. Crown thinning (all forms of crown thinning can be observed; see description under Norway spruce) results from needle losses of green, yellowed, or necrotic needles. Frequently the older needles in the shadowed crown area are shed first. The current shoots are lost only shortly before the tree dies.

"Stork Nest" Development. The stork nest development is a normal phenomenon in older firs. Owing to reduced height growth the tree top becomes flatter and flatter. Simultaneously a dense growth of lateral branches in the crown top takes place, and thus the crown top looks like a "stork nest." This development was also recently observed in younger firs and addressed as a damage symptom. However, this is controversial.

Development of Secondary Branches. The development of secondary branches along the tree stem is not a new symptom. But in declining firs exhibiting pronounced needle losses, increased secondary branching is observed. As these branches are developed from adventitious buds below the normal crown, timber quality is remarkably reduced.

Various recent investigations eliminated a primary cause of biotic factors in silver fir decline (Eichhorn 1981, 1985; Moosmayer 1984; Schwenke 1982, 1985), but insect infestations may accelerate the process of dieback. Sierpinsky (1984) suggested that increased infestations of *Dreyfusia nordmannianae* are correlated with increased air pollution. Furthermore, declining firs often reveal a reduced fine rooting system, which might be caused by various root fungi (Schoenhar 1985).

Damages in European Beech

As indicated, decline symptoms were first observed in fir and spruce trees. But since the early 1980s symptoms have also been evident in deciduous trees, particularly in beech (the most common deciduous tree species in central Europe). Crown thinning of the upper outer crown area, dieback of branches, insufficient shoot development, premature foliage discolorations and losses, bark necroses along the stems and branches, growth reductions, and root damages were registered (Eckstein et al. 1984; Flueckiger 1986; Moehring 1982a, 1982b; Schuett and Summerer 1983).

As a result of these observations deciduous trees were included in the annual terrestrial forest damage inventories. But as little is known about the morphology of healthy beech crowns, foliage density as a parameter of various damage classes is even more critical in beech than in spruce and fir. Thus, in addition to crown morphology, vitality classes of beech are based on parameters that look at the life history of the trees over various decades. It is therefore possible to distinguish short-term influences from long-term impacts (Roloff 1984a, 1984b, 1985a, 1985b, 1985c, 1985d).

According to Roloff, four damage classes can be differentiated due to the morphology of the branching system. The main criterion is the differentiation of long and short shoots. Along with an abnormal branching morphology, premature leaf and shoot shedding as well as unusual leaf morphology are seen at least to some degree as symptoms of new type forest damages. In this context it is important to note that beech leaves vary in both size and shape more than in any other tree species.

Furthermore, various pathogens (insects, fungi) are having an impact on European beech and influencing, at least regionally, the vitality of beeches significantly. However, Flueckiger (1986) suggested that increased biotic infestations in beech might be correlated with air pollution impacts. Again, this is seen in relation to nutritional disturbances such as overoptimal supply of nitrogen due to high atmospheric N deposition and nutrient deficiencies like Mg and K.

In beech, as in spruce, damages increase with increasing altitude as well as with decreasing distance to the coast. Beech damages are particularly pronounced in the Harz Mountains, where no healthy beeches can be found anymore. Severe damages also occur in the Black Forest, in various regions of Switzerland, and in the Vosges Mountains.

During recent years damages were observed not only in older beeches but also in natural beech regenerations. These damages were particularly pronounced at acidic sites. Under these conditions beech seedlings showed root damages, reduced shoot elongation, leaf necroses, chlorotic foliar discolorations, and in severe cases leaf losses. Of special concern is the impaired root development. Roots are primarily distributed in the humus layer. Deeper roots into the mineral soil frequently die off (Huettermann and Gehrmann 1982, Huettermann and Ulrich 1982). The observed root damages would, however, indicate that the dieback of seedlings is caused by soil-borne stresses.

Various investigations into older beech stands revealed that the chemical soil condition close to the stem is strongly influenced by stemflow, showing greatly reduced pH values as well as low basic cation contents (Ca^{2+} and Mg^{2+}) and increased contents of H^+, Al^{3+}, and Fe^{3+}. In addition, an accumulation of heavy metals was found (Glatzel et al. 1983, Glavac et al. 1985, Schulte and Spiteller 1985). The soil chemical changes are also reflected in soil vegetation.

The nutritional status of declining beeches (Kaufunger Forest) exhibiting different degrees of damages were characterized by Bredow et al. (1986). They found that increased damages were correlated with decreased foliar contents of Mg, K, and Ca. Heavily damaged beeches were clearly marked by K deficiencies. Similar observations were made by Flueckiger (1986) in Switzerland.

Forest Decline and Tree/Stand Growth

Numerous investigations have been carried out to illustrate the growth patterns of damaged and undamaged trees and stands. Before we consider the detailed information, it must be emphasized that growth of forest trees and stands is influenced by many factors and is thus very complex.

Of particular importance in this context are climatic factors. Furthermore, it is known that forest stands affected by classical fumigation damages can experience reduced growth long before symptoms such as needle loss, crown thinning, and crown structure changes become obvious. These growth declines are probably related to disturbances of metabolic processes and production of assimilates. On the other hand, however, needle losses particuarly of older needle-year classes may not necessarily be related to decreased increment growth. This is especially so for affected trees that are relatively vital and are thus able to counteract a probable water stress by shedding older, not very productive needles (Pollanschuetz 1986).

It is well known that trees affected by phytotoxic air pollutant concentrations react to this stress with decreased radial and volume growth, particularly in the lower and middle stem area. These changes follow the damaging influence relatively rapidly. A further characteristic feature is that growth variations due to climatic conditions are reflected much less in damaged trees than in undamaged trees. It has also been found that these reactions are reflected adequately only in dominant or predominant trees. Thus, for valid proof of the impacts of gaseous air pollutants on the physiological function of forest trees and thus on their growth productivity, it is necessary to take corings or carry out stem analysis only in dominant or predominant trees with adequate crown development.

Further difficulties arise when we try to relate growth patterns to specific impacts. It is impossible to find forests in central Europe that have not been affected by air pollution. Yet remarkable concentration gradients of various air pollutants exist. As no "normal" control can be found, trend extrapolation of year-ring widths are used whenever possible to establish a baseline before tree/stand growth was affected by air pollution (Neumann and Pollanschuetz 1982, Pollanschuetz 1966). Finally, Abetz (1985) suggested the use of reference trees. This method is based on the theoretical construction of dominant trees that follow a specific growth pattern optimal for a particular species and a specific site. Recent research on growth patterns was focused primarily on single tree investigations. Very few data are available that describe growth patterns on a stand level.

The reactions of forest trees to the negative and positive factors in their environment are as varied as the reports of new type forest damages in central Europe. The specific physiological reactions and metabolic processes are influenced not only by various exogenic but also by many endogenic influences, and thus they result in many different growth and increment reactions of single trees and stands. Growth is hence a very important parameter to mark the vitality of trees and/or stands. Indeed, growth-related parameters might be good indicators of vitality changes in forest stands. But, as outlined, they are extremely difficult to establish.

Norway Spruce

Franz and Roehle (1985) reported a remarkable reduction of increment growth in Norway spruce trees since about 1970 in heavily damaged trees of the Fichtel Mountains. They detected similar growth reductions for the last 15 years in damaged stands of the Bavarian Forest. In areas where damages have occurred only since 1982, such as the Bavarian Alps, no significant growth modifications of damaged and undamaged trees were detected.

Franz and Roehle (1985) also indicated that some spruces with needle losses greater than 60 percent did not reveal any growth decline. Johann (1986) reported on trees of the Bavarian Alps that were already marked by heavy damages (classes 3 to 4) in 1975 and that were reinvestigated in 1985. Year-ring analyses proved that these trees, in spite of their very poor vitality, had increased growth since 1950.

But in other areas, such as the Bavarian Forest and the Fichtel Mountains, growth decline is evident in spruce stands since 1970. There, as in the classical fumigation damage areas, in addition to growth reductions in heavily damaged spruces the well-known increment translocations from lower to upper stem parts have also been detected. A significant correlation between degree of damage symptoms and growth reduction was found in northeast Bavaria by Greve et al. (1986). They attributed these damages to the impact of air pollution.

According to Kenk (1985), the development of volume growth based on a comprehensive long-term study in spruce trial stands in southwestern Germany was found to be much higher than would be expected from well-established site-specific yield tables. This is especially true for older Norway spruce stands. Increased volume growth began in the late 1950s/early 1960s and implies a growth increase of generally 20–50 percent. Kenk names as causes for this curious phenomenon increased N deposition, higher carbon dioxide (CO_2) contents in the atmosphere, favorable precipitation regimes in the 1960s, and improved forest management practices.

In summarizing his findings, Kenk reported that no clear trend can be established between amount of damage (needle losses) and growth patterns in Norway spruce. This is true for all stands where needle losses do not exceed 40–50 percent. Only in cases of extreme foliar losses or of yellowing can pronounced increment growth reductions be detected that might finally lead to the complete breakdown of a stand.

Silver Fir

Various authors (Aldinger 1987, Aldinger and Kremer 1985, Kenk et al. 1984), have found that in older silver fir stands of the Black Forest—one of the major silver fir habitats—growth had declined by up to 40 percent over a long time period. Particularly on poorer soils, firs exhibiting distinct needle losses (greater than 25 percent) were marked by significant growth reductions when compared with healthy control trees. Spruces at the same sites revealed similar reduction only in case of heavy needle losses (greater than 40–50 percent) or pronounced yellowing. However, recent terrestrial forest damage inventories indicated that fir stands in the Black Forest revealed growth increases for the most recent years following a period of significant growth decreases.

But this improvement process was not very pronounced when the declining firs had reached a rather low productivity level. Similar positive growth reactions were detected in fir stands of Switzerland. Schmid-Haas (1986) found that from 1983 through 1985 the increment of various declining fir stands there had doubled and that the year-ring widths had reached values of those found 30 years ago. Comparable observations were also made by Pollanschuetz (1986) in various states of Austria when height growth of firs at different declining stages was investigated.

Concluding Remarks

Distinct differences exist in the growth patterns of damaged spruces and firs. In spruce trees, no general correlation between needle loss and growth can be found, whereas silver firs are marked by significant growth decreases when needle losses exceed 25–30 percent. Growth reductions in spruce and fir trees are frequently much more evident in the lower stem area than in the upper stem part. In general, 1972/1973 to about 1978/1979 was characterized by a pronounced growth depression. This growth depression, which peaked in the dry, warm year of 1976, was followed beginning in 1982 and (occasionally only) in 1984 by increased increment growth. How long this period of improved growth will last remains to be seen (Pollanschuetz 1986).

Atmospheric Deposition Load

The effects of air pollutants on forest ecosystems are seen as one major factor causing forest declines (Manion 1981, Rat von Sachverstaendigen fuer Umweltfragen 1983). In general two major pathways are differentiated. Direct damages may occur due to phytotoxic concentrations of gaseous air pollutants. Examples of this type of impact are the dramatic SO_2 fumigation damages, particularly in Norway spruce stands of the Ore Mountains in Czechoslovakia (Materna 1986) and the well-described SO_2 fumigation damages in forest trees close to the Sudburry Smelter in Canada. Acute damages related to ozone (O_3) have been reported for pine trees of the San Bernardino Mountains in southern California (Miller 1987) and for forest stands in the vicinity of Mexico City (Bauer et al. 1987). Phytotoxic NH_3/NH_4 air concentrations are causing direct damages in forests of the Netherlands, West Germany, and probably also Belgium (Boxman et al., Knabe et al. 1985).

Indirect damages are provoked by the negative impacts of wet and dry deposition of air pollutants via soil-mediated processes. Considerable soil degradation may be caused by acidic precipitation resulting in accelerated soil acidification and thus, for example, in enhanced nutrient losses from the rooted soil substrate (Tamm and Haellbaecken 1986, Ulrich 1986). Increased nitrogen deposition might produce nutrient imbalances in the soil and thus disturb nutrient uptake via tree roots (Huettl 1986).

Finally, interactions between the direct and indirect pathways may occur. For example, O_3 may damage the foliar tissue of forest trees and in combination with acid precipitation, particularly acid fog, nutrient leaching and losses of organic compounds may be increased, leading to nutrient deficiencies in forest trees and stands (Arndt et al. 1982, Krause et al. 1983, Prinz et al. 1982, Zoettl and Huettl 1986).

Proton concentrations higher than would be expected due to the carbonic acid equilibrium in the precipitation result primarily from the oxidation of SO_2 and nitrogen oxides (NO_x) to sulfuric and nitric acids. Whereas on a global level the natural sulfur and nitrogen emissions account for 30–40 percent and 30–70 percent, respectively, of the total emissions, they contribute less than 10 percent to West Germany's emissions (Umweltbundesamt 1986). Because of the increased anthropogenic emissions, a rain pH of 4.0–4.5 can be estimated that is today considered "normal" in central Europe (Winkler 1983).

According to Winkler (1985), pH measurements indicate that the acidic load of rain has not much changed during the last four or five decades (cf. also Clarke and Lambert 1987, Lacaux et al. 1987). However, owing to the drastic reduction of alkaline particulate emissions, the atmospheric buffering capacity has been much decreased. Schenck (1988) computed that today only 5.0 percent of the amount of Mg is deposited as earlier. The primary source of ammonium (NH_4) in precipitation is ammonia (NH_3) emissions, which are generally very high in areas of high livestock density. It is important to note that precipitation pH is elevated due to the partial neutralization of protons by ammonia ($NH_3 + H^+ \rightarrow NH_4^+$). Data from various regions of West Germany are presented here to characterize atmospheric deposition loads, concentrations, and amounts as well as regional variation.

Gaseous Air Pollutants

Sulfur dioxide is a much discussed air pollutant in the context of the new types of forest damages. But owing to the spatial distribution of SO_2 deposition it cannot be seen as a single causative agent. Concentrations of SO_2 are highest in the industrial and densely populated areas, such as the Ruhr area of Northrhine-Westphalia. But the most pronounced forest damages are found in remote areas. In addition, Kandler (1985) showed that SO_2 concentrations in urban areas have decreased significantly by describing the recolonization of SO_2-sensitive lichen species in the city of Munich. The overall decreasing SO_2 emission trend can be seen in figure 2.1. Recent German federal regulations should reduce SO_2 emissions by 70 percent in 1995, with 1982 as the baseline (Umweltbundesamt 1986).

Measurements of annual mean ambient SO_2 concentrations range from 60 micrograms per cubic meter ($\mu g\ m^{-3}$) in industrial areas (Ruhr area: Ixfeld et al. 1985) to 35 $\mu g\ m^{-3}$ in central mountains influenced by industry, for example,

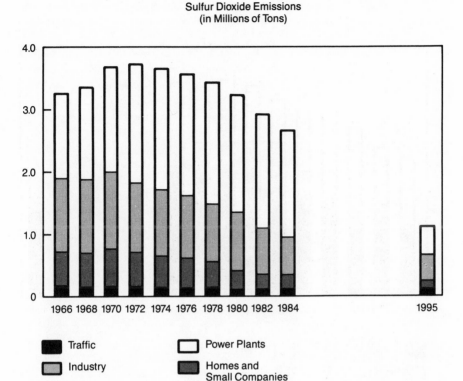

Figure 2.1. SO_4 emissions in West Germany by emission sources (1966–95). Source: Umweltbundesamt 1986.

Egge Mountains (Pfeffer 1985) and the northern and middle Black Forest, to less than 10 µg m^{-3} in the southern Black Forest and the southern Bavarian Forest (both remote, so-called clean air areas) (cf. Oblaender and Hanss 1985). But the remote Fichtel Mountains are heavily influenced by SO_2 emissions from Czechoslovakia. There, annual mean SO_2 concentrations of 60 to 80 µg m^{-3} are reported, with peaks (greater than 500 µg m^{-3}) particularly during wintertime.

Owing to the high density of automobile traffic, NO_x emissions are more evenly distributed than SO_2. Nevertheless, because of the dilution factor and atmospheric transformation processes during transmission from highly populated and industrialized areas to the forest damage areas, the deposition load is relatively low. This is especially true for nitric oxide (NO), which is readily oxidized to nitrogen dioxide (NO_2). The ambient NO_x concentrations are generally about 50–65 percent that of the SO_2 concentrations. However, longer measurement periods are missing. As figure 2.2 indicates, the overall increasing trend of NO_x in West Germany has leveled off since about 1980. The scheduled reduction of NO_2 should reach 40 percent by 1995, again with 1982 as the

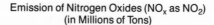

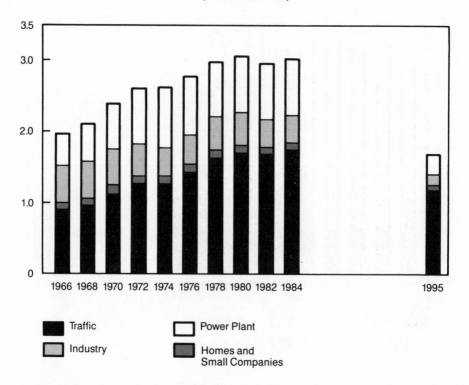

Figure 2.2. NO$_x$ emissions in West Germany by emission sources (1966–95).
Source: Umweltbundesamt 1986.

baseline (Umweltbudesamt 1986). Overall NO$_x$ emissions have increased since
the 1970s in central Europe (Sartorius 1984).

No real trend is clear for emission data of VOC (volatile organic compounds)
(figure 2.3). It must be emphasized that the measurement of VOC is rather complicated and the data presented might be rather low (Umweltbundesamt 1986).
Furthermore, only scant data are available for forest decline areas.

The only investigated gaseous air pollutant that shows a clear trend of
increasing concentrations with increasing distance from emission sources as
well as with increasing concentrations at higher geographical altitudes is O$_3$.
This is indicated by measurements of Oblaender and Hanss (1985) in southwestern Germany, as can be seen in figure 2.4. Ozone is a secondary air pollutant
that is formed under the influence of ultraviolet light in photochemical reactions
involving NO$_x$ and reactive hydrocarbons. In some regions of southern Germany,
O$_3$ concentrations increased during the last two decades (Attmannspacher et al.

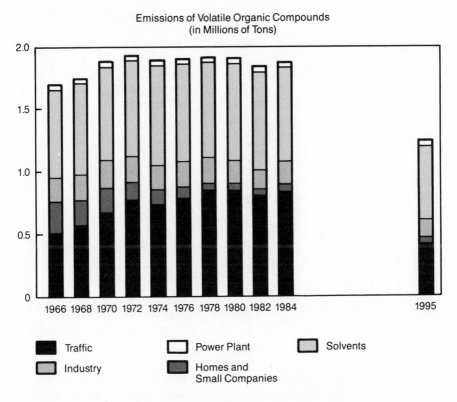

Figure 2.3. VOC (volatile organic compound) emissions in West Germany by emission sources (1966–95). Source: Umweltbundesamt 1986.

1984, Warmbt 1979). Over the past 25 years, O_3 concentrations have been found to increase even during winter (Feister and Warmbt 1984). In 1980 the monthly mean O_3 concentration at the Schauinsland station in the southern Black Forest was, for example, 113 $\mu g \ m^{-3}$ and thus in the range of concentrations measured in the United States (Ashmore et al. 1985). In severely affected mountainous areas of Northrhine-Westphalia, O_3 concentrations are generally higher by a factor of 2.0 to 2.4 compared with the Ruhr area, and can reach monthly means during the summer of 100 $\mu g \ m^{-3}$ (Pfeffer 1985). High O_3 concentrations with peak values greater than 600 $\mu g \ m^{-3}$ have been registered under favorable weather conditions in 1976 (Becker et al. 1983). According to Krause et al. (1986), O_3 maximum concentrations in 1976, 1980, 1982, and 1983 coincide to some degree with the increase in forest damages observed in most areas of West Germany (cf. Arndt et al. 1982, Prinz et al. 1982).

Krause et al. (1986) also indicated that at higher elevations O_3 concentrations are high enough to cause injury to O_3-sensitive plant species. Similar

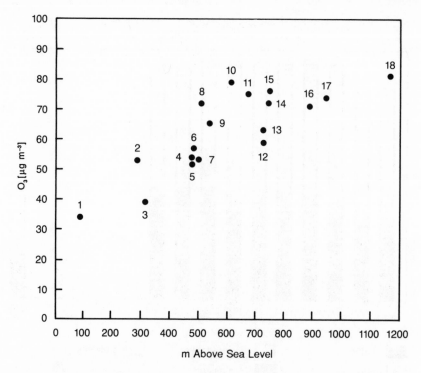

Figure 2.4. O_3 concentration and altitude in southwestern Germany. Source: Forschungsbeirat für Waldschaeden/Luftverunreinigungen der Bundesregierung und der Laender 1986.

findings have been reported from other parts of Europe by active monitoring techniques (Posthumus and Tonneijk 1982, Skaerby and Sellden 1984). Besides the acute injury symptoms suffered by sensitive plants, even more resistant tree species such as spruce and fir may be affected on the cellular level and in combination with other pollutants as well as with various site factors (Krause et al. 1985).

Wet and Dry Deposition

"Acid rain"—often used as a synonym for wet and dry atmospheric deposition—has been hypothesized to cause damages to forest ecosystems. Dry and wet deposition are the two major sinks for emissions from natural and anthropogenic sources. Rain analyses show the presence of cations such as H^+, NH_4^+, Ca^{2+}, Mg^{2+}, K^+, and Na^+ as well as anions such as SO_4^{2-}, NO_3^-, and Cl^-. As can be seen from table 2.2, bulk deposition in open-land precipitation varies to a great extent in West Germany.

From a historical viewpoint, three major changes have taken place during

Table 2.2. Element input by open-land bulk precipitation in West Germany.

Location (region)	Elevation (a.s.l.) (m)	Rainfall (mm)	H⁺	Na	K	Mg	Ca	SO₄-S	NH₄-N	NO₃-N	Cl
							$kg \cdot ha^{-1} \cdot a$				
Baden-Württemberg (Southwestern Germany)											
Schauinsland (Black Forest)	1200	1330	0.51	2.8	1.6	0.6	4.4	10.8	5.7	6.1	5.5
Kaelbelescheuer (Black Forest)	900	1560	0.83	4.1	2.8	0.9	5.4	15.6	9.5	7.8	5.2
Rotenfels (Black Forest)	710	1170	0.54	—	2.6	1.0	4.2	12.8	6.4	6.9	18.4
Schoenbuch (Swabian Mountains)	450	810	0.30	—	3.3	0.8	2.6	8.1	5.6	4.8	13.3
Bavaria (Southeastern Germany)											
Bodenmais (Bavarian Mountains)	1215	1490	0.46	7.3	7.2	1.6	8.9	28.2	11.5	9.1	—
Oberwarmensteinach (Fichtel Mountains)	760	1420	0.94	—	13.7	4.1	12.2	37.4	13.9	3.4	25.6
Wuelfersreuth (Fichtel Mountains)	680	1110	0.58	—	8.0	2.1	9.0	27.4	8.5	2.3	8.1
Northrhine-Westphalia (Northwestern Germany)											
Elberndorf	570	1480	0.74	11.8	4.4	1.5	10.4	20.7	7.0	8.4	20.7
Monschau	515	990	0.50	8.9	4.0	2.0	7.9	15.5	7.7	5.6	15.9
Haard	70	750	0.68	8.9	3.7	1.4	7.4	18.7	7.2	5.4	24.2
Lower Saxony (Northeastern Germany)											
Lange Bramke (Harz Mountains)	600	1270	0.73	9.1	3.2	1.7	6.9	25.0	14.8	8.9	16.6
Solling	500	940	0.81	7.7	3.3	1.9	10.4	23.8	10.9	7.7	16.9
Lueneburger Heide	80	770	0.54	7.6	2.2	1.1	4.5	19.1	7.4	5.4	15.0

Source: Mies 1987.

the last two decades. To some extent sulfate was replaced by nitrate in rainwater (figure 2.5). Particulate emissions such as Mg^{2+} and Ca^{2+} have been much reduced (see figure 2.6). Nitrogen deposition, both NO_3^-N and NH_4^-N, has increased remarkably.

Although often anticipated, a significant change in rain pH has not occurred in the past four to five decades, having reached a point of "acid saturation" in 1940 at a pH of about 4.2 (Winkler 1982). As demonstrated already, forest decline has developed rapidly during recent years, although few spatial differences in rain pH values throughout central Europe are observed (Winkler 1983).

When one discusses atmospheric deposition in the context of forest ecosystems, various important factors have to be taken into account. Forests, particularly coniferous stands, act as filters that can absorb relatively large amounts of dry deposition. Various chemical reactions may take place between deposited substances and the tree surfaces, for example, foliar leaching and uptake of nutrient ions. Thus, ion content of throughfall might be quite different from open-land bulk precipitation (see table 2.3). Owing to the large relative surface area and long residence time of fog droplets, this form of wet deposition has a much greater ionic load in comparison to rain (Schrimpf et al. 1984), and pH values below 3.5 are observed frequently (Fuzzi et al. 1983, Kaminski and

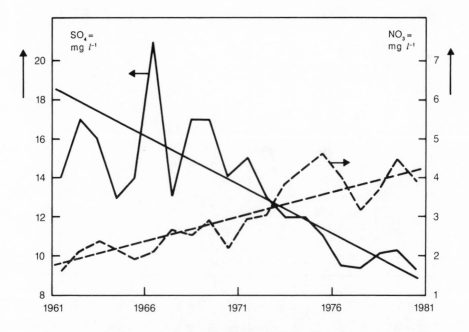

Figure 2.5. Temporal development of SO_4 and NO_3 contents in rainwater of 1961 to 1981 at Delft, Netherlands. Source: Forschungsbeirat für Waldschaeden/Luftverunreinigungen der Bundesregierung und der Laender 1986.

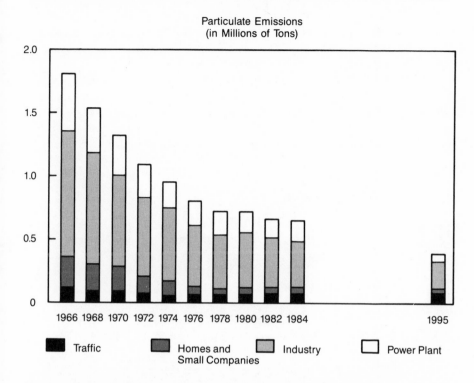

Figure 2.6. Particulate emissions in West Germany (1966–95). Source: Umweltbundesamt 1986.

Winkler 1983). Since fog episodes are much more frequent at higher elevations, impacts on vegetation and soil have been anticipated (Flueckiger 1986, Ulrich and Matzner 1983, Zoettl and Huettl 1986).

Increased input of N compounds into forest ecosystems has become the subject of considerable concern (Nihlgard 1985), although higher N deposition might also have positive effects on forest growth (cf. Kenk 1985).

Finally, it must be mentioned that forest ecosystems act also as a sink for heavy metal deposition. Many authors (for example, Mies 1987) have indicated that heavy metal input is accumulated at different levels in the forest ecosystem. For example, lead (and copper) are primarily stored in the forest canopy, whereas manganese (and zinc) contents are much higher in throughfall as compared with open-land precipitation, indicating high foliar leaching rates. However, heavy metals such as lead and cadmium are also accumulated in the organic matter of the forest floor. Table 2.4 shows that heavy metal deposition in German forests is moderate and varies to a great extent (Zoettl 1985a). No correlation exists between forest damages and heavy metal deposition, but the accumulation of heavy metals in the forest ecosystem implies a potential hazard.

Table 2.3. *Element input by throughfall and input differences of throughfall versus open-land bulk precipitation in West Germany.*

Tree species (ID)	Rainfall (mm)	H+	Na	K	Mg	Ca	SO$_4$-S	NH$_4$-N	NO$_3$-N	Cl
						kg · ha^{-1} · a				
Norway spruce	840	1.0 (60)	—	30.3 (244)	3.6 (45)	15.8 (80)	37.5 (40)	7.3 (−21)	2.3 (−4)	13.9 (26)
Norway spruce	690	1.9 (197)	16.5 (54)	25.7 (424)	3.9 (160)	22.8 (171)	61.0 (222)	19.1 (79)	15.1 (110)	45.4 (103)
Norway spruce	750	3.2 (295)	16.9 (117)	29.6 (722)	4.8 (153)	32.8 (212)	87.0 (267)	14.8 (24)	14.8 (85)	38.4 (125)
Norway spruce	510	0.5 (108)	—	15.5 (384)	2.5 (218)	15.8 (365)	23.1 (178)	8.0 (51)	10.3 (129)	18.8 (40)
European beech	870	0.9 (10)	10.4 (33)	19.0 (428)	3.1 (63)	19.7 (88)	34.0 (43)	10.3 (−13)	10.3 (9)	23.9 (40)
European beech	680	0.5 (−27)	12.4 (16)	17.9 (265)	2.6 (73)	14.2 (69)	25.0 (32)	11.6 (8)	8.3 (15)	28.5 (27)
Mixed spruce, fir, beech stand	1060	1.1 (61)	10.3 (167)	24.4 (874)	3.5 (331)	12.8 (177)	25.5 (97)	7.2 (−12)	13.1 (89)	13.8 (123)

Source: Mies 1987.
Note: ID = input difference in percent.

Table 2.4. *Heavy metal input by throughfall and input differences of throughfall versus open-land bulk precipitation at various sites in West Germany.*

Tree species (ID)	Rainfall (mm)	Mn	Zn	Cu	Pb	Cd
		$g \cdot ha^{-1} \cdot a^{-1}$				
Norway spruce	1460	1200	302	24	77	11.4
		(1100)	(74)	(7)	(−40)	(165)
Norway spruce	1000	555	117	36	42	0.8
		(620)	(125)	(−21)	(50)	(86)
Norway spruce	690	8110	—	368	108	7.9
		(3490)		(17)	(−27)	(49)
European beech	680	5140	—	354	110	4.9
		(2170)		(13)	(−25)	(−8)
Mixed spruce, fir, beech stand	1060	910	300	43	100	3.1
		(570)	(63)	(−3)	(7)	(44)

Source: Mies 1987.
Note: ID = input difference in percent.

Hypotheses and Some Research Results

Numerous hypotheses have been put forward on the new type forest damages. Only the most important and substantial hypotheses are briefly discussed here.

Acid Rain and Soil Acidification

According to Ulrich (1986)—the pioneer in analyzing the new type forest damage problem—acid deposition can be noticed in the soil in two different ways. It can provoke a decrease in soil alkalinity, such as through a reduction of exchangeable nutrient contents such as Ca, Mg, and K. This was demonstrated, for example, by Zezschwitz (1985). Acid deposition can also cause an increase of acidity in the soil. The findings of the Solling project (since 1969) and of the watershed Lange-Bramke study in the Harz Mountains (since 1977) indicate that the present soil acidification partially due to the impact of acid deposition is marked by both mechanisms. Soil acidification results in decreased pH values only when the acids pass over into the soil solution via solution or cation exchange. Therefore, the base content of the soil may decrease without having an impact on the pH values.

The load of strong acids (pH less than 5) will lead to a decrease of base saturation and to an accumulation of strong acids when the rate of acid loading exceeds the rate of proton consumption due to mobilization of basic cations during the process of silicate weathering. It has been shown that the rate of proton consumption by silicate weathering is in many cases already surpassed by the proton production due to forest growth. Hence, any additional acid loading via acid deposition must lead to soil acidification. Indeed, numerous authors (for example, Tamm and Hallbaecken 1986) report that increased acid deposition causes accelerated soil acidification. The situation is intensified when acid deposition results in decomposition of the nitrogen storage in the soil.

The extent of soil acidification depends on the deposition rate and duration. As mentioned, Winkler (1982) indicated that over the last 40 to 50 years no trend of pH values in rainwater was recorded for central Europe. Thus, precipitation deposition implies an annual input of 0.8 kmol of H^+ per hectare at least since 1930. Of course, in areas with particulate contamination this rate would be considerably decreased.

From throughfall data available since 1965 for the Kaufunger Forest and since 1968 for the Solling, Ulrich postulates that the current acid deposition load has existed at least for the last two decades. It has increased ever since industrialization started. Thus, forest ecosystems might have experienced increasing rates of acid deposition for a long time in central Europe.

In spite of decreasing amounts of alkaline cations such as Mg^{2+}, Ca^{2+}, and K^+, enhanced soil acidification leads to the mobilization of phytotoxic metal ions, for example, Al^{3+}. Thus, soil acidification in forest ecosystems— particularly when acidification pulses occur with high concentrations of acids and low Ca:Al, Mg:Al ratios—may lead to root damages. Stienen et al. (1984) showed that the Al concentration in fine roots of damaged spruces increased. Simultaneously the uptake of Ca and especially of Mg was much reduced. Thus, acid toxicity may lead to nutrient disorders such as Mg deficiency. Also, water conductivity was reduced, which may lower drought resistance.

Decreased root biomass, particularly of fine roots due to accelerated soil acidification, may lead to the same impairments. Comprehensive root studies were carried out by Murach (1984), Jorns and Hecht-Buchholz (1985), Stienen et al. (1984), Puhe et al. (1986), and Schulte-Bisping and Murach (1984) indicating that declining trees were marked by root damages such as reduced fine root growth, reduced mycorrhiza symbiosis, and overall reduced root biomass. In addition, root damages were also found on a histological level. However, the hypothesis that soil acidification causes root damage due to high concentrations of H^+ and/or Al^{3+} is controversial (Bauch 1983, Rehfuess 1983a, Zoettl 1983, Zoettl and Huettl 1986).

So far there has been no experimental proof of acid toxicity related to forest tree species under realistic site conditions (Krause 1987, Makkonen-Spiecker 1984). But it was clearly demonstrated that Al^{3+} has an antagonistic effect on Mg and Ca uptake. Furthermore, observations of forest damages on soils where the chemical soil reaction makes excessive mobilization of toxic metal ions impossible show that acid toxicity cannot explain new type forest damages at all sites. Nevertheless, high Al^{3+} concentrations are definitely an important negative factor in acidic soils.

Direct Impacts of SO_2 and NO_x

Sulfur dioxide heads the list of classic air pollutants with a potential effect on vegetation. As indicated, SO_2 emissions have been decreasing for some years in central Europe. Owing to the demonstrated SO_2 concentration gradient from urban to remote areas, no correlation exists between SO_2 concentrations in ambient air and new type forest damages. In most areas where forest damages are observed, SO_2 concentrations are below the well-established International Union of Forest Research Organizations standards, thus implying values for protection of trees (IUFRO 1983). Thus, direct SO_2 damages can be eliminated as an overall cause for new type forest damages. This conclusion is underlined by histological research findings. Fink (1986) demonstrated that phytotoxic SO_2 concentrations cause a destruction of the mesophyll tissue close to the stomata, where all gaseous air pollutants enter the forest tree foliage. But no such tissue damages were found in declining forest trees when researchers investigated all important forest damage areas in West Germany.

Furthermore, needle analysis carried out in areas where new type forest damages are present indicate that S contents rarely exceed normal concentrations of 800–1200 µg per gram dry matter (Forschungsbeirat Waldschaeden/ Luftverunreinigungen der Bundesregierung und der Laender 1986).

As discussed already, NO_x emissions have increased markedly during the 1970s in most of Europe. However, gaseous N compounds have a rather low phytotoxic effect (Taylor et al. 1975). Again, as ambient NO_x concentrations decrease from urban to remote areas, NO_x can be eliminated as a possible single cause of new type forest damages.

It can therefore be concluded that SO_2 and NO_x cannot be seen as single causes of the observed new type of forest damages. However, SO_2 and NO_x might cause chronic damages in forest trees that are not yet fully understood. Furthermore, SO_2, NO_x, and other gaseous air pollutants are much more phytotoxic when they occur in combination than as single agents (Krause 1987).

The Ozone Hypothesis

Ozone is a secondary air pollutant. Ultraviolet light, NO_x, O_2, and reactive hydrocarbons are necessary to form ozone. Owing to increased NO_x and VOC emissions, various authors have suggested that O_3 together with other photooxidants might have reached ambient air concentrations high enough to injure both highly sensitive and less sensitive plant species (Arndt et al. 1982, Bosch et al. 1983, Guderian et al. 1985, Krause et al. 1983, Prinz et al. 1982, Skeffington and Roberts 1985). In addition, more resistant tree species such as spruce and fir may be damaged on a cellular level by ozone, particularly in combination with other air-pollution-related factors such as acid precipitation, especially acid fog.

The primary site of action for O_3 is cell membranes (Krause et al. 1986). As O_3 might cause higher membrane permeability, leaching rates of inorganic ions that are rather mobile in the cellular liquid might be increased (Keitel and Arndt 1983). However, it must be emphasized that symptoms comparable to those found in the field have not yet been reproduced in laboratory experiments. But enhanced nutrient leaching was frequently observed in young trees when O_3 fumigation was combined with misting experiments (Bosch et al. 1986, Krause et al. 1983, Krause et al. 1986). These mechanisms might finally lead to nutrient deficiencies on rather poor and/or acidic substrates. Miller et al. (1969), when investigating the well-known O_3 impacts in pine trees of the San Bernardino Mountains in southern California, found reduced production of assimilates associated with the impact of O_3. This may consequently lead to reduced growth of fine roots before the leaf area is affected.

Histological investigations into this problem showed that O_3, when causing acute damages to forest tree foliage, also leads to the destruction of the mesophyll tissue rather than to damages within the vascular bundle (Fink 1986). Because damages of the vascular bundle are frequently observed in yellowed needles of damaged spruces, acute O_3 injuries were ruled out by Fink (1986) when he considered new type forest damages in West Germany. Since O_3 concentrations are not elevated in other countries of central Europe, these findings can be transferred to all forest damage areas in this region. But based on these findings, chronic O_3 damages cannot be ruled out. Indeed, Bosch et al. (1986) showed that O_3 in combination with acid fog, extreme frost events, and poor, acid soil enhances tree damages substantially.

Of all investigated gaseous air pollutants in areas of new type forest damages, only O_3 reveals a positive correlation with the spatial and temporal development of damages. In addition, O_3 has the highest degree of phytotoxicity based on concentration and duration of exposure. It is therefore reasonable to assume that photooxidants may play a role, at least regionally, in causing this phenomenon.

The Nitrogen Hypothesis

Nihlgard (1985) suggested that increased nitrogen deposition might result in forest damages. The increased combustion of fossil fuels, particularly at high temperatures, leads to increased emissions of NO_x-N. Input of nitrate in forest ecosystems can lead to the depletion of the base saturation in the soil, to reduced microbial activity (Haertl and Cerny 1981), and to increased ammonium fixation (Huettermann and Ulrich 1984). Agricultural activities such as N fertilization and high livestock densities result in increased deposition of NH_3-N/NH_4-N. In forest ecosystems, this may lead to unbalanced nutrient supply and nutrient deficiencies (Bosman et al. 1987, Flueckiger 1986, Friedland et al. 1984, Huettl 1986, Roelofs et al. 1985). High nitrogren soil contents may negatively influence the fine root mycorrhiza symbiosis and thus tree vitality might be affected (Meyer 1985, Mohr 1986, Nihlgard 1985).

Schulze et al. (1987) showed a negative correlation between Mg and N needle contents for declining spruce forests in the Fichtel Mountains. The highest N contents were found in spruces with lowest Mg contents. These trees were marked by pronounced tip-yellowing. Occasionally spruces were found with low Mg and N contents. These trees did not exhibit any decline symptoms but appeared green and healthy.

Huettl (1987) simulated increased nitrogen deposition by adding N fertilizers, both NO_3-N and NH_4-N, to younger Norway spruce trees in stands marked by low but still sufficient Mg needle contents. As expected, the increased nitrogen supply induced a pronounced Mg deficiency coinciding with foliar analysis results and tip-yellowing symptoms. These trials indicate that increased nitrogen deposition might provoke nutrient deficiencies. Of course, it is important to realize that increased nitrogen supply generally leads to increased growth of trees. Therefore only when one or more specific nutrients are in limited supply might nutrient deficiencies be induced. In the case of increased NH_4 deposition, this might be closely related to nutrient uptake antagonisms such as $K:NH_4$, $Mg:NH_4$, and $Ca:NH_4$.

As nitrogen deposition rates vary between 15 and 200 kg per hectare and year in central Europe, further research is needed to determine potential links between nitrogen deposition and forest decline development.

The Virus Hypothesis

The observation that decline symptoms (particularly needle discolorations in coniferous trees) would occasionally spread from one tree or a small group of trees into larger areas within a stand led to the hypothesis that this phenomenon might be caused by a disease due to infections by viruses or microorganisms (Ebraim-Nesbat and Heitefuss 1985, Frenzel 1983, Kandler 1983, Nienhaus 1985a). It was also suggested that the phloem damages detected in yellow needles via histological investigations could be caused by the infection of phloem-specific biotic factors (Fink 1983, Fink and Braun 1978, Parameswaran et al. 1985). So far, research results indicate that viruses or virus-like particles were detected in forest soils as well as in water samples obtained from forest sites. In addition, viruses and virus-like structures were found in needles of declining spruces, firs, and pines. Also, in leaves as well as in the rooting area of older beeches, virus-like particles have been isolated (Nienhaus 1985a, 1985b). According to Nienhaus (1986), microplasm-like organisms have not been found in conifers or beeches. Ebraim-Nesbat and Heitefuss (1985) isolated rickettsia-like organisms in the woody tissue of fine roots of spruces.

The detected microorganisms and viruses are known to cause various diseases in herbaceous and woody plants. However, no evidence is available indicating that viruses or microorganisms might be primary factors related to the observed forest damages. Also, according to the spatial and temporal development of the damages, this type of biotic disease is probably not a cause of this phenomenon (Nienhaus 1985a, 1985b).

These findings were substantially underlined by grafting trials carried out by Mehne (1987). Grafting is a classic method to test whether disease symptoms are caused by biotic factors such as viruses or microorganisms. In the case of virus diseases, transference from the symptomatic tissue of the scion to the healthy tissue of the plant should occur after successful development of mutual tissue. If the symptoms of the scions are primarily caused by infectious agents, the healthy tissue should develop the same symptoms. Since spring 1984 about

3000 graftings have been applied for a variety of different decline symptoms, as scions were sampled from all major decline areas in West Germany. In not one single case was transference of symptoms observed. But whenever graftings of scions marked by discoloration symptoms were successful, the damaged scions regreened and stayed healthy for all subsequent growth periods. Thus, classic grafting experiments also indicate that the frequently observed discoloration symptoms in stands revealing new type forest damages are due to nutritional disturbances, as shown by Huettl and Mehne (1988).

Of course, it is not possible at present to rule out a predisposing effect of virus and microorganism infestations completely. But further assessments of the potential importance of viruses or microorganisms in relation to forest damages are possible only when the results of successful retransference trials are available (cf. Forschungsbeirat für Waldschaeden/Luftverunreinigungen der Bundesregierung und der Laender 1986).

Site-Specific Declines due to Varying Sets of Stress Factors

The following hypothesis is primarily based on the observation of recent disturbed nutrient supply in many forest ecosystems marked by varying site and stand characteristics. The most common nutrient disorder is the magnesium deficiency, which is a truly new phenomenon in older Norway spruce stands of middle- and high-elevation forests in central Europe. This damage type has been described by many authors, as detailed already. Other nutritional disturbances of concern are calcium, potassium, manganese, and zinc. Huettl and Zoettl (1985), Zoettl and Huettl (1985), Bonneau and Landmann (1986), Reemtsma (1986), and others use historical needle analysis comparisons to show that many of the observed nutrient deficiencies have developed within a rather short time period within large forest areas in central Europe. The following complex hypothesis has been suggested to explain these historical changes of the nutritional status, and it is probably the most accepted explanation of the real new type forest damages in central Europe.

Owing to air pollutants and acid precipitation, particularly acid fog, the foliar tissue of forest trees might be damaged, resulting in enhanced leaching of nutrient ions from the canopy. The nutrient elements of concern are Mg, Ca, K, Zn, and Mn because these elements are mobile in the phloem or easily exchangeable. In reaction, the affected trees try to compensate for these higher leaching losses by increased nutrient uptake from the soil. Depending on the geological parent material and the element availability, this effort will be more or less successful.

In addition, soil acidification occurs naturally under humid climatic conditions and is enhanced by forestry practices applied in central Europe, particularly in coniferous forest stands. Soil acidification, however, is probably accelerated by long-term increased anthropogenic acid deposition (also due to the recent heavy reduction of alkaline particulate emissions), causing enhanced nutrient leaching from the rooted soil horizons. Special attention must be drawn to the rhizosphere. A stimulated nutrient turnover of cations in the forest ecosystem will lead to higher rhizosphere acidification because the tree has to release, at least to some degree, protons when taking up increased amounts of Mg^{2+}, Ca^{2+}, and K^+ in order to keep the internal anion and cation balance. Depending on the chemical reaction of the substrate, well-known antagonisms will impede nutrient uptake. On acidified substrates Mg and Ca uptake might be affected by high Al concentrations. On more alkaline soils a higher release of protons will immediately be buffered, leading to enhanced Ca concentrations in the nutrient solution of the rhizosphere and provoking a less favorable K:Ca ratio, which then might cause reduced K uptake.

Another important factor in this context is increased nitrogen input into declining forest ecosystems. In particular, NH_4-N is a strong antagonist for all other nutrient cations. In the case of nitrogen nutrition being shifted from the NO_3-N source to the more favorable NH_4-N source, nutrient imbalances might be induced. To some degree this problem might be enhanced by stimulated tree growth because of higher N availability on substrates where Mg, K, or Ca supply is limited. If we take all these considerations into account, this complex hypothesis is applicable to all "recent" nutrient deficiencies or imbalances, whether they occur in forest trees/stands on more alkaline substrates or more acidic soils.

A tree's ability to compensate for the increased nutrient losses from the phyllosphere pedosphere by higher uptake from the soil depends on the intensity of the leaching losses from both the canopy and the soil as well as from the litogenic weathering supply and the atmospheric deposition rates of the discussed nutrient elements. It is therefore to be expected that a Mg deficiency will occur in parent materials that are, for example, primarily poor in plant-available Mg. When the upper soil is poor in K and/or the K uptake is impeded by competing cations (such as NH_4^+, Ca^{2+}), acute K deficiency situations will develop.

Thus, the frequently poor nutrient supply of forest soils is an important predisposing stress factor. As causes of the observed new type forest damages, the direct as well as primarily the indirect impacts of air pollutants on forest ecosystems are seen. Biotic infestations and climatic extremes are considered as accompanying factors (Zoettl and Huettl 1986; cf. Bosch et al. 1983, Kaupenjohann et al. 1987, Rehfuess 1983a).

This hypothesis was tested under field and laboratory conditions (Bosch et al. 1986, Huettl 1985). Very promising results were derived from diagnostic fertilization and liming trials. Results of these investigations are presented in the section on revitalization and restabilization of declining forest ecosystems.

Further Causal Aspects

In addition to these major hypotheses on the causes of new type forest damages, a number of further causal aspects have been suggested. Cramer and Cramer-Middendorf (1984) and Rehfuess and Bosch (1987) as well as various other authors have stated that climatic factors—such as a series of rather dry and warm years since 1976 or frost episodes (for example, drastic temperature changes)—might account at least on a regional scale as inciting factors for new type forest damages. Again, these climatic stresses are considered as interactive processes associated with other negative impacts on forest ecosystems.

Besides the virus hypothesis, other biotic diseases have been mentioned to explain part of the observed damages. As described earlier, the observed needle necroses in older stands in southern Germany are correlated with the infestation of various needle cast fungi. However, Butin and Wagner (1985) showed that a considerable portion of investigated necrotic needles taken from spruces exhibiting this decline type was not affected by fungal diseases. Thus in this case, too, part of the necrotic needle biomass might be due to abiotic factors.

Rehfuess has repeatedly stated (for example, in 1987) that due to the extensive planting of Norway spruce over about 200 years in many regions where new type forest damages are now most pronounced, planting stock has been used that is unsuited for specific site requirements. Off-site planting stock is generally much less resistant to many natural stress factors, and less vital trees should be much less resistant to air pollution impacts. Indeed, Rehfuess proved this relationship in the Bavarian Forest by comparing naturally regenerated spruce stands that were next to planted spruce stands where off-site planting stock was utilized. The planted stands frequently showed severe damage symptoms, whereas the site-specific regenerated trees were healthy or revealed only slight injuries. On the other hand, however, other tree species such as European beech do not follow this pattern so clearly. For example, over about the last four years natural beeches in the Black Forest have also been heavily declining.

Isermann (1985), Zoettl (1985b), and others believe that former and recent forestry practices might at least in places have led to a pronounced destabilization of forest ecosystems. Forest management practices such as long-term litter racking, forest pasture, monoculture, establishment of off-site species, intensive harvesting, heavy forest road construction, and soil compaction due to heavy logging machinery were and often continue to be common in many forests of central Europe. But it must also be stated that damages do occur in forest ecosystems that can be considered undisturbed.

Hunting in central Europe, particularly in West Germany, Austria, and Switzerland, has a very long, highly respected history and is a rather prestigious pleasure in these countries. Hunting also represents an economic factor in land management business. It is therefore not surprising that wild game populations are high. Owing to high population density and the many anthropogenic disturbances of wildlife, wild game damages (for example, deer browsing) represent an important factor related to reduced vitality of forest trees and stands. A striking example of this is the impeded natural regeneration in alpine forests (Forschungsbeirat für Waldschaeden/Luftverunreinigungen der Bundesregierung und der Laender 1986).

Another hypothesis discussed in recent years suggests that radioactive emissions of nuclear power plants could be responsible for new type forest damages (Metzner 1985, Reichelt 1984). However, various studies (for instance,

Huettermann 1987, Schoepfer 1986) proved that no correlation exists between these potential emission sources and the temporal and spatial distribution of new type forest damages.

Schenck (1985) has put forward a hypothesis that has gained widespread recognition. He postulated that forest damages associated with nutrient deficiencies (particularly Mg) are caused by photodynamic processes in which primarily hydrocarbons, photooxidants, and radiation are involved. Another important factor in his hypothesis is reduced emissions of alkaline particles such as Mg and Ca. As mentioned earlier, Schenck computed that today only 5 percent of the amount of Mg is deposited in forest ecosystems of central Europe as previously. Besides heavy reductions in dust emissions, this is related to a historical change in the use of coal as a primary fuel source.

Concluding Remarks

The new type forest damages are characterized by many symptoms, of which only few are truly new. The recent damages cannot be related to one specific causal mechanism. The problem is complex and is probably caused by regionally varying sets of stress factors, both natural and unnatural in origin. However, most of the scientists involved in this research in central Europe consider anthropogenic air pollution an important factor in the cause of this serious phenomenon.

Means to Counteract This Phenomenon

Because "acid rain"—respectively, the related air pollutants (particularly SO_2 and NO_x)—is regarded as a major factor in provoking forest declines, it was considered essential to reduce emission of these substances.

Reduction of Emissions

As acid rain may cause extensive damages not only in forest ecosystems but also in other terrestrial ecosystems, in aquatic ecosystems, and on buildings (Wisniewski et al. 1987), this section documents some methods and plans related to the reduction of SO_2 and NO_x emissions in West Germany—the most progressive nation in this context. Reduction plans have also been established in Switzerland, Austria, France, and the Benelux countries, as well as in many other industrialized countries that experience the same or similar environmental problems.

The original Clean Air Act in West Germany was passed in 1964, but it had no real enforcement provisions. In 1974 this law was amended and all new stationary sources were required to install the best available control technology for SO_2 emissions. The current law, requiring installation of the best available control technology for SO_2 and NO_x at existing sources, was passed in 1983. The forest damages phenomenon was the primary motivating force behind the current law.

The final compliance date in the statute is April 1, 1993, with several interim steps. The primary targets for reduction are coal-fired power plants of 50 MW or larger. Additional requirements concern smaller units and other industries. With 1982 as the baseline, emission reductions are targeted at 70 percent for SO_2 (from 3.2 to 1.0 million tons) and at 40 percent for NO_x (from 2.9 to 1.7 million tons).

These reductions will be achieved by retrofitting all existing coal-fired plants that will continue in operation after April 1, 1993, with wet limestone scrubbers and either low-NO_x burners or selective catalytic reduction (SCR) equipment. The scrubbers will produce a usable gypsum end-product.

Current plans call for plants producing about 40,000 MW to be retrofitted and for those generating about 12,000 MW to be operated through March 31, 1993, and then retired. This is all the coal-fired capacity in West Germany not regulated under new source standards established in 1974.

The plan is to market the ash and gypsum produced. Currently all bottom ash and about 80 percent of the fly ash are used in the domestic construction industry. The fly ash not used is to a large extent shipped to either France or East Germany for disposal. West Germany currently produces 1 million tons of gypsum annually. This production will increase to about 4 million tons per year in 1993, with disposal of this excess gypsum a potential problem.

Although most plants will control NO_x by burner modification, some SCR equipment is being installed. Although this removes NO_x efficiently, it is extremely expensive and the ammonia by-products are difficult to sell or distribute.

Implementation of the 1983 law has been and will continue to be expensive. Using 1986 figures, the West German utility industry forecasts a capital cost of 11 billion DM (5 billion US$) for NO_x control and 17–20 billion DM (7.7–9 billion US$) for SO_2 control by April 1, 1993. Other industries anticipate a cost of 10 billion DM (4.5 billion US$) for combined SO_2/NO_x reductions. The utility industry alone will spend about 7700 DM (3500 US$) for every ton of SO_2 removed from the stack gases and about 9150 DM (4160 US$) per ton for NO_x removal.

To ensure the best use of the investments, all power plant retrofits are being conducted as part of repowering and/or life extension programs. Other plants not scheduled for such work will be retired.

There is no cost-sharing program to assist industry in paying for capital costs associated with the controls. The federal government has, however, been the principal source of funding for research on the type of equipment being installed. In a program similar to the Clean Coal Technology Reserve in the United States, the government provides the majority of funding for over fifty NO_x control projects. These projects are full-sized demonstrations in addition to laboratory bench studies or pilot plant projects. A similar effort was mounted for flue gas desulfurization (FGD), but only a small number of technologies were investigated due to the focus on yielding a usable by-product. This government program has enabled the industry to accelerate installation of control equipment by a decade or more.

As approximately 50 percent of the total NO_x emissions in West Germany are caused by car exhausts, efforts have been undertaken to reduce these emissions. Recent federal regulations grant considerable tax reimbursements to car buyers who purchase an automobile equipped with a catalytic converter because this equipment may reduce NO_x emissions by up to 90 percent. In addition, the unleaded gas used by cars with catalytic converters is less expensive than leaded gasoline. However, the catalytic converter has not (yet) been very well accepted by car buyers in central Europe (Umweltbundesamt 1986).

Revitalization and Restabilization of Declining Forest Ecosystems

To complement the reduction of air pollutants, some tools are available to directly alleviate various types of forest damages and to improve the vitality of forest trees and stands.

Fertilization and Liming. Owing to the frequently poor nutrient supply of forest sites, atmospheric deposition may result in a predisposition for damages. Therefore, in cases where acute nutritional disturbances are induced by air pollution, the improvement of the nutrient supply should reduce the damages. Under favorable conditions the symptoms should even disappear. A quick improvement of the nutrient supply is possible by the application of fast-soluble fertilizers. The application of lime appears useful (Wentzel 1959, Ulrich 1986, Ulrich 1972, Gussone 1984) to mitigate the effects of acid precipitation on the soil as well as to improve the chemical and biological soil conditions.

Diagnostic Fertilization Trials. To test whether appropriate fertilization with fast-soluble fertilizers can mitigate or remove the nutrient deficiencies attributed to air pollution, diagnostic fertilization experiments have been carried out since 1979 in West Germany (Huettl 1985, Huettl and Zoettl 1986, Isermann 1987, Kaupenjohann et al. 1987, Zech 1983). Three typical trials are presented in this section.

Mg Deficiency Trials: Diagnostic Fertilization Trial Elzach 10 (Huettl 1987). In fall 1983 the 12-year-old Norway spruce stand in the forestry district of Elzach (crystalline Black Forest) exhibited needle discoloration symptoms that indicated Mg deficiency (tip-yellowing in older needle-year classes). The spruces grow on an acidic brown earth derived from solifluction debris over granite. The humus form is mull. The slope is northwesternly exposed at 900 m above sea level and receives about 1500 mm precipitation annually.

The needle analytical results (table 2.5) are clearly correlated with the observed damages. They show a dramatic Mg deficiency. In the still green needles of the first whorl, relatively low Ca and Zn values were also found. All other

Table 2.5. Diagnostic fertilization trial in 12-yr-old Norway spruce stand: Elzach 10.

Elzach 10		Diagnostic fertilization trials							
	Deficiency threshold range	Control				To reduce Mg deficiency MgSO₄·H₂O (27% MgO)		To enhance N:Mg Imbalance (NH₄)₂SO₄ (21% N)
Element (mg g⁻¹ d.m.)		1983	1984	1985	1986	1984	1986	1986
N	12.0–13.0	16.0	16.2	17.8	14.6	15.1	14.2	15.2
P	1.1–1.2	2.6	2.5	2.3	2.4	2.6	2.3	1.9
K	4.0–4.5	7.5	8.6	7.9	6.4	10.1	6.2	6.6
Ca	1.0–2.0	2.4	3.0	3.2	1.7	2.4	1.6	1.5
Mg	0.7–0.8	0.20	0.32	0.31	0.28	0.56	0.56	0.18
Mn	0.020–0.080	0.640	0.69	0.83	0.75	0.74	0.43	0.52
Zn	<0.013	0.018	0.018	0.022	0.217	0.025	0.020	0.015
Element ratios	Balanced nutrition (range)							
N:P	6–12	6.2	6.5	7.7	6.1	5.8	6.2	8
N:K	1–3	2.1	1.8	2.3	2.3	1.5	2.3	2.3
N:Ca	2–7	6.7	5.4	5.6	8.6	6.3	8.9	10.1
N:Mg	8–30	80	51	57	52	27	25	84

Source: Author.

Note: Soil application of kieserite in spring 1984 and ammonium sulfate in spring 1985, 150 g per tree (600 kg ha⁻¹ of each fertilizer). Nutrient element contents in current needles (fall sampling; n = 10).

nutrients showed at least a sufficient supply. The soil analysis data indicated a poor Ca and Mg supply related to very low base saturation and extremely low Mg:Al and Ca:Al ratios.

To counteract the acute Mg deficiency, a part of this young spruce stand was fertilized with Kieserite (150 grams per tree, 160 kilograms of MgO per hectare) in spring 1984. Within 4 to 5 months, in fall 1984, the fertilized plot looked much more vital when compared with the control trees. To a certain extent the yellowed spruces had regreened, needle loss was reduced, and the current shoots were well developed and lushly green. The general visual improvement was verified by the needle analysis data.

In comparison with the control (the analytical data of the four sampling dates vary within the normal range), the Mg content of the fertilized trees almost doubled. The supply of all other investigated nutrient elements was optimal except for Ca. The positive fertilization effect continued in 1985 and 1986 (cf. table 2.5).

Even though the Mg supply improved remarkably, the Mg values were still below the deficiency threshold range. This finding coincides well with the incomplete disappearance of the discoloration symptoms in the trial trees.

As discussed earlier, increased anthropogenic N input, particularly NH_4-N, was thought to induce nutrient imbalances and thus provoke deficiencies. To test this hypothesis, ammonium sulfate (150 grams per tree, 600 kilograms per hectare) was applied to one of the trial plots in spring 1985. As indicated in table 2.5, the NH_4 application caused a further depression of the Mg uptake. This becomes especially obvious when we look at the foliar N:Mg ratio.

Diagnostic Fertilization Trial *Oberwarmensteinach* (Kaupenjohann et al. 1987). Based on successful revitalization of declining silver firs marked by Mg and K deficiency (Zech 1983), a 40-year-old Norway spruce stand with Mg deficiency symptoms in the forestry district of Oberwarmensteinach (Fichtel Mountains) was fertilized in spring 1983 with Kieserite (plot 1: 270 kg MgO per hectare) and with half-oxidized dolomite (plot 2: 3,000 kg MgO per hectare). The productive spruces grow on a brown-earth podzol derived from base poor parent material ("phyllite"). Needle analysis data (table 2.6) indicated a sufficient to good N, P, and K supply for the control spruces. The Ca contents were low and the Mg values clearly in the deficiency range.

As in Elzach, the Kieserite fertilization provoked within one vegetative period a pronounced improvement of the Mg contents as well as a much improved phenotype (table 2.6). Four years after the fertilizer application the Mg values of the fertilized trees had increased by more than 100 percent when compared with the control spruces.

In the beginning the dolomite plot showed a less pronounced increase of the Mg contents. Even though in comparison with the Kieserite plot the dolomite plot had received 11 times as much MgO, the delayed effect was probably due to the considerably slower solubility of the applied limestone.

Owing to the increase of the Mg contents into the range of optimal supply, a

Table 2.6. *Diagnostic fertilization trial in 40-yr-old*
Norway spruce: Oberwarmensteinach.

Element ($mg \cdot g^{-1}$ d.m.)	Control plot	Plot 1 (kieserite)	Plot 2 (dolomite)
N	14.0	14.4	16.2
P	1.9	1.8	1.7
K	6.1	6.6	7.1
Ca	1.7	2.8	1.7
Mg	0.6	1.0	0.8

Source: Kaupenjohann 1987.
Note: Soil application (spring 1983): plot 1: kieserite ($MgSO_4 \cdot H_2O$); 100 kg ha^{-1}. Plot 2: dolomite-limestone (MgO, $CaCO_3$; 30% MgO); 10.000 kg ha^{-1}. Nutrient element contents in current needles (fall sampling).

pronounced regeneration was observed in both plots. Four years after fertilization the revitalization effect was clearly visible and analytically evident.

Investigations of the mineral soil over a 1-year period showed a clear increase of Mg contents in the soil solution for the second year after fertilization and lime application. The liming resulted in an enhanced NO_3 leaching, whereas the Kieserite fertilization led to a small decrease of NO_3 contents in the soil solution. The application of Kieserite more than doubled the plant-available Mg content in the upper soil (0–30 cm). At this soil depth, 33 percent of the applied Mg was adsorbed in exchangeable form 18 months after Kieserite fertilization. These results coincide well with the investigations of Huettl and Zoettl (1986), who had found higher Mg contents and reduced Al^{3+} and H^+ adsorption at the exchange sites in the upper mineral soil due to Mg fertilization. Because of the higher dosage the exchangeable Mg content was increased even more obviously on the dolomitic lime plot. But here only 29 percent of the applied Mg was adsorbed in plant-available form.

K Deficiency Trial; Diagnostic Fertilization Trial Saulgau 2 (Huettl 1987). A further diagnostic fertilization trial in spruce was established in the forestry district of Saulgau in southwestern Germany. The 15-year-old Norway spruces grow on an organic soil. In the very dry and warm summer of 1983, typical K deficiency symptoms were observed in this stand. The visible symptoms were also correlated with data of chemical needle and soil analyses (K content of current needles, fall 1983: 2.6 mg per gram dry matter; cf. Huettl and Wisniewski 1987). Particularly striking was the extremely low K:Ca ratio of the exchangeable cations in the soil. Kalimagnesia (240 kg K_2O per hectare) was applied in single-tree fertilization in spring 1984 to improve nutritional status. Only a few months later the yellowed needles of almost all treated trees had remarkably regreened. Compared with the control plot, the K content of the fertilized spruces almost tripled in fall 1984 (table 2.7). The revitalization trend

Table 2.7. *Diagnostic fertilization in 15-yr-old Norway spruce: Saulgau 2.*

Plot	Sampling year	N	P	K	Ca	Mg	Mn	Zn
		mg · g⁻¹ d.m.					µg g⁻¹ d.m.	
Control	1984	18	3.9	2.4	3.2	0.7	70	12
Control	1986	16	3.1	2.2	3.7	0.9	66	12
Fertilized	1984	18	3.2	6.9	3.4	0.6	100	24
Fertilized	1985	21	3.0	7.3	5.4	0.8	130	31
Fertilized	1986	19	3.7	6.5	5.4	0.8	130	31

Source: Author.

Note: Soil application of Kalimagnesia (K_2SO_4, $MgSO_4$; 30% K_2O, 10% MgO) in spring 1984, 800 kg · ha⁻¹. Nutrient element contents in current needles (fall sampling; n = 10).

continued in 1985 and in 1986 (table 2.7). Thus, in this trial the vitality of declining trees was also remarkably improved by appropriate fertilization.

To investigate the character of the needle discoloration symptoms frequently associated with new type forest damages, diagnostic fertilization trials were recently initiated in France by Bonneau; in the Netherlands by van Diest, van den Burg, and Janssen; in Switzerland by Flueckiger; in Austria by Glatzel and Huettl; in Belgium by Delecour, van Praag, and Weissen; in Sweden by Andersson and Popović; in Canada by Tomlinson and Jones, and in the United States by Skelly, Pye, Zoettl, and Huettl.

Some Further Research Results. Isermann (1987) demonstrated that in addition to younger spruce stands, older Norway spruce stands suffering from Mg deficiency can also be revitalized with appropriate fertilization (cf. Kaupenjohann 1987). Comparable findings were obtained by Huettl and Zoettl (1986) from fertilization trials in older Norway spruce stands that were not sufficiently supplied with K.

Huettl and Fink (1988) indicated via microscopic investigations that the improvement of the phenotype of fertilized spruces coincides with a pronounced regeneration on a histological level. In contrast to the observed mesophyll tissue damages of SO_2-fumigated trees as well as of O_3-damaged conifers, Mg-deficient tip-yellowed needles of declining Norway spruce trees revealed damages within the vascular bundle. Above all the phloem-cells collapsed (see section on the ozone hypothesis).

To verify these observations, Fink grew young Norway spruce plants in a Mg-free nutrient solution to induce a Mg deficiency. By the end of the first vegetative period after transplanting, the trees showed the typical tip-yellowing. The histological investigation of these needles revealed the same anatomical damages as found for the tip-yellowed needles of declining trees at field sites. Electron microscopy showed for previously yellowed needles that had regreened due to the application of Mg fertilizers that new phloem cells adjacent to the necrotic phloem tissue had been formed by cambium activity, indicating the well-known regeneration potential of plant tissue. Thus, besides diagnostic fertilization trials the histological approach is a valuable diagnostic tool to differentiate direct air pollution impacts and tissue damages due to nutrient deficiencies.

In forest ecosystems affected by atmospheric deposition, root and mycorrhiza damages are frequently discussed, particularly in heavily acidified soils (for example, Murach 1984). The rapid uptake of the fertilized nutrients and the quick improvement of the nutritional status provide some evidence that the root systems of the trial trees were still able to function. The same must therefore be assumed for the mycorrhiza at the investigated sites (cf. Zoettl 1985b). However, the trials of Huettl (1985) also indicated that once a critical degree of damage is reached, no revitalization is possible.

An interesting nutritional aspect was produced via grafting trials by Huettl and Mehne (1987). By means of the classic method of grafting, Mehne tried to test whether the new type damages in spruce and other tree species are caused by microorganisms or viruses. Scions marked by discoloration symptoms typical of new type damages were harvested at various sites in West Germany and grafted onto healthy plants in order to examine a potential transference of viruses or microorganisms. These investigations indicated that the yellowed scions became green after the scions and the young healthy Norway spruce trees that were well supplied with nutrients had developed mutual tissue. This observation shows, as already stated in the section on the virus hypothesis, that the discoloration symptoms are probably not caused by viruses or microorganisms but are due to nutritional disturbances such as Mg or K deficiencies.

Experiences from Former Liming and Fertilization Trials. Finally the question arises whether the vitality of forest stands that have been fertilized or limed before the appearance of the new type forest damages would indeed be more vital today compared with the control plots. Generally, former fertilization trials were aimed at increasing wood production of forest stands. Therefore Mg fertilization was not of any major concern. Furthermore, because both Mg deficiency and poor Ca supply in high-elevation forests and K deficiency at certian moraine sediment sites must be recognized as new, the number of eligible trials is very limited.

Former Liming Trials. The evaluation of older liming trials produced varying results. Aldinger and Kremer (1985) did not find significant growth increases or improved vitality in older Norway spruce stands at sandstone sites of the Black Forest that were limed between 1964 and 1975. However, the application of lime containing Mg in this area, which since the late 1970s has been marked by Mg deficiency symptoms, impeded the appearance of tip-yellowing. Furthermore, growth of these stands compared with the control trees was improved.

In addition, Bauch et al. (1985) demonstrated better growth and smaller damages in different Norway spruce stands that had been treated with limes containing Mg between 1953 and 1980.

Kreutzer's investigations (1984) of older liming trials (to a certain extent lime with Mg had been applied)—in some cases combined with P, K, and Mg fertilization—indicated for spruce and pine stands in Lower Franconia an overall significant positive fertilization and liming effect in relation to new type forest damages. These stands were limed and/or fertilized between 1973 and 1979.

Former Fertilization Trials. Evers (1984) evaluated a fertilization trial in an older Norway spruce stand that had been established at a sandstone site in the Oden Forest at the end of the 1950s. More than 20 years after the application of Kalimagnesia, needle analysis data in the fall of 1982 still showed a positive fertilizer response. In contrast to the control spruces, tip-yellowing observed in this area since about 1980 did not occur in the KMg fertilizer plot. In addition, volume growth was increased due to Mg application. At comparable sites Evers found similar results in beech stands.

Baule (1984) demonstrated a remarkable vitality effect in a historical Norway spruce fertilization trial that was established on a sandstone site in Lower Saxony in 1964. An inventory of the crown biomass revealed high needle losses for the control spruces as well as for the NP plot. The NP+KMg plot was obviously (owing to the beneficial K and Mg fertilizer effects) almost undamaged. In this case, the visible vitality differences of the investigated spruces also coincided with chemical needle analysis data.

Based on the experiences from older liming and fertilization trials and on the results of the recently established diagnostic fertilization trials, it can be concluded that in all cases where air pollutants have a negative impact upon forest ecosystems an optimal and balanced nutrient supply is an inevitable necessity for the best vitality and resistance of forest trees and stands for any type of stress factor.

From Diagnosis to Therapy. The overall positive results of fertilizer and lime applications to remove and mitigate, respectively, forest damages probably caused by atmospheric deposition have led to a new federal regulation in West Germany. Forest fertilization and liming in stands revealing new type forest damages are subsidized by 80 percent of total costs (for fertilizers and application).

In practice, a distinction is drawn between compensative and meliorative fertilization. Compensative fertilization implies the application of lime with Mg (generally 3000 kilograms per hectare) to neutralize further acidic deposition in forest soils. Meliorative fertilization is focused on the specific application of nutrients, particularly in stands exhibiting acute nutritional disturbances. This is accomplished with fast-soluble Mg and K fertilizers applied in amounts of

500–1000 kilograms per hectare. In cases of insufficient P supply or to activate soil microorganisms, P fertilizers are utilized. Especially during meliorative fertilization, foliar and soil analyses are recommended.

Silvicultural Tools. In addition to the appropriate selection of site-specific planting stock and the establishment of mixed stands wherever possible, early stand management aimed at the production of vital trees seems essential. Furthermore, the stabilization of stand edges and the preservation of stands that have protection functions must be given special consideration. Another important practice is the establishment of seedlings under the shelter of older trees. This practice implies that tree species that are rather shade tolerant are planted in gaps that have been caused by new type forest damages. Erosion problems in former monocultural stands can be minimized by appropriate reforestations. Besides site-specific reforestation the planting of more resistent tree species, especially in areas where high concentrations of air pollutants are to be expected, seems advisable. Various research projects are therefore focused on breeding of resistant plants or plant provenances. As mentioned earlier, forest damages due to high game population densities must be drastically reduced, especially in alpine forests (Forschungsbeirat für Waldschaeden/Luftverunreinigungen der Bundesregierung und der Laender 1986).

Conclusions

This review of findings on the so-called new type forest damages observed in central Europe since the mid–1970s indicates that primary and secondary gaseous air pollutants, acid precipitation, nitrogen, heavy metals, and other substances deposited in forest ecosystems all may play more or less important roles in the cause of this phenomenon. Futhermore, a considerable number of other natural and unnatural stress factors have been suggested to explain causal mechanisms responsible for the observed damages. However, the dramatic increases of the damages have leveled off since 1985. Scientists generally agree that the damages are site specific and related to regional sets of stress factors. No single overall causal mechanism can be named. Thus, new type forest damages present a complex problem.

A substantial portion of research on this issue has been focused on the nutritional status of declining forest ecosystems. These findings clearly indicate that many damage types are associated with nutritional disturbances. In this respect it was found that fertilization and liming are valuable tools to counteract damages correlated with nutrient disorders. Furthermore, various silvicultural practices can be utilized to stabilize declining forest ecosystems. Nevertheless, all air pollutants that affected forest ecosystems negatively must be reduced.

Finally it can be concluded that the forests of central Europe are damaged significantly. But if all activities that have been initiated to counteract this problem proceed and/or are enhanced, these forests have a fair chance to survive and provide their many vital functions for future human generations.

References

Abetz, P. 1985. Ein Vorschlag zur Durchfuehrung von Wachstumsanalysen im Rahmen der Ursachenforschung von Waldschaeden in Suedwestdeutschland. *Allgemeine Forst- und Jagdzeitschrift*, 156:177–187.

Aldinger, E. 1987. Elementgehalte im Boden und in Nadeln verschieden stark geschaedigter Fichten-Tannen-Bestaende auf Praxiskalkungsflaechen im Buntsandstein-Schwarzwald. *Freiburger Bodenkundliche Abhandlungen* 19:226.

Aldinger, E., and W. L. Kremer. 1985. Zuwachsuntersuchungen an gesunden und geschaedigten Fichten und Tannen auf Praxiskalkungsflaechen. *Forstwissenschaftliches Centralblatt* 104:360–373.

Arndt, U., G. Seuffert, and W. Nobel. 1982. Die Beteiligung von Ozon an der Komplexkrankheit der Tanne (*Abies alba* Mill.)—eine pruefenswerte Hypothese. *Luft* 42:243–247.

Ashmore, M., N. Bell, and J. Rotter. 1985. The role of ozone in forest damages in West Germany. *Ambio* 14:81–87.

Attmannspacher, W., R. Hartmannsgruber, and P. Lang. 1984. Langzeit-Tendenzen des Ozons in der Atmosphaere aufgrund der 1967 begonnenen Ozon-Messreihen am Meteorologischen Observatorium Hohenpeissenberg. *Meteorologische Rundschau* 37:193–199.

Bauch, J. 1983. Biological alterations in the stem and root of fir and spruce due to pollution influence. In *Effects of Accumulation of Air Pollutants in Forest Ecosystems*, B. Ulrich, and J. Pankrath, ed. 377–386.

Bauch, J., H. Stienen, B. Ulrich, and E. Matzner. 1985. Einfluss einer Kalkung bzw. Duengung auf den Elementgehalt in Feinwurzeln und das Dickenwachstum von Fichten aus Waldschadensgebieten. *Allgemeine Forstzeitschrift (AFZ)* 40:1148–1150.

Bauer, de L. I., T. Hernandez, and D. Alvarado. 1987. Forest decline in southern areas of Mexico City. Fourteenth International Botanical Congress, Berlin, July 24–August 1.

Baule, H. 1984. Zusammenhaenge zwischen Naehrstoffversorgung und Walderkrankungen. *AFZ* 39:775–778.

Baule, H., and C. Fricker. 1967. *Die Duengung von Waldbaeumen* 259 pp.

Becker, K. H., J. Loebel, and U. Schurath. 1983. Bildung, Transport und Kontrolle von Photooxidantien. *Umweltbundesamt Berichte* 5:3–132.

Bonneau, M. 1986. "Analyses folières." In *Recherches sur le dépérissement des forêts attribué á la pollution atmosphérique* (Report).

Bonneau, M. 1987. Effects of air pollutants through the soil. In *Les recherches en France sur le dépérissement des forêts*. DEFORPA program, First report, 64–70.

Bonneau, M., and P. Joliot. 1987. Conclusion. In *Les recherches en France dépérissement des forêts*, FORPA program, First report, 79–82.

Bonneau, M., and G. Landmann. 1986. Dépérissement et état nutritionnel des peuplements de sapin pectine et d'épicea commun dans le Massif Vosgien. Collected papers of January 22–24, Center for Forest Research, INRA. Nancy, France.

Bosch, C., E. Pfannkuch, U. Baum, and K. E. Rehfuess. 1983. Ueber die Erkrankung der Fichte (*Picea abies* Karst.) in den Hochlagen des Bayerischen Waldes. *Forstwissenschaftliches Centrablatt* 102:167–181.

Bosch, C., E. Pfannkuch, K. E. Rehfuess, K. H. Runkel, P. Schramel, and M. Senser. 1986. Einfluss einer Duengung mit Magnesium und Kalzium von Ozon und saurem Nebel auf Frosthaerte, Ernaehrungszustand und Biomasse-Produktion junger Fichten (*Picea abies* (L.) Karst.). *Forstwissenschaftliches Centralblatt* 105:218–229.

Boxman, A. W., H. F. G. van Dijk, and J. G. M. Roelofs. 1987. Some effects of ammonium sulphate deposition on pine and deciduous forests in the Netherlands. In *Acid Rain: Scientific and Technical Advances*, ed. R. Perry, R. M. Harrison, J. N. B. Bell, and J. N. Lester, 680–687.

Brandl, H. 1985. Zur Bedeutung bestandesgeschichtlicher Untersuchungen in der Forstgeschichte am Beispiel des 'Tannensterbens' im Schwarzwald. *Allgemeine Forst- und Jagdzeitschrift* 156:142–146.

Bredow, B. v., A. Buggert, A. Eckhoff, B. Hollstein, M. Neumann, R. Schindel, A. Weber, S. Zech, and V. Glavac. 1986. Vergleichende Untersuchungen der Boden-, Wurzel- und Blattmineralstoffgehalte von Baeumen verschiedener Schadstufen in einem immissionsbelasteten Altbuchenbestand. *AFZ* 41:551–554.

Breloh, P., and G. Dieterle. 1986. Ergebnisse der Waldschadenserhebung 1985. *AFZ* 41:1377–1380.

Bruck, R. I. 1985. Observations of boreal mountain forests decline in the Southern Appalachian Mountains. International Symposium on Acidic Precipitation, Muskoka, Ontario, Canada. September 15–20.

Bundesamt für Forstwesen und Landschaftsschutz. 1986. *Sanasilva—Waldschadensbericht 1986*. 27 pp.

Burg, van den J. 1987. Beobachtungen ueber Einfluesse von Stickstoff und saurer Deposition in niederlaendischen Forsten. Zur Duengung kranker Waldbestaende. *AFZ* 42:285–287.

Butin, H., and C. Wagner. 1985. Mykologische Untersuchungen zur "Nadelroete" der Fichte. *Forstwissenschaftliches Centralblatt* 104:178–186.

Clarke, A. G., and D. R. Lambert. 1987. Local factors affecting the chemistry of precipitation. In *Acid Rain: Scientific and Technical Advances*, ed. R. Perry, R. M. Harrison, J. N. B. Bell, and J. N. Lester, 252–259.

Cramer, H. H., and M. Cramer-Middendorf. 1984. Untersuchungen ueber Zusammenhaenge zwischen Schadensperioden und Klimafaktoren in mitteleuropaeischen Forsten seit 1858. *Pflanzenschutz-Nachrichten Bayer* 37:208–334.

Ebel, B., and J. Rosenkranz. 1984. Untersuchungen an Fichtennadeln aus den Abteilungen 103 und 65 im Hils (noerdlich von Goettingen). *Berichte des Forschungszentrums Wald-Oekosysteme/Waldsterben* 3:69–129.

Ebel, B., and J. Rosenkranz. 1985. Ultrastrukturuntersuchungen und Elementmikroanalyse von geschaedigten und ungeschaedigten Fichtennadeln. *Exkursionsfuehrer des Forschungszentrums Wald-Oekosysteme/Waldsterben* (Summer) 35–38.

Ebraim-Nesbat, F., and R. Heitefuss. 1985. Rickettsien-aehnliche Bakterien (RLO) in Feinwurzeln erkrankter Fichten unterschiedlichen Alters. *European Journal of Forest Pathology* 15:182–187.

Eckstein, D., K. Richter, R. W. Aniol, and F. Quiehl. 1984. Dendroklimatologische Untersuchungen zum Buchensterben im suedwestlichen Vogelsberg. *Forstwissenschaftliches Centralblatt* 103:274–290.

Eichhorn, O. 1981. Zoologische Aspekte des Tannensterbens. *Forstwissenschaftliches Centralblatt* 100:270–275.

Eichhorn, O. 1985. Zoologische und ertragskundliche Aspekte des Tannensterbens. *Der Forst- und Holzwirt* 40:415–419.

Evers, F. H. 1984. Welche Erfahrungen liegen bei Kalium- und Magnesium-Grossduengungsversuchen auf verschiedenen Standorten in Baden-Wuerttemberg vor? *AFZ* 39:767–768.

Feig, R., and A. Huettermann. 1985. Untersuchungen von stress-physiologisch relevanten Parametern an Fichten-Jungbestaenden. *Exkursionsfuehrer des Forschungszentrums Wald-Oekosysteme/Waldsterben* (Summer) 260–269.

Feister, U., and W. Warmbt. 1984. Long-term surface ozone increase at Arkona." In Proceedings of Quadr. International Ozone Symposium, Chalkidiki, Greece. September 3–7.

Fink, S. 1983. Histologische und histochemische Untersuchungen an Nadeln erkrankter Tannen und Fichten im Suedschwarzwald. *AFZ* 38:660–663.

Fink, S. 1986. *Pathologische und regenerative Anatomie der Holzpflanzen. Habilitationsschrift*, University of Freiburg. 279 pp.

Fink, S., and H. J. Braun. 1978. Zur epidemischen Erkrankung der Weisstanne (*Abies alba* Mill.). I. Untersuchungen zur Sympotomatik und Formulierung einer Virus-Hypothese. *Allgemeine Forst- und Jagdzeitschrift* 149:145–150, 184–195.

Flueckiger, W. 1986. Untersuchungen ueber den Ernaehrungszustand von Buchen (*Fagus sylvatica* L.) und Fichten (*Picea abies* Karst.) und den Naehrstoffgehalt im Boden in festen Beobachtungsflaechen in der Schweiz. In *Moeglichkeiten und Grenzen der Sanierung immissionsgeschaedigter Wald-Oekosysteme*, ed. G. Glantzel, 65–81.

Flueckiger, W., H. Flueckiger-Keller, and S. Braun. 1984. Untersuchungen ueber Waldschaeden in der Nordwestschweiz. *Schweizerische Zeitschrift für das Forstwesen* 135:389–444.

Forschungsbeirat für Waldschaeden/Luftverunreinigungen der Bundesregierung und der Laender. 1986. 2. *Bericht*. 229 pp.

Franz, F., and H. Roehle. 1985. Zum Wuchsverhalten geschaedigter Waldbestaende in Bayern. In *Was wir ueber das Waldsterben wissen*, ed. E. Nieslein and G. Voss, 234–246.

Frenzel, B. 1983. Beobachtungen eines Botanikers zur Koniferenerkrankung. *AFZ* 38:743–747.

Friedland, A. J., R. A. Gregory, L. Karenlampi, and A. H. Johnson. 1984. Winter damage to foliage as a factor in red spruce decline. *Canadian Journal of Forest Research* 14:963–965.

Fuzzi, S., G. Orsi, and M. Mariotti. 1983. Radiation fog liquid water acidity as a field station in the Po Valley. *Journal of Aerosol Science* 14:135–138.

Gerriets, M., and H. Schulte-Bisping. 1985. Vergleich eines geschaedigten und eines symptomfreien Fichten-Jungbestandes im Hils: Deposition, Boden- und Bestandesinventur. *Exkursionsfuehrer des Forschungszentrums Wald-Oekosysteme/Waldsterben* (Summer) 129–134.

Glatzel, G., E. Sonderegger, M. Kazda, and H. Puxbaum. 1983. Bodenveraenderungen durch schadstoffangereicherte Stammablauf-Niederschlaege in Buchenbestaenden des Wienerwaldes. *AFZ* 38:693–694.

Glavac, V., H. Jochaim, H. Koenies, R. Rheinstaedter, and H. Schaefer. 1985. Einfluss des Stammablaufwassers auf den Boden im Stammfussbereich von Altbuchen in unterschiedlich immissionsbelasteten Gebieten. *AFZ* 40:1297–1398.

Greve, U., D. Eckstein, W. Aniol, and F. Scholz. 1986. Dendroklimatische Untersuchungen an Fichten unterschiedlicher Immissionsbelastung in Nordostbayern. *Allgemeine Forst- und Jagdzeitschrift* 157:1986.

Guderian, R., K. Kueppers, and R. Six. 1985. Reaktionen von Fichte und Pappel auf Schwefeldioxid- und Ozon-Einwirkung bei unterschiedlicher Versorgung mit Kalzium und Magnesium. *VDI-Berichte* 560:657–701.

Gussone, H. A. 1984. Welche neueren Versuchsergebnisse liegen bei der Kalkduengung in Norddeutschland vor? *AFZ* 39:779–789.

Gussone, H. A. 1986. Die Waldschaeden 1986 in der Bundesrepublik Deutschland. *Der Forst- und Holzwirt* 20:540.

Haertl, O., and M. Cerny. 1981. Veraenderungen in Fichtenwaldboeden durch Langzeitwirkung von Schwefeldioxid. *Mitteilungen der Forstlichen Bundesversuchsanstalt Wien* 137/II:233–240.

Hartig, R. 1889. *Lehrbuch der Baumkrankheiten*.

Hauhs, M. 1985. Wasser- und Stoffhaushalt im Einzugsgebiet der Langen Bramke (Harz).

Berichte des Forschungszentrums Wald-Oekosysteme/Waldsterben der Universitaet Goettingen, 17.

Horras, C. 1986. Beziehungen zwischen den Makronaehrstoffen Magnesium, Kalzium, Kalium und Schwefel in Fichtennadeln und Waldschadenssymptomen im Saarland. Diplomarbeit (Master thesis). Subject 6.6 biogeography, Universitaet des Saarlandes, Saarbruecken, Federal Republic of Germany. 157 pp.

Hoshizaki, T., S. K. S. Wong, and B. N. Rock. 1987. Pigment Analyses of Spruce Trees undergoing Forest Decline in the Northeastern United States and Germany, XIV. International Botanical Congress, Berlin, West Germany. July 24–August 1.

Huettermann, A. 1987. Zur Frage einer moeglichen Beteiligung von elektromagetischen Strahlen an der neuartigen Schaedigung des Waldes. Der Forst- und Holzwirt 42:567–573.

Huettermann, A., and J. Gehrmann. 1982. Auswirkungen von Luftverunreinigungen auf eine Buchen-Naturverjuengung in immissionsexponierter Lage. Der Forst- und Holzwirt 37:406–410.

Huettermann, A., and B. Urlich. 1982. Solid phase–soil–solution–root interactions in soil subjected to acid deposition. Philosophical Transactionf of Royal Society B 305:353–363.

Huettl, R. F. 1985. "Neuartige" Waldschaeden und Naehrelementversorgung von Fichtenbestaenden (Picea abies Karst.) in Suedwestdeutschland. Freiburger Bodenkdliche Abhundlungen 16:195.

Huettl, R. F. 1986. Forest Decline and Nutritional Disturbances. Eighteenth IUFRO World Congress, Ljubljana, Yougoslavia. September 7–21. Division I, vol. 2:774.

Huettl, R. F. 1987. "Neuartige" Waldschaeden, Ernaehrungsstoerungen und Duengung. AFZ 42:289–299.

Huettl, R. F., and S. Fink. 1988. Diagnostische Duengungsversuche zur Revitalisierung geschaedigter Fichtenbestaende (Picea abies Karst.) in Suedwestdeutschland. Forstwissenschaftliches Centralblatt 107:173–183.

Huettl, R. F., and B. M. Mehne. 1988. "New type" of forest decline, nutrient deficiences and the "virus"-hypothesis. In Air Pollution and Ecosystems, ed. P. Mathy, 870–873.

Huettl, R. F., and J. Wisniewski. 1987. A Critique of Forest Fertilization Efforts in "New Type" Decline Forests Associated with Nutrient Deficiencies: The West German and United States Experience. (in press).

Huettl, R. F., and H. W. Zoettl. 1985. Ernaehrungszustand von Tannenbestaenden in Sueddeutschland—ein historischer Vergleich. Allgemeine Forstzeitschrift 40:1011–1013.

Huettl, R. F., and H. W. Zoettl. 1986. Diagnostische Duengungsversuche in geschaedigten Nadelbaumbestaenden Suedwestdeutschlands. IMA-Querschnittsseminar "Restabilisierungsmassnahmen—Duengung." Karlsruhe, Federal Republic of Germany. April 15–16:3–14.

Innes, J. L. 1987. The interpretation of international forest health data. In Acid Rain: Scientific and Technical Advances, ed. R. Perry, R. M. Harrison, J. N. B. Bell, and J. N. Lester, 633–640.

Isermann, K. 1985. Diagnose und Therapie der "neuartigen Waldschaeden" aus der Sicht der Waldernaehrung. VDI-Berichte 560:897–920.

Isermann, K. 1987. Revitalisierung geschaedigter Fichten-Altbestaende durch Mineralduengung. AFZ 42:997–1000.

IUFRO. 1983. Aquilo Series on Botany 19:175.

Ixfeld, H., K. Ellermann, and M. Buck. 1985. Bericht ueber die Ergebnisse der diskontinuierlichen Schwefeldioxid- und Mehrkomponenten-Messungen im Rhein-Ruhr-Gebiet fuer die Zeit vom 01. Januar 1984 bis 31. Dezember 1984. LIS-Berichte 63:73–126.

Jagels, R. 1986. Acid fog, ozone and low elevation spruce decline. *IAWA Bulletin* 7:299–307.

Johann, K. 1986. *Institutsinterner Bericht ueber die Jahrestagung der Sektion Ertragskunde*, May 1986 Schwangau, Federal Republic of Germany.

Jorns, A., and C. Hecht-Buchholz. 1985. Aluminium-induzierter Magnesium-und Kalziummangel im Laborversuch bei Fichtensaemlingen. *AFZ* 41:1248–1252.

Kaminski, O., and P. Winkler. 1983. Saure Aerosolpartikel und Nebel und ihre Wirkung auf die Biosphaere. *Annals of Meteorology* 20:149–150.

Kandler, O. 1983. Waldsterben: Immissions- oder Epidemie-Hypothese? *Naturwissenschaftliche Rundschau* 36:488–490.

Kandler, O. 1985. Immissions- versus Epidemie-Hypothesen. In Kortzfleisch, G. *Waldschaeden—Theorie und Praxis auf der Suche nach Antworten*, ed. G. Kortzfleish, 19–59.

Kandler, O., W. Miller, and R. Ostner. 1987. Dynamik der "akuten Vergilbung" der Fichte. *AFZ* 42:715–723.

Kaupenjohann, M. 1987. Waldduengung und neuartige Waldschaeden: Ergebnisse aus Duengungs- und Kalkungsversuchen. In *Moeglichkeiten und Grenzen der Sanierung immissionsgeschaedigter Wald-Oekosysteme*, ed. G. Glatzel, 82–98.

Kaupenjohann, M., W. Zech, R. Hantschel, and R. Horn. 1987. Ergebnisse von Duengungsversuchen mit Magnesium an vermutlich immissionsgeschaedigten Fichten (*Picea abies* (L.) Karst.) im Fichtelgebirge. *Forstwissenschaftliche Centralblatt* 106:78–84.

Keitel, A., and U. Arndt. 1983. Ozoninduzierte Turgeszenzverluste bei Tabak (*Nicotiana tabacum* var. Bel W3)—ein Hinweis auf schnelle Permeabilitaetsveraenderungen der Zellmembranen. *Angewandte Botanik* 57:193.

Kenk, G. 1985. Zuwachs und oekonomische Bewertung. Goettingen, Federal Republic of Germany. Seminar, January 30.

Kenk, G., W. Kremer, D. Bonaventura, and M. Gallus. 1984. Jahrring- und zuwachsanalytische Untersuchungen in erkrankten Tannenbestaenden des Landes Bades-Wuerttemberg. *Mittelungen der Forstlichen Versuchsanstalt Baden-Wuerttemberg* 112:1–38.

Knabe, W., J. Block, and G. Cousen. 1985. Die Beziehungen zwischen Immissionsbelastung, Bodenzustand und Waldgesundheit an Standorten des Pilot-Messprogrammes "Saure Niederschlaege" des Landes Nordrhein-Westfalen. *VDI-Berichte* 560:599–626.

Koehl, M. 1987. Gegenwaertiger Stand der Methoden zur Inventur und Ueberwachung gefaehrdeter Waelder. *AFZ* 42:573–577.

Koenig, E. 1983. Gegenwaertige Forstschutzsituation in Suedwestdeutschland. *AFZ* 38:284–289.

Kozlowski, T. T., and H. A. Constantinidou. 1986. Responses of woody plants to environmental pollution. *Forestry Abstracts* 47:5–51.

Kozlowski, T. T., and J. B. Mudd. 1975. *Responses of Plants to Air Pollution*. New York: Academic Press. pp. 1–8.

Kowalski, T., and K. J. Lang. 1984. Die Pilzflora von Nadeltrieben und Aesten unterschiedlich alter Fichten (*Picea abies* (L.) Karst.) mit besonderer Beruecksichtigung vom Fichtensterben betroffener Altbaeume. *Forstwissenschaftliche Centralblatt* 103:349–360.

Krause, G. H. M. 1987. Forest decline and the role of air pollutants. In *Acid Rain: Scientific and Technical Advances*, ed. R. Perry, R. M. Harrison, J. N. B. Bell, and J. N. Lester, 621–632.

Krause, G. H. M., U. Arndt, C. J. Brandt, J. Bucher, G. Kenk, and E. Matzner. 1986. Forest decline in Europe: Development and possible causes. *Water, Air and Soil Pollution* 31:647–668.

Krause, G. H. M., K. D. Jung, and B. Prinz. 1983. Neuere Untersuchungen zur Aufklaerung immissionsbedingter Waldschaeden. *VDI-Berichte* 500:257–266.

Krause, G. H. M., K. D. Jung, and B. Prinz. 1985. Experimentelle Untersuchungen zur Aufklaerung der neuartigen Waldschaeden in der Bundesrepublik Deutschland. *VDI-Berichte* 560:627–656.

Kreutzer, K. 1970. *Manganmangel der Fichte.* Habiltationsschrift: Universitaet Muenchen.

Kreutzer, K. 1984. Mindern Duengungsmassnahmen die Waldschaeden? *AFZ* 39:771–773.

Kroemer-Butz, S. 1986. Waldschaeden: Wie sieht die Situation im uebrigen Europa aus? *Unser Wald* 38:77–79.

Lacaux, J. P., J. Servant, and J. G. R. Baudet. 1987. Acid rain water in the tropical forest of Western Africa. In *Acid Rain: Scientific and Technical Advances*, ed. R. Perry, R. M. Harrison, J. N. B. Bell, and J. N. Lester, 264–269.

Makkonen-Spiecker, K. 1984. *Auswirkungen des Aluminiums auf junge Fichten (Picea Abies Karst.) verschiedener Provenienzen.* Monograph Series, Waldbau Institute, University of Freiburg, 2, 129 pp.

Manion, P. D. 1981. *Tree Disease Concepts.* 399 pp.

Materna, J. 1986. Erfahrungen bei der Aufforstung extremer Waldschadensflaechen im Erzgebirge, CSSR. Symposium Moeglichkeiten und Grenzen der Sanierung immissionsgeschaedigter Wald-Oekosysteme. Vienna, Austria. November 6–7.

Mehne, B. M. 1987. Vergleichende Untersuchungen an Pfropfungen mit Reisern gruener und vergilbter Fichten (*Picea abies* Karst.) aus Walschadensgebieten. Ph.D. diss. University of Freiburg, Federal Republic of Germany.

Metzner, H. 1985. Waldschaeden durch Kerntechnische Anlagen? *Literaturstudie im Auftrag des Ministers fuer Ernaehrung, Landwirtschaft, Umwelt und Forsten, Baden-Wuerttemberg.* September.

Meyer, F. H. 1985. Einfluss eines Stickstoff-Factors auf den Mykorrhizabesatz von Fichtensaemlingen im Humus einer Waldschadensflaeche. *AFZ* 40:208–219.

Michael, G. 1941. Ueber die Aufnahme und Verteilung des Magnesiums und dessen Rolle in der hoeheren gruenen Pflanze. *Zeitschrift für Pflanzenernaehrung Bodenkunde* 29:65–120.

Mies, E. 1987. Elementeintraege in tannenreiche Mischbestaende des Suedschwarzwaldes. *Freiburger Bodenkundliche Abhandlungen* 18:247.

Mies E., and H. W. Zoettl. 1985. Zeitliche Aenderung der Chlorophyll- und Elementgehalte in den Nadeln eines gelbchlorotischen Fichtenbestandes. *Forstwissenschaftliche Centralblatt* 104:1–8.

Miller, P. R. 1987. Root and shoot growth during early development of *sequoiadendron giganteum* seedlings stressed by ozone. Fourteenth International Botanical Congress, Berlin, Federal Republic of Germany. July 24–August 1.

Miller, P. R., J. R. Parmeter, B. H. Flick, and C. W. Martinuz. 1969. Ozone dosage response of ponderosa pine. *Journal of Air Pollution Control Association* 19:435–438.

Moehring, K. 1982a. Zum Erkennen von Schaeden an Fichte und Buche. *AFZ* 37:464–465.

Moehring, K. 1982b. Erfahren wir die Antwort der Natur? Beobachtungen und Gedanken zu einer noch namenlosen Waldkrankheit. *Forstarchiv* 53:123–128.

Mohr, H. 1986. Die Erforschung der neuartigen Waldschaeden—eine Zwischenbilanz. *Biologie in unserer Zeit* 16:83–89.

Moosmayer, H.-U. 1984. Erkenntnisse ueber die Walderkrankungen. Dargestellt an Projekten der Forstlichen Versuchs- und Forschungsanstalt Baden-Wuerttemberg. *Forstwissenschaftliche Centralblatt* 103:1–16.

Murach, D. 1984. Die Reaktion der Feinwurzeln von Fichten (*Picea abies* Karst.) auf zunehmende Bodenversauerung. *Goettinger Bodenkundliche Berichte* 77:126.

Neger, F. W. 1919. *Die Krankheiten unserer Waldbaeume und wichtigsten Gartengehoelze.*

Neger, F. W. 1924. *Die Krankheiten unserer Waldbaeume.*

Neumann, M., and J. Pollanschuetz. 1982. Untersuchungen ueber Auswirkungen gasfoermiger Immissionen auf Waldbestaende im Raum Gailitz-Arnoldstein. *Carinthia II* Special issue. Klagenfurt, Austria, 39pp.

Nienhaus, F. 1985a. Zur Frage der parasitaeren Verseuchung von Forstgehoelzen durch Viren und primitive Mikroorganismen. *AFZ* 40:119–124.

Nienhaus, F. 1985b. Nachweis von Viren und primitiven Prokaryonten in Wald-Oekosystemen. *VDI-Berichte* 560:961–971.

Nienhaus, F. 1986. Zusammenfassung der Ergebnisse des IMA-Statusseminars *"Biotische Schadfaktoren—Epidemiologie."* Merkenheim, Federal Republic of Germany. April 24–25.

Nihlgard, B. 1985. The ammonia hypothesis—an additional explanation of the forest dieback in Europe. *Ambio* 14:2–8.

Nobbe, F. 1893. Ueber die Fichtennadelroete und ihre Verbreitung in den saechsischen Forsten. *Taranther Forstliches Jahrbuch* 43:39–55.

Oblaender, W., and A. Hanss. 1985. Zwischenbericht ueber Schadstoffmessungen in Waldgebieten Baden-Wuerttemberg. *LFU-Bericht* 97:1–18.

Oesterreichische Duengerberatungsstelle (OeDB). 1986. *Duengung—ein Weg zu gesunden Waeldern.* 23 pp.

Parameswaran, N., S. Fink, and W. Liese. 1985. Feinstrukturelle Untersuchungen an Nadeln geschaedigter Tannen und Fichten aus Waldschadensgebieten im Schwarzwald. *European Journal of Forest Pathology* 15:168–182.

Pfeffer, H.-U. 1985. Immissionsmessungen in Waldgebieten des Eggegebirges und der Eifel. *LIS-Berichte* 57:43–49.

Pollanschuetz, J. 1966. Methodik der Rauchschadensfeststellung, wie sie gegenwaertig von der Forstlichen Bundesversuchsanstalt angewandt wird. *Mitteilungen der Forstlichen Bundesversuchsanstalt Wien* 73:81–89.

Pollanschuetz, J. 1986. Auswirkungen auf Zuwachs und Holzproduktion. Interforest Congress, Munich, Federal Republic of Germany. 1–4 July.

Pollanschuetz, J. 1987. Verfahren der Waldzustands-Inventur in Oesterreich und Ergebnisse 1984–86. In *Waldschaeden-Holzwirtschaft,* ed. H. P. Rossmanith, 39–51.

Posthumus, A. C., and A. E. G. Tonneijk. 1982. Monitoring of Effects of photo-oxidants on plants. In *Monitoring of Air Pollutants by Plants,* ed. L. Steubing and H. J. Jaeger, 115–120.

Prinz, B., G. H. M. Krause, and H. Strathmann. 1982. Waldschaeden in der Bundesrepublik Deutschland. *LIS-Berichte* 28:1–154.

Puhe, J., H. Persson, and I. Boerjesson. 1986. Wurzelwachstum und Wurzelschaeden in skandinavischen Nadelwaeldern. *AFZ* 42:488–492.

Rat von Sachverstaendigen fuer Umweltfragen. 1983. Waldschaeden und Luftverunreinigungen. *Sondergutachten* March. 338 pp.

Reemtsma, J. B. 1986. Der Magnesiumgehalt von Nadeln niedersaechsischer Fichtenbestaende und seine Beurteilung. *Allgemeine Forst- und Jagdzeitschrift* 157:196–200.

Rehfuess, K. E. 1983a. Walderkrankungen und Immissionen—eine Zwischenbilanz. *AFZ* 38:601–610.

Rehfuess, K. E. 1983b. Ersatztriebe an Fichten. *AFZ* 38:1111.

Rehfuess, K. E. 1987. Einfuehrung und Ueberblick ueber den Themenbereich F "Bodenchemie Naehrstoffhaushalt" (einschliesslich Meliorationsmassnahmen). *Tagungsbericht zum Statusseminar "Ursachenforschung zu Waldschaeden."* Jülich, Federal Republic of Germany. March 30-April 3, 272–291.

Rehfuess, K. E., and C. Bosch. 1987. Experimentelle Pruefung der Auswirkungen eines Witterungsstresses in Expositionskammern. *GSF-Berichte* 10:175–184.

Rehfuess, K. E., and H. Rodenkirchen. 1984. Ueber die Nadelroete-Erkrankung der Fichte (*Picea abies* Karst.) in Sueddeutschland. *Forstwissenschaftliches Centralblatt* 103:245–262.

Reichelt, G. 1984. Zur Frage des Zusammenhanges zwischen Waldschaeden und dem Betrieb von Atomanlagen. *Forstwissenschaftliches Centralblatt* 103:290–297.

Roehle, H. 1985. Ertragskundliche Aspekte der Walderkrankungen. *Forstwissenschaftliches Centralblatt* 104:225–242.

Roelofs, J. G. M., A. J. Kempers, A. L. F. M. Hondijk, and J. Jansen. 1985. The effect of airborn ammonium sulphate on *Pinus var. maritima* in the Netherlands. *Plant and Soil* 102:45–56.

Roloff, A. 1984a. Nur zwei trockene Sommer oder mehr? Zum Bestimmen der Ursache derzeitiger Kronenduerre und Absterbeerscheinungen in Buchenbestaenden. *Forst- und Holzwirt* 39:364–366.

Roloff, A. 1984b. Morphologie der Verzweigung von *Fagus sylvatica* L. (Rotbuche) als Grundlage zur Beurteilung von Triebanomalien und Kronenschaeden. *Berichte des Forschungszentrums Wald-Oekosysteme/Waldsterben* 3:1–25.

Roloff, A. 1985a. Schadstufen bei der Buche—Vorschlag fuer eine bundeseinheitliche Einordnung der Buche in vier Schadstufen bei terrestrischen Aufnahmen. *Forst- und Holzwirt* 40:131–134.

Roloff, A. 1985b. Untersuchungen zum vorzeitigen Laubfall und zur Diagnose von Trockenschaeden in Buchenbestaenden. *AFZ* 40:157–160.

Roloff, A. 1985c. Ein schleichendes Absterben: Auswirkungen von Immissionsschaeden in Buchenbestaenden. *AFZ* 40:905–908.

Roloff, A. 1985d. Morphologie der Kronenentwicklung von *Fagus sylvatica* L. (Rotbuche) unter besonderer Beruecksichtigung moeglicherweise neuartiger Veraenderungen. Ph.D. diss., Forestry Science, University of Goettingen.

Sartorius, R. 1984. In *Kolloquium Waldsterben—Diagnose und Therapie*. Berlin, Federal Republic of Germany. April 2–3.

Scheifele, M. 1985. Waldschadenserhebung auch in Frankreich. *AFZ* 40:340.

Schenck, G. O. 1985. Is the new forest decline a photodynamic disease caused by light and peroxyacetylnitrate, ozone, halocarbons and others? *EPA-Newsletter* (April/June 1985):15–41.

Schenck, G. O. 1988. Zur Beteiligung fotochemischer Prozesse an den fotodynamischen Lichtkrankheiten der Pflanzen (Waldsterben). *Rheinisch-Westfaelische Akademie der Wissenschaften (Vortraege)* N 360:25–112.

Schmid-Haas, P. 1985. Der Gesundheitszustand des Schweizer Waldes. *Schweizerische Zeitschrift für das Forstwesen* 251–273.

Schmid-Haas, P. 1986. In *Auswirkungen auf Zuwachs und Holzproduktion*, ed. J. Pollanschuetz, Interforest Symposium, Munich, Federal Republic of Germany July 1–4.

Schoenhar, S. 1985. Untersuchungen ueber das Vorkommen pilzlicher Parasiten an Feinwurzeln der Tanne (*Abies alba* Mill.). *Allgemeine Forst- und Jagdzeitschrift* 156:247–251.

Schoepfer, W. 1986. Verstaerkte Waldschaeden durch Radioaktivitaet? *AFZ* 41:95–98.

Schrimpf, D., O. Klemm, R. Eiden, T. Frevert, and R. Herrmann. 1984. Anwendung eines Grunow-Nebelfaengers zur Bestimmung von Schadstoffgehalten in Nebel-Niederschlaegen. *Staub-Reinh. Luft* 44:72–75.

Schroeter, H., and E. Aldinger. 1985. Beurteilung des Gesundheitszustandes von Fichte und Tanne nach der Benadelungsdichte. *AFZ* 40:438–442.

Schuett, P., and H. Summerer. 1983. Waldsterben-Symptome an Buche. *Forstwissenschaftliches Centralblatt* 102:210–206.

Schulte, A., and M. Spiteller. 1985. Veraenderungen bodenchemischer Parameter im Stammabflussbereich von Buchenwald-Oekosystemen auf Kalk und Basalt. *Exkursionsfuehrer des Forschungszentrums Wald-Oekosysteme/Waldsterben* (Summer):115–117.

Schulte-Bisping, H., and D. Murach. 1984. Inventur der Biomasse und ausgewaehlter chemischer Elemente in zwei unterschiedlich stark versauerten Fichtenbestaenden im Hils. *Berichte des Forschungszentrums Wald-Oekosysteme/Waldsterben* 2:207–265.

Schulze, E. B., M. Fuchs, and M. J. Fuchs. 1977. Special distribution of photosynthetic capacity and performance in a mountain spruce forest of northern Germany. *Oecologia* 29:43–61, 30:239–248.

Schulze, E.-D., R. Oren, and R. Zimmermann. 1987. Die Wirkung von Immissionen auf 30 jaehrige Fichten in mittleren Hoehenlagen des Fichtelgebirges auf Phyllit. *AFZ* 42:725–730.

Schwenke, W. 1982. Moegliche Beziehung tierischer Schaedlinge zu derzeitigen Krankheitserscheinungen bei Tanne und Fichte. *AFZ* 37:446.

Schwenke, W. 1985. Beziehungen zwischen tierischen Schaedlingen und Baumerkrankungen. *Forstwissenschaftliches Centralblatt* 104:220–225.

Schwenke, W., G. Braun, and E. Maschning. 1983. Situation und Prognose des Schaedlingsbefalls in Bayern 1982–1983. *AFZ* 38:292–294.

Seitschek, O. 1981. Verbreitung und Bedeutung der Tannenerkrankung in Bayern. *Forstwissenschaftliches Centralblatt* 100:138–148.

Sierpinsky, Z. 1984. Ueber den Einfluss von Luftverunreinigungen auf Schadinsekten in polnischen Nadelbaum-Bestaenden. *Forstwissenschaftliches Centralblatt* 103:83–92.

Skaerby, L., and G. Sellden. 1984. The effects of ozone on crops and forests. *Ambio* 13:68–72.

Skeffington, R. A., and T. M. Roberts. 1985. Effect of ozone and acid mist on Scots pine and Norway spruce—an experimental study. *VDI-Berichte* 560:747–760.

Stienen, H., R. Barckhausen, R. Schaub, and J. Bauch. 1984. Mikroskopische und roentgenenergiedispersive Untersuchungen an Feinwurzeln gesunder und erkrankter Fichten verschiedener Standorte. *Forstwissenschaftliches Centralblatt* 103:262–274.

Tamm, C. O., and L. Hallbaecken. 1986. Changes in soil pH over a 50-yr-period under different forest canopies in SW Sweden. *Water, Air and Soil Pollution* 31:337–342.

Taylor, O. C., C. R. Thompson, D. T. Tingey, and R. A. Reinert. 1975. Oxides of nitrogen. In *Responses of Plants to Air Pollution*, ed. T. T. Kozlowski, 121–139.

Ulrich, B. 1972. Forstduengung und Umweltschutz. *AFZ* 27:147–148.

Ulrich, B. 1986. Die Rolle der Bodenversauerung beim Waldsterben: Langfristige Konsequenzen und forstliche Moeglichkeiten. *Forstwissenschaftliches Centralblatt* 105:421–435.

Ulrich, B., and E. Matzner. 1983. Raten der oekosysteminternen H^+-Produktion und der sauren Deposition und ihre Wirkung auf Stabilitaet—Elastizitaet von Wald-Oekosystemen. *VDI-Berichte* 500:289–300.

Umweltbundesamt. 1986. *Daten zur Umwelt 1986/87.* 550 pp.

Wachter, A. 1978. Deutschsprachige Literatur zum Weisstannensterben (1830–1978). *Zeitschrift für Pflanzenkrankheiten und Pflanzenschutz* 85:361–381.

Warmbt, W. 1979. Ergebnisse langjaehriger Messungen des bodennahen Ozons in der DDR. *Zeitschrift für Meteorologie* 29:24–31.

Weissen, F., and van H. J. Praag. 1984. Sulfur transfer in oligitropic forest soil and spruce dieback by atmospheric pollutants, related to sulphur content of needles (Ardennes). *Proceedings "Acid Deposition and Sulphur Cycle."* Brussels (June):173–182.

Wentzel, K. F. 1959. Zur Bodenbeeinflussung durch industrielle Luftverunreinigungen und Duengung in Rauchschadenslagen, insobesondere mit Kalk. *Der Forst und Holzwirt* 14:6.

Winkler, P. 1982. Zur Trendentwicklung der pH-Werte des Niederschlages in Mitteleuropa. *Zeitschrift für Planzenernaehrung und Bodenkunde* 145:576–585.

Winkler, P. 1983. Der Saeuregehalt von Aerosol, Nebel und Niederschlaegen. *VDI-Berichte* 500:141–147.

Winkler, P. 1985. Die messtechnische Erfassung der trockenen und nassen Deposition. *VDI-Berichte* 560:289–291.

Wisniewski, J., R. Beck, and R. F. Huettl. 1987. *SO$_2$ and NO$_x$ Emissions and Control Strategies: A Tale of Three Countries (West Germany, Canada and the United States).* Fourth Annual Pittsburgh Coal Conference, Pittsburgh, Pa. September 28–October 2.

Zech, W. 1983. Kann Magnesium immissionsgeschaedigte Tannen retten? *AFZ* 38:237.

Zech, W., and E. Popp. 1983. Magnesiummangel, einer der Gruende fuer das Fichten- und Tannensterben in NO-Bayern. *Forstwissenschaftliches Centralblatt* 102:50–55.

Zezschwitz, E. v. 1985. Qualitaetsaenderungen des Waldhumus. *Forstwissenschaftliches Centralblatt* 104:205–220.

Zoettl, H. W. 1983. Zur Frage der toxischen Wirkung von Aluminium auf Pflanzen. *AFZ* 38:206–208.

Zoettl, H. W. 1985a. Heavy metal levels and cycling in forest ecosystems. *Experientia* 41:1104–1113.

Zoettl, H. W. 1985b. Waldschaeden und Naehrelementversorgung. *Duesseldorfer Geobotanisches Kolloquium* 2:31–41.

Zoettl, H. W., and R. F. Huettl. 1985. Schadsymptome und Ernaehrungszustand von Fichtenbestaenden im Suedwestdeutschen Alpenvorland. *AFZ* 40:197–199.

Zoettl, H. W., and R. F. Huettl. 1986. Nutrient supply and forest decline in Southwest Germany. *Water, Air and Soil Pollution* 31:449–462.

Zoettl, H. W., and E. Mies. 1983. Naehrelementversorgung und Schadstoffbelastung von Fichten-Oekosystemen im Suedschwarzwald unter Immissionseinfluss. *Mitteilungen der Deutschen Bodenkundlichen Gesellschaft* 38:429–434.

3

Concept of Forest Decline in Relation to Western U.S. Forests

PAUL R. MILLER

Forested lands of the Pacific Coast and Rocky Mountains cover 352 million acres, of which 128 million acres are categorized as commercial forest; this represents 27 percent of the total commercial forest of the nation (USDA 1982). Forest stands in the West typically contain all ages of trees, but compared with the more intensively managed forests in the East they contain more old-growth stands, that is, composed of old, large trees. Tree species in these stands are subject to disease, insect attack, damaging climatic extremes, and an assortment of abuses caused by humans. Air pollution has damaged nearby forests in the past, but now the wider distribution of urban areas and certain industries has raised the possibility of chronic exposure to low levels of several atmospheric pollutants.

In several forested regions of the United States, including the Pacific Coast and Rocky Mountains, the National Acid Precipitation Assessment Program is sponsoring research to investigate the current health of forests in relation to air pollutants. This chapter reviews the present state of health of western forests in an attempt to identify tree and stand conditions that either are explainable by the expected behavior of well-known forest pest and disease problems and abiotic stresses or are characterized by a reduction of vigor that cannot easily be assigned to a single causal agent, and that may involve complex interactions of background levels of pollutants with known forest pests and abiotic stresses.

Overview of Forest Resources in the Western United States

It is not within the scope of this report to discuss the present health of all western forest types. Several methods of classifying these forests will be presented, however, in order to provide an appreciation for the great extent and diversity of the resource.

Forest resources in the western United States, including Alaska, are grouped by silviculturists into 21 forest types (USDA 1973). These include 8 forest types in the Pacific Northwest, 4 in the Pacific Southwest, 5 in the northern Rocky Mountains, and 4 in the central and southern Rocky Mountains (see table 3.1). Forest type names are derived from the dominant and codominant tree species that commonly occur together in a particular region. The name assigned to the type is usually the common name of one or more of the species most abundant in the mixture, such as western hemlock–Sitka spruce. This indicates that other species may be present in smaller numbers but that western hemlock and Sitka spruce are the most abundant. Sometimes only one species is used to define a type, as in the case of coastal Douglas fir.

The Society of American Foresters (Eyre 1980) groups a total of 54 western forest types according to biogeoclimatic zones, including Northern Interior (boreal); High Elevations; Middle Elevations, Interior; North Pacific; Low Elevations, Interior; and South Pacific (except for high mountains). The color-coded map (Eyre 1980) of forest types (not included here) for the contiguous western states represents the 10 most important types: Douglas fir, hemlock–Sitka

Table 3.1. *Forest types as recognized for distinct silvicultural treatments in the western United States.*

Region	Forest types
Pacific Northwest	Western hemlock–Sitka spruce
	Coastal douglas fir
	Mixed conifers of sw Oregon
	True fir–mountain hemlock
	Mixed pine–fir of e Oregon and Washington
	Northwestern ponderosa pine
	Interior Alaska white spruce
	Interior Alaska hardwoods
Pacific Southwest	Redwood
	Red fir–white fir
	California mixed conifers
	Pacific ponderosa pine
Northern Rocky Mountain	Ponderosa pine and Rocky Mtn. Douglas fir
	Western larch
	Western white pine and associated species
	Engelmann spruce
	Lodgepole pine
Central and southern	Southwestern ponderosa pine
	Southwestern mixed conifers
	Rocky Mountain aspen
	Black Hills ponderosa pine

Source: Forest Service, *Agricultural Handbook, No. 445* (Washington, D.C.: U.S. Department of Agriculture, 1973).

spruce, redwood, ponderosa pine, white pine, lodgepole pine, larch, fir-spruce, hardwoods, and pinyon juniper. Alaskan coastal forests, hemlock–Sitka spruce, and Alaskan interior spruce-hardwoods are the most important types in that region. Species included on the Society of American Foresters map (Eyre 1980) are essentially those of greatest commercial importance.

Distribution of Commercial Forest Cover Types

Another simple classification used for describing the area of productive forestland is based on the commercial importance of single or associated species (see table 3.2). This classification first subdivides species into softwoods and hardwoods and follows with a list of those under management for commercial timber production. Douglas fir, ponderosa pine, and fir-spruce are those occupying the largest acreages in the West, followed by lodgepole pine and hemlock–Sitka spruce.

Productivity Statistics

Commercial timberland is further classified in terms of cubic feet of wood produced per acre per year into one of the following four productivity classes

Table 3.2. *Area of commercial timberland in the United States, by forest type, 1977.*

Forest type	Total area	Proportion
	(million acres)	(percent)
Western types		
Softwood types		
Douglas fir	30.9	6.4
Ponderosa pine	26.6	5.5
Fir-spruce	19.9	4.1
Lodgepole pine	12.7	2.7
Hemlock–Sitka spruce	12.9	2.7
Larch	2.4	0.5
White pine	0.4	0.1
Redwood	0.7	0.1
Other western softwoods	0.5	0.1
Total	107.0	22.1
Western hardwood types	14.9	3.1
Nonstocked	6.4	1.3
Total West	128.3	.26.6
Eastern Types		
Softwood Types		
Total	96.9	19.1
Hardwood types		
Total	248.0	51.4
Nonstocked	10.0	2.1
Total East	354.2	73.4
United States	482.5	100.0

Source: Forest Service, *Forest Resource Report No. 23* (Washington, D.C.: U.S. Department of Agriculture, 1982).

(USDA 1982): 120 or more cubic feet, 85–120 cubic feet, 50–85 cubic feet, or 20–50 cubic feet. Sites producing less than 20 cubic feet per year are not considered commercial forest. Furthermore, on 5 or 6 million acres of nonstocked forestlands in the West, the tree cover is not sufficient to designate a forest type accurately. Forests capable of producing more than 120 cubic feet per year constitute less than 10 percent of all commercial timberland in the United States. About half this is located in the Pacific Northwest on the west side of the Cascade Range. The forest types included here are Douglas fir, hemlock–Sitka Spruce, and Western Hardwoods. About half the commercial forest acreage in the Rocky Mountains is in the productivity class of 20–50 cubic feet per year.

Softwood timber volume in the West is about seven times larger than that of hardwoods. Some 59 percent of the softwood growing stock inventory of the United States is located in the Pacific Coast region, even though a much greater acreage of commercial forests is found in the eastern states. The explanation is that the Pacific Coast has a high proportion of old-growth stands with high volumes per acre.

Timber inventories have decreased in the Pacific Coast region as a result of harvesting old-growth stands. In the Rocky Mountains, harvests have been at a relatively slower pace and timber inventories have not changed substantially since 1952. In the West, mortality in old-growth stands offsets much of the total annual growth. Therefore, net annual growth of stock in the West was 5.2 billion cubic feet, which is less than one-quarter of the national total.

Timber inventories also reflect the mortality losses resulting from within-stand competition, insects, diseases, fire, and blowdowns. Competition is one of the most important causes of mortality. Bark beetles also cause great losses in old-growth stands of ponderosa pine and lodgepole pine. Losses from both these mortality agents are exacerbated by drought. Diseases such as root rots, white pine blister rust, and both true and dwarf mistletoes can contribute to tree mortality and also reduce annual growth. Stem rots do not contribute to mortality but do reduce the amount of merchantable wood in harvested stems. Diseases and some insects extend the time required to grow trees to salable size.

Losses from fire appear to be spectacular, as in the summer of 1987 throughout the West, but on the average fires do not contribute a large amount to the total mortality. Mortality agents will be discussed in greater detail later in the chapter.

Projected Demand versus Anticipated Timber Supply

The supply-demand balance for western forests from 1952 to 2030 shows two projections, depending on assumptions about price trends. The first assumption is that the trends established from the late 1950s to the mid–1970s will continue at the same rate. In this case, the supply-demand balance will remain negative in the Rocky Mountain, Pacific Northwest, and Pacific Southwest regions from 1990 to 2030 (USDA 1982)—that is, demand will outstrip supply (see table 3.3). Under the second assumption, prices are allowed to increase sufficiently to maintain a balance between supply and demand. Consumers can

Table 3.3. *Softwood timber demand on and supply from forests in the Pacific Northwest, Pacific Southwest, and Rocky Mountain regions between 1952 and 2030.*

Region	Base period	Projections					
		Base level price trends			Equilibrium price trends		
	1952	1990	2010	2030	1990	2010	2030
	(billion cubic feet)						
Pacific NW							
Douglas fir							
Subregion							
Demand	2.09	2.58	2.40	2.26	2.44	2.20	2.10
Supply	2.09	2.26	2.14	2.05	2.44	2.20	2.10
Balance	0	−.32	−.26	−.21	0	0	0
Ponderosa pine							
Subregion							
Demand	.33	.65	.83	.92	.57	.67	.76
Supply	.33	.54	.63	.72	.57	.67	.76
Balance	0	−.11	−.20	−.20	0	0	0
Pacific sw region							
Demand	.76	1.00	1.05	1.06	.89	.89	.91
Supply	.76	.77	.80	.88	.89	.89	.91
Balance	0	−.23	−.25	−ı18	0	0	0
Rocky Mountain region							
Demand	.42	1.14	1.37	1.54	1.03	1.17	1.36
Supply	.42	.91	1.08	1.13	1.03	1.17	1.36
Balance	0	−.23	−.29	−.41	0	0	0

Source: Forest Service, *Forest Resource Report No. 23* (Washington, D.C.: U.S. Departmemt of Agriculture, 1982).

expect shortages under the first assumption, while under the second they will pay increasingly higher prices.

Summary

In summary, the forests of the West contain more than half the nation's reserve of softwood (USDA 1982). Old-growth trees that are already in a natural decline of growth rate and younger trees suffering severe competition from companion trees are sensitive to additional stress agents, such as air pollution. From the viewpoint of commercial forestry practice it is essential to harvest old-growth stands while trees with large volumes of wood are still alive. It is much less attractive to attempt only the salvage of dead and dying trees. On the other hand, market demand cannot handle a wholesale sell-off of older timber just to shift from management of old stands to the management of young ones.

The management of young stands requires considerable thinning, which in turn could result in an excess of certain timber products derived from small-diameter trees. In addition, these old-growth forests represent an inheritance that provides other amenity values, such as recreation and a high-quality water-shed, for many generations to come. Older trees already under stress from other biotic and abiotic agents have a poorer capacity to survive additional stress. It therefore becomes even more important to avoid the degradation of air quality in older stands.

Review of Past and Current Status of Forest Health

The health of each western forest species is threatened by different specific diseases, insects, and environmental extremes. The symptoms (appearances of leaf, stem, and root injuries) and signs (actual structures produced by pathogens) are usually sufficient to assign the cause of a disease to a specific agent. Injury caused by insects attacking leaves and stems can often be recognized by the general appearance of the affected tree and by observing the evidence for past or present activity of insects. Damage caused by abiotic agents such as wind, snow, fire, drought, and most air pollutants is also easily distinguished by the experienced forester, forest pathologist, or forest entomologist.

Several biotic and abiotic agents often act simultaneously or sequentially on individual trees, causing sufficient injury to kill them. Examples include the successful attack by bark beetles after prolonged drought, chronic root rot, or dwarf mistletoe infections on conifers. Experienced diagnosticians have a high rate of success in explaining the cause or causes of poor tree appearance and tree death. Yet there may still be baffling cases that require following many avenues of investigation for a prolonged period.

The Concept of Tree Decline

The gradual decline in health and vigor of a tree species involving young as well as old trees has been termed a decline disease. There is a need to be more precise with terminology. A more restricted definition of decline refers to a deterioration of vigor and death of trees caused by a combination of biological and nonbiological stress factors (Houston 1981). No single factor is capable of causing death. Manion (1981) interprets tree decline as a succession of events in time—beginning with predisposing factors (poor site quality or tree age) that make trees more sensitive to inciting factors (insects, drought, air pollution), and ending with contributing factors (bark beetles, root diseases) that are the actual cause of death. The term *forest decline* has been used frequently in recent times when really only a single species is involved; tree decline would be a more accurate designation.

In the Pacific region forests (Hawaii), decline may be an ecological process in which a particular age cohort of plants is predisposed by environmental events as the plants age together, leading to synchronous senescence of the entire cohort (Mueller-Dombois et al. 1983).

Pole blight of western white pine may be the best example of a decline disease known in the West. However, it does not necessarily include all the variables that Manion (1981) associates with a "classic" tree decline. The pole blight disease of western white pine in the northern Rocky Mountains, studied for more than 20 years starting in the 1940s, could not be explained on the basis of current knowledge when it was first observed (Leapheart and Wicker 1966). Terminal growth of affected pole-sized trees decreased year after year, and eventually trees died from the top down. Rootlet mortality and overall deterioration of the lateral root system occurred in trees showing early symptoms of pole blight (McMinn 1956). Insects and fungus pathogens did not initiate the decline of growth. Another puzzling aspect of the problem was that only trees 40–100 years old (pole size) were affected. Companion species, including Douglas fir, grand fir, western larch, and western red-cedar were not affected.

Tree mortality was most frequent on soils with a low moisture-holding capacity. Climatological records showed that a particularly severe drought had culminated in the 1930s. The magnitude of this drought was greater than had ever been experienced by even the oldest white pine stands. The hypothesis for pole blight was that the pole-size cohort of white pines suffered injury to roots on poor sites, leading to an unfavorable balance of roots and tops. This condition occurred at a stage of development when the physiological demands of trees were at their peak and recovery was very difficult. This may not be a complete explanation of the cause of pole blight but it illustrates that a severe abiotic stress, acting alone, can cause an irreversible process of decline in tree health. It was reported, however, that some pole-size trees began to recover when precipitation increased. Another important lesson from this example is that a long period of investigation may be required to solve a problem of this complexity.

Pole blight was confined to one age class of a single species occurring on soils low in moisture-holding capacity. This situation is simple in comparison with the widespread occurrence of insects and disease organisms that often affect more than one species on the same site.

Prominent Infectious Diseases and Insect Pests

Most tree mortality in the West is the result of drought followed by bark beetle infestations (USDA 1982). Based on statistics for 1976, most softwood mortality occurred in western states, particularly in the Pacific Coast region, where there is such a huge concentration of timber volume. In many cases, infectious diseases weaken trees and make them more susceptible to bark beetle attack. Diseases alone cause volume loss because tree growth is suppressed over many years and stems may be malformed or decayed.

The category of disease responsible for the largest volume losses in the West is stem decay followed by dwarf mistletoe (Davidson and Prentice 1967). Root diseases, needle casts, and rusts play an important role over wide areas of the West (see table 3.4). It is not unusual for more than one disease to be present in the same tree. For example, a single ponderosa pine may have Annosus root rot, dwarf mistletoe, and *Elytroderma* needle cast. The same tree could be infested by flat-headed borers that persist in a tree for years before it is killed by a more

Table 3.4. *Prominent insect pests and diseases causing growth reduction and mortality of Douglas fir and ponderosa pine.*

Host	Principal insect pests	Principal diseases
Douglas fir Pacific region (P) Rocky Mountain Region (R)	Douglas fir beetle (P, R) *Dendroctonus pseudotsugae* Flat-headed borer (P) *Melanophila drommondi* Spruce budworm (P, R) *Choristoneura fumiferana*	Laminated root rot (P) *Poria weirii* Heartwood decays (P, R) Red ring rot *Fomes pini* Red-brown butt rot *Polyporus schweinitzii* Dwarf mistletoe (P, R) *Arceuthobium douglasii*
Ponderosa pine Pacific region (P) Rocky Mountain Region (R)	Mountain pine beetle (P, R) *Dendroctonus monticolae* Western pine beetle (P, R) *Dendroctonus brevicomis* Cal. flat-headed borer (P, R) *Melanophila californica*	Annosus root rot (P, R) *Fomes annosus* Heartwood decays (P, R) Red rot *Polyporus anceps* White pocket rot *F. pini* Needle cast (P, R) *Elytroderma deformans* Comandra rust (P, R) *Cronartium comandrae* Limb rust (P, R) *Peridermium filamentosum* Dwarf mistletoe (P, R) *Arceuthobium campylopodum* var. *campylopodum*

Sources: Forest Service, *Agricultural Handbook, No. 271* (Washington, D.C.: U.S. Department of Agriculture, 1965); A. G. Davidson and R. M. Prentice, *Important Forest Insects and Diseases of Mutual Concern to Canada, the United States, and Mexico* (Ottawa, Canada: The Queen's Printer, 1967); R. V. Bega, "Diseases of Pacific Coast Conifers," *Agricultural Handbook No. 521* (Washington, D.C.: Forest Service, U.S. Department of Agriculture, 1978).

aggressive insect pest. As indicated, each tree species has its own unique insect pest and disease complexes. The survival of the tree depends on year-to-year changes in climatic conditions that either weaken the tree or help it to maintain an equilibrium with the pathogens. Poor growing sites also interact with pest and disease complexes to hasten mortality. Another variable, discussed in greater detail later, is the effect that air pollutants may have on resistance to pest complexes.

In addition to damage from insects and diseases there may also be significant damage to trees by animals native to the West, including Alaska (Hiratsuka 1987). The amount of injury is not necessarily in proportion to the size of the animal. Bark may be chewed away by the snowshoe hare, several species of squirrel, or porcupine. Large patches of bark are rubbed away by deer, elk, and moose; these animals may also feed heavily on the young shoots of smaller trees.

Bears scrape and tear bark from trees. Beavers cut down significant numbers of trees in the process of dam building. Squirrels and other small mammals feed on the seeds of conifers by cutting green cones or seeking seeds out on the ground. One consequence of damage to tree bark and the breakage of branches is the establishment of an environment where spores of heartwood decay fungi can germinate and initiate infection.

Role of Natural Abiotic Stresses

Natural abiotic stresses are the result of incidents of extremes in the climatic environment of a tree. (Abiotic stresses can also be caused by humans—for example, mechanical injury to the roots or stems during a construction project or the misapplication of toxic chemicals.) The most common incidents of natural abiotic stress result from temperature extremes and drought (Bega 1978).

Extremely low winter temperatures can kill needles, buds, twigs, and inner bark of conifers. In eastern Washington, low winter temperatures have been observed to injure both ponderosa pine and Douglas fir. Ponderosa pine was injured more under similar circumstances. Although injured trees can recover, top kill and frost crack can remain as a permanent result of such injury. Fall and spring frosts occur at a time when current year foliage has insufficient cold hardiness. A partial list of western conifers in order of increasing sensitivity to frost includes lodgepole pine, Jeffrey pine, ponderosa pine, incense cedar, sugar pine, and white fir. Recovery is not possible until the following spring.

The interaction of alternating low night temperatures and unusually high day temperatures in winter results in a condition called red belt. Variable lengths of tree crowns from the tip downward display dead, reddish-colored needles. The trees are distributed in well-defined bands varying from less than 60 feet to 3000 feet wide on mountain slopes and sloping benches. The belt is formed because of a warm temperature inversion layer that may be present at varying elevations along the slope, depending on time of day. Nearby valley bottoms are constantly filled with extremely cold air, but the sudden occurrence of warm dry foehn winds forms a thin layer of warm air on the top that does not mix downward. At night, normal cold air drainage fills the valley to a greater depth and raises the level of cold air into the zone occupied by warm air during the day.

This alternate exposure to warm and cold air when the soil is usually frozen causes dessication of the foliage because high daytime transpiration removes moisture from the foliage more rapidly than it can be replaced by roots in frozen soil. Red belt can extend for many miles along major mountain ranges in the Rocky Mountains. Trees produce normal foliage in the spring unless buds are killed. Similar tree injury called sun scorch and parch blight have been reported to be caused by desiccation under winter conditions. Winter yellows (Perry and Baldwin 1966) and winter fleck (Miller and Evans 1974) are also attributed to winter conditions.

Soil drought conditions during years of low rainfall develop first on gravels and sands with low moisture-holding capacity and on shallow soils that overlay rock or gravel. Drought-damaged trees therefore usually occur in groups. Symptoms include premature needle loss and small twig dieback. These symptoms progress from the top to the base of the tree. The symptoms remaining on surviving trees in subsequent years include dead tops, fewer needle whorls, and shortened needles in whorls formed during the drought years. During severe drought conditions in California in 1929 and 1960, Douglas fir and incense cedar were killed or injured on poorer sites but ponderosa pine, sugar pine, and white fir on similar sites were less seriously affected.

Other natural abiotic factors that cause injury to western conifers include lightning, breakage by snow and ice loads, hail damage to foliage and twigs, and blowdowns during severe winds (Bega 1978).

Historic Incidents of Air Pollutant Damage

The typical source of air pollution damage to forests during the first half of the twentieth century was a nonferrous-ore-processing facility releasing a plume of pollutants into the surrounding landscape. Emission controls were absent or too rudimentary to be effective. Since the enactment of the Clean Air Act and its amendments, there has been more satisfactory control of emissions. In addition, economic constraints have meant that operations with a long history of unsatisfactory abatement have closed. The following accounts illustrate the effects on forest species of both acute and chronic exposures to sulfur dioxide or hydrogen fluoride. Other incidents of sulfur dioxide injury to forest vegetation in the West have been reported from Redding, California (Haywood 1905) and Anyox, British Colombia (Errington and Thirgood 1971). Similarly, incidents of fluoride injury to forest species have also been reported at The Dalles, Oregon (Compton et al. 1961), Georgetown Canyon, Idaho (Treshow et al. 1967), and Columbia Falls, Montana (Carlson and Dewey 1971).

Copper Smelter at Trail, British Columbia. The Trail smelter is located 11 miles north of the United States border in the gorge of the Columbia River. It began operations in 1896. The emission of sulfur dioxide continued to increase, to nearly 10,000 tons per month in 1930. Following the installation of stack gas scrubbing equipment in 1931 and the institution of voluntary cutbacks during unfavorable weather, sulfur dioxide emissions were reduced by half (Scheffer and Hedgecock 1955). During the period of highest emissions, a zone in which 30 percent of the trees were dead or in morbid condition extended 52 miles southward, along the course of the Columbia River. A zone in which 60 percent of the trees were dead or severely damaged reached 33 miles south of the smelter. At some points the damaged forested area was up to 10 miles wide. The species composition of the affected zones varied with elevation.

At lower elevations, ponderosa pine, Rocky Mountain Douglas fir, western larch, and lodgepole pine were all present; at higher elevations the principal species were Douglas fir and western larch. Douglas fir was the most sensitive to sulfur dioxide, followed by ponderosa pine, lodgepole pine, and western larch.

Growth reduction was documented for ponderosa pine. Other evidences of injury included lack of cone production, and small numbers of seedlings and saplings in damaged stands (Scheffer and Hedgecock 1955).

Copper Smelter at Anaconda, Montana. The smelter at Anaconda was located on a hill (6600 feet elevation) near the southern end of a narrow valley about 35 miles in length. The surrounding upper slopes and ridge tops were the principal forested areas affected by smelter emissions. Sulfur dioxide emitted from the stack at 7200 feet diffused into tributary drainages and across surrounding ridges to reach other forested areas beyond. Field surveys in 1908 found symptoms on Douglas fir as far as 19 miles from the smelter. Injury to lodgepole pine was not seen at distances greater than 10 miles from the source. Further observations in 1910 determined the relative susceptibility of native species on the basis of frequency and amounts of foliage thinning of trees at varying distances from the smelter. The order of decreasing susceptibility was subalpine fir, Douglas fir, lodgepole pine, Engelmann spruce, ponderosa pine, and limber pine (Scheffer and Hedgecock 1955).

Aluminum Ore Reduction Plant at Spokane, Washington. Severe injury to the foliage of ponderosa pine was first observed north of Spokane in 1943 (Adams et al. 1956). By 1952 nearly all trees in a 3-square-mile area near the aluminum ore reduction plant were dead and significant foliar injury was detectable in a 50-square-mile area. The possible biotic causes of injury were systematically eliminated, and findings of up to 600 parts per million (ppm) fluoride in foliage strongly implicated the ore reduction plant as the source of the problem (Shaw et al. 1951).

Confirming evidence was obtained by fumigating ponderosa pine with anhydrous hydrogen fluoride, by comparing symptoms on field and fumigated foliage, and by analyzing fumigated needles for fluoride content (Adams et al. 1956). Foliage was most sensitive during the period of elongation; older foliage was less sensitive. Injured foliage was shed prematurely. Extreme reductions were observed in diameter growth of injured trees (Shaw et al. 1951).

Ozone Injury to Mixed Conifer Forests in Southern California. In the early 1950s a problem described as "x-disease" was observed on ponderosa pines in the San Bernardino Mountains of southern California. Early investigators described the symptoms as a chlorotic mottle or chlorosis of older needles, leading to premature senescense. Not all ponderosa pines growing at the same site were equally affected. Healthy trees and trees with definite symptoms of injury grew with branches intermingled. Reciprocal grafts between such pairs of trees showed that the original condition of the scion persisted after graft establishment. Therefore, a graft transmissible agent was rejected as the possible cause of symptoms (Parmeter et al. 1962).

The precipitation record showed some interesting extremes. Between 1911 and 1960 (the total record available at that time), the years 1946 and 1953 were the driest ever recorded, and the average deviation from the annual mean of 41.24 inches was always negative between 1946 and 1960. These dry years were preceded by a 5-year span (1941–45) when precipitation was higher than the average. Could the 5-year span of favorable precipitation have made trees less able to withstand subsequent dry years?

The role of drought as a principal cause of the observed symptoms was doubtful because injury symptoms progressed from the bottom to the top of the tree and from the inside to the outside of the crown. Drought symptoms progress in the opposite way, and one would also expect trees growing side by side to be similarly affected if drought were the cause. Furthermore, symptomatic trees continued to decline in growth while healthy trees responded to years of more favorable moisture (Parmeter et al. 1962). Other species known to be more drought sensitive than ponderosa pine, particularly incense cedar, were not affected. It was necessary to investigate other possible causes of x-disease.

Since 1953 air pollution (smog) had been recognized as the cause of damage to crops and ornamental plants in the nearby Los Angeles basin (Middleton et al. 1950). The transport of polluted air to mountain areas by the afternoon sea breeze had been described (Coffin 1959). Air pollution (primarily ozone in smog) was suspected as the possible cause of foliage injury. Branches of nearby ponderosa pines, one with typical symptoms of injury and the other with little evidence of injury, were enclosed in branch chambers and treated with carbon-filtered air, ambient air, and ozone added to filtered air. Changes in visible symptoms and chlorophyll content revealed improvements in filtered air treatments, and new or increased injury in ambient air ozone treatments (Miller et al. 1963). This field experiment and later ozone fumigations of seedlings (Richards et al. 1968) confirmed that ozone was responsible for injury to ponderosa pine foliage.

If ozone was responsible for primary injury to foliage and subsequent reduction of foliage surface area, how did the good moisture years from 1941 to 1945 and the consistently dry years thereafter, particularly 1946 and 1953, interact to affect tree vigor? At best we can only suggest that the two stresses are additive and that ozone continues to cause injury even in years with abundant precipitation.

Aerial photo surveys, based on stratified random sampling, were used in the San Bernardino and San Gabriel mountains to establish the extent of injury in relation to the Los Angeles basin (Wert 1969). Ground surveys in other parts of California in 1970 established evidence of ozone injury symptoms to conifers in the Laguna Mountains east of San Diego, in the Sierra Nevada in the Sequoia National Forest (east of Fresno), and at scattered locations in the San Francisco and Monterey bay areas (Miller and Millecan 1971).

Beginning in the early 1970s, an expanded research effort was begun in the San Bernardino Mountains by cooperating agencies. Between 1973 and 1978 an interdisciplinary study team was assembled to study the effects of 25 years of exposure to ozone on selected ecological processes in these mixed conifer forests. The project was funded for the most part by the U.S. Environmental Protection Agency and carried out by researchers from the Forest Service, U.S. Department of Agriculture, and the University of California Riverside and Berkeley campuses.

Two questions were investigated:

• What are the responses of organisms and processes in the conifer forest to different levels of chronic oxidant exposure?
• Can the observed responses be interpreted in an ecosystem context?

The records for temperatures, relative humidity, and ozone concentrations were assembled from existing records and measured concurrently with the study. The system components selected for study included water availability as a function of soil and site attributes, foliage injury along a gradient of decreasing ozone dose, leaf litter accumulation, litter decay and partitioning of selected nutrients, tree seed production, seedling establishment, tree growth, insect and disease components responsible for tree mortality, tree population dynamics, and small mammal population dynamics.

Data required to study each system component were obtained from 18 vegetation plots located in west-to-east transects (see figure 3.1) along which a distinct gradient of decreasing ozone concentration was measured (see figure 3.2). The California mixed conifer forest type was the general designation for the forest cover type but five subtypes were identified whose locations were a function of elevation and distance eastward. In a general sense the following subtypes were identified in a west-to-east sequence: ponderosa pine, ponderosa pine–white fir, ponderosa pine–Jeffrey pine, Jeffrey pine–white fir, and Jeffrey pine. The common unit of vegetation at all plots was a minimum of 50 ponderosa or Jeffrey pines larger than 30 centimeters in diameter at breast height. All other species and size classes were included that were in the area occupied by the 50 pines (Miller 1983).

The ecosystem components most affected by ozone were the tree species, the fungal microflora present on living needles, and the foliose lichens growing on tree bark. Ponderosa pine and Jeffrey pine were equally the most sensitive to chronic ozone exposure, with white fir, California black oak, incense cedar, and sugar pine following in decreasing order of sensitivity. Foliar injury on sensitive ponderosa and Jeffrey pines was present when the 24-hour-average ozone concentration was 0.05–0.06 ppm during May through September. Thus, the eastern half of the study area did not have ozone concentrations sufficient to cause more than very slight visible injury to foliage (Miller 1983). In the area with higher average ozone concentrations (namely 0.10–0.12 ppm), the decrease in tree vigor precipitated a series of other events that together had the effect of changing stand dynamics.

The relative effects of ozone on different species throughout the study area can also be expressed by dividing the 50 kilometers from the heavily ozone-polluted western end to the slightly polluted eastern end into zones. Stem

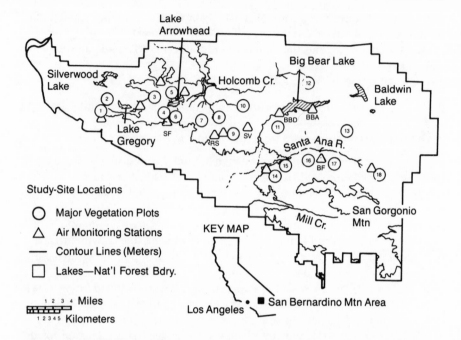

Figure 3.1. Eighteen vegetation study sites and thirteen semipermanent air monitoring stations in the San Bernardino National Forest. Source: P. R. Miller, "Ozone effects in the San Bernardino Mountains," *Symposium on Air Pollution and Productivity of the Forest* (Washington, D.C.: Izaak Walton League/Penn State University, 1983).

growth, of ponderosa pine in particular, was limited as much as 50 percent in the first (western) 20-kilometer zone studied (Gemmill et al. 1982), and accumulated mortality during 1973 to 1978 reached as high as 10 percent. White fir growth was reduced in this first zone, and mortality was increased. Foliage injury and up to 25 percent stem growth decrease were observed on California black oak in the first 20-kilometer zone (Miller et al. 1980). Stands containing Jeffrey pine were located 25 kilometers or more from the sites where the severest injury to ponderosa pine was observed. However, foliage retention was reduced by one annual whorl, and mortality rates were higher in the 30–40-kilometer zone than in the zone between 40 and 50 kilometers.

Injured ponderosa and Jeffrey pines older than 130 years produced significantly fewer cones per tree than uninjured trees of the same age (Luck 1980). The smaller number of trees that did have abundant cone crops were favored by grey squirrels that feed on cones that were still green.

Dahlsten and Rowney (1980) concluded that ozone-damaged trees, compared with healthier trees, produced about the same total brood of new beetles

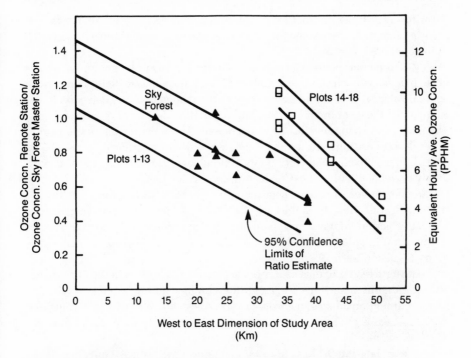

Figure 3.2. Estimated gradients of ozone concentrations along two west-to-east axes in the San Bernardino Mountains (expressed as a ratio obtained by dividing doses for all comparable times at semipermanent monitoring stations by the dose at Sky Forest).

with fewer initial attacks. It was therefore proposed that a given population of western pine beetles killed more trees and increased at a greater rate in stands with a high proportion of ozone-injured trees. A pest survey completed on private lands in the area of severe ozone injury revealed the interplay of other factors in the mortality process. In addition to ozone injury, trees were subjected to disturbance from urban construction activities, breakage from wind and snow or ice, and further damage from fire and lightning. The ultimate cause of death for 53 percent of pine trees was bark beetles alone, followed by a combination of bark beetles and construction (20 percent), root diseases and bark beetles (10 percent), and other combinations of biotic and abiotic factors (the remaining 17 percent) (Miller 1983).

There was some evidence that disease development was hastened by chronic ozone injury where trees were infected with root disease caused by *Fomes annosus* (James et al. 1980). Perhaps the most important relationship between ozone injury and the root rot was that both led to tree death (usually in conjunction with bark beetles), which resulted in new clearings. However, those clearings resulting from root rot could not support regeneration of native conifers because of the persistence of the disease organism in dead roots (from which it infected healthy roots).

Measurements of litter accumulation indicated excessive amounts of needle litter in areas with chronic ozone injury (Arkley and Glauser 1980). The greatest annual litter fall was found under trees in the moderate injury category. It was shown that the taxonomic diversity and population density of fungi growing on the surface of living needles were decreased. These organisms participate in litter decomposition, and therefore the rate of litter decomposition may be affected. In fact, the decomposition rate of ponderosa pine needle litter was positively correlated with increasing leaf injury (Bruhn 1980).

A comparison of lichen species found on conifers during the years 1976 to 1979 with collections from the early 1900s showed 50 percent fewer species in the latter period. Marked morphological deterioration of the common species *Hypnogymnia enteromorpha* was common in areas of high ozone concentrations (Sigal and Nash 1983).

The entire suite of data sets collected from 1973 to 1978 in the San Bernardino Mountains made some progress toward understanding responses of some of the prominent organisms to chronic ozone stress, but only a little was learned about processes such as the flux of water, nutrients, and carbon under different levels of ozone.

Newer data (McBride et al. 1985) combined with those collected earlier (1973–78) have been useful in gaining a better understanding of the process of forest succession under ozone stress. Eight of the original 18 plots dominated by ponderosa pine (numbers 1, 2, 4, 5, 6, and 7 [severe injury] and numbers 3 and 14 [slight to moderate injury]) were resurveyed in 1982–83 to determine changes in species composition and regeneration. In the 6 severe class plots the accumulated mortality noted in the young mature (50–99 years) age class was 33 percent since 1973; only 7 percent of this class died on the slight-to-moderate injury plots. In the same age-class group, 24 percent of the white fir died on the severe plots, while no mortality was observed in this age class in the other two plots (see table 3.5).

High mortality occurring in younger age classes in both categories of plots is presumed to be the result of greater competition for space, light, water, and nutrients among smaller trees. The lower mortality in the mature age class in the severely injured plots suggests that remaining mature trees are individuals with more tolerance to ozone. These trees survived the entire period of ozone stress from the mid–1950s to 1983. Although data are not available, one possible hypothesis is that mortality rates and losses to sanitation salvage logging among mature trees in the early period of stress were at least equal to those now documented for young mature trees.

For nearly two decades during the early phases of ozone injury in the San Bernardino Mountains there was intensive sanitation salvage logging to remove trees with weakened crowns that were judged to be more susceptible to bark

Table 3.5. *Mortality by age class group and tree species in areas of severe and slight foliar ozone injury in the San Bernardino Mountains, 1974–84.*

Age class group	Species 1	Foliar ozone injury	
		Severe	Slight
		(Percent)	
Seedling	PP	54.5	41.2
(0–9 years)	WF	57.5	44.2
	IC	41.4	—
	SP	51.1	0*a*
	BO	63.5	68.5
Saplings and poles	PP	67.8	85.7
(10–49 years)	WF	42.4	45.8
	IC	19.4	—
	SP	30.9	100*a*
	BO	56.5	69.7
Young mature	PP	33.2	6.9*a*
(50–99 years)	WF	24.2	0
	IC	4.2	—
	SP	35.6	20
	BO	3.3	0*a*
Mature	PP	14.4	22.5
(100 + years)	WF	0	50*a*
	IC	—	—
	SP	—	0
	BO	0	50*a*

Source: J. R. McBride, P. R. Miller, and R. D. Laven, Effects of oxidant air pollutants on forest succession in the mixed conifer type of southern California. In *Proceedings of Conference on Air Pollutants Effects on Forest Ecosystems* (St. Paul, Minn.: Acid Rain Foundation, 1985).

Note: PP = *Pinus ponderosa*; WF = *Abies concolor*; IC = *Libocedrus decurrens*; SP = *Pinus lambertiana*; BO = *Quercus kelloggii*.

*a*Difference significant at 0.01 percent.

beetle attack. Even casual observation of older trees that have survived to the present day with moderate ozone injury suggests that many high-volume trees may have been removed that might otherwise have survived. This outcome is more meaningful in the context of the San Bernardino National Forest, where the main objectives are to provide recreation, protect wildlife, and maintain a high-quality watershed. The decision to remove a tree was based mainly on the general appearance, such as foliage density, color, and length. Salvage logging may have influenced the potential for seedling establishment of some species.

The zero-to-nine-year age class (see table 3.6) showed lower numbers of seedlings of all species, except incense cedar, established in severe-injury plots in 1975–84 than in 1965–74. Projections from these data suggest a decline of

Table 3.6. *Comparison of ten-year seedling establishment in areas of severe and slight foliar ozone injury in the San Bernadino Mountains, 1965–74 and 1975–84 (average number of seedings established per hectare).*

Species 1	Severe injury		Slight injury	
	1965–74	1975–84	1965–74	1975–84
PP	92	94	180	131
WF	1321	226[a]	172	94[a]
IC	193	890[a]	0	0
SP	0	14[a]	5	5
BO	327	175	265	83[a]

Source: J. R. McBride, P. R. Miller, and R. D. Laven, Effects of oxidant air pollutants in the mixed conifer type of southern California. In *Proceedings of Conference on Air Pollutants Effects on Forest Ecosystems* (St. Paul, Minn.: Acid Rain Foundation, 1985).

Note: PP = *Pinus ponderosa*, WF = *abies concolor*, IC = *Calocedrus decurrens*, SP = *Pinus lambertiana*, BO = *Quercus kelloggii*.

[a]Difference significant at 0.01 percent.

ponderosa pine and an eventual dominance of incense cedar on the severe-injury plots (McBride et al. 1985).

Finally, the worst-case scenario for forest succession involves the frequency and intensity of wildfires. The stands historically dominated by ponderosa and Jeffrey pines were the so-called fire climax, a species mixture that was the product of frequent fires—a completely natural phenomenon before the influence of European settlers. The shorter interval between fires in the past limited accumulation of ground fuel and eliminated species in the understory easily killed by fire because of thin bark or an abundance of branches near the ground, which resulted in the whole tree being ignited. Fire also provided a suitable seedbed for ponderosa and Jeffrey pines so there was a fairly continuous establishment of young trees, resulting in an uneven-aged stand dominated by these species. Because in recent years the interval between fires has lengthened because of fire control policies, shade-tolerant species that successfully establish in the presence of litter (incense cedar and white fir) have filled in the understory. This is particularly true of incense cedar.

Now the consequences of fire are likely to be more drastic. Wildfires that enter stands with heavy litter accumulations and an understory of white fir or incense cedar are so intense that even the largest remaining pine overstory trees are likely to be destroyed. Without costly intervention to reestablish conifers, large areas would probably be converted to a self-perpetuating cover of shrubs and California black oak that are capable of resprouting from the roots after fires. This cover type produces crown closure rapidly, thus excluding the establishment of a conifer forest cover. The burn-resprout cycle may continue indefinitely.

As to the two questions posed at the outset of the interdisciplinary study between 1973 and 1978, the first—regarding identification of components and processes affected by ozone—was answered to a large extent. The second asked for interpretation of effects in an ecosystem context. The limited time available did not allow sufficient data analysis and the preparation or adaptation of the models needed to examine this question. Nevertheless, word models were developed for selected systems, which led to a more cohesive understanding of the problem (Miller and Elderman 1977, Kickert and Gemmill 1980).

Ozone Injury to the Mixed Conifer Forests in the Sierra Nevada. The initial investigations of ozone effects in the Sierra Nevada have consisted mainly of repeated surveys for ozone injury symptoms by the U.S. Forest Service in the Sequoia National Forest and by the National Park Service (NPS) in Sequoia and Kings Canyon national parks. Both agencies have maintained ozone-monitoring stations. The first evidence of ozone injury to ponderosa pine was found at Happy Gap, located in the Sequoia National Forest just a few miles west of the Big Stump entrance to Sequoia and Kings Canyon national parks (Miller and Millecan 1971). Subsequently, a short period of combined ground and aircraft measurements of ozone in August 1970 confirmed the transport of ozone from the Central Valley eastward to the Mineral King Valley, located at 2287 meters in the Sequoia National Forest (now a part of Sequoia National Park). The ozone concentration peaked at 1800 hours, with a 10-day average peak value of 0.08 ppm total oxidant, as a result of afternoon up-canyon transport (Miller et al. 1972).

An initial survey of part of the Sequoia National Forest was completed in 1974–75, including the establishment of four permanent plots with 25, 19, 21, and 35 trees (Williams et al. 1977). The plots were reexamined in 1983 to determine if changes in foliage symptoms could be detected (Williams and Williams 1986). Chlorotic mottle symptoms appeared on current-year needles of 23.8 percent of the trees in 1983, compared with 14.5 percent in 1975. For 1-year-old needles, symptoms were present on 60.7 percent of the trees in 1983 compared with 44.2 percent in 1975. Inferences were made about possible growth reductions resulting from chronic injury to foliage. These conclusions are subject to challenge, however, since abbreviated procedures were employed to interpret the sources of variation contributing to trends in annual ring width. Peterson et al. (1987) completed a tree ring analysis of Jeffrey pines growing in areas of Sequoia and Kings Canyon national parks with and without ozone exposure. A reduction of ring width index was noted in recent years for large trees growing on harsh sites in the ozone-exposed area. This work is discussed more thoroughly later in this chapter.

Additional surveys were carried out by Forest Pest Management (FPM) of the Forest Service's Region 5 (Pronos and Vogler 1981) over a broader area in the Sequoia National Forest. This work resulted in the establishment of a large number of semipermanent, 10-tree ponderosa or Jeffrey pine plots extending from points east of Bakersfield northward to points east of Fresno. The amount of foliage injury in these plots was categorized as mostly none to slight, with occasional moderate injury, compared with the severest levels of injury known to occur in the San Bernardino National Forest. Some of these plots have been revisited in subsequent years and survey activities have been extended northward to the Sierra, Stanislaus, Eldorado, and Tahoe national forests. Some evidence of slight damage was detected in these forests, but the most significant injury continues to be evident in the Sierra and Sequoia national forests.

By 1982 the Resources Management Division of Sequoia and Kings Canyon national parks had established 54 plots spread throughout the parks at points where 5000-, 6000-, 7000-, and 8000-foot contour intervals intersected with roads or trails. The FPM scoring method was used for these plots. Branches were pruned from the lower crown of each tree and after close inspection the youngest needle whorl showing evidence of chlorotic mottle was recorded. The number code assigned zero to the youngest needle whorl and higher numbers to each successivly older whorl; a low number therefore meant more chronic ozone injury. In 1984 the scoring method was made more descriptive by recording the percentage of chlorotic mottle on each needle whorl present and the percentage of other injuries (insect injury, winter flecks) on each whorl. Other variables recorded were the percentage of the original number of needles retained per annual whorl, needle length, and some other tree and site information. Further method modifications by the NPS include an increase from 10 to 15 trees per plot and an independent evaluation on 5 branches per tree.

The results of the FPM and NPS surveys showed that ozone injury symptoms were severest on the slopes of mountains facing the Central Valley (front range) in drainages and tributaries of the Kaweah, Kern, and Kings rivers (Wallner and Fong 1982). Surveys were repeated in the front-range plots in 1984–85. The general trend revealed from both FPM and NPS surveys is an increase in injury based on FPM scores (lower numbers mean more injury). For example, the NPS survey showed scores of 2.8 in 1980–82, 2.6 in 1984, and 1.8 in 1985. In effect, the mean injury progressed from the fourth to the third whorl. Furthermore, 48 percent of all trees in plots were injured in 1980–82, and in 1985 an average of 87 percent of the trees had some level of injury (National Park Service, Air Quality Division, Internal Report).

In 1985, twenty 15-tree pine plots were established in Yosemite National Park along the western border of the park and in a west-east transect along Tioga Road. Data from a single year show injury to 58 percent of all trees. Injury is located mainly in the 5000- to 6000-foot zone, as it is in Sequoia and Kings Canyon national parks (National Park Service, Air Quality Division, Internal Report).

In 1977–78 an air quality survey in the Bakersfield area (southwest of Sequoia National Park) revealed that sulfate and sulfur dioxide concentrations were among the highest recorded in California (Duckworth and Crowe 1979). Ozone concentrations in the Bakersfield area are routinely elevated during the summer months and the pollutant mixture is advected eastward into the Sequoia National Forest by up-slope and up-canyon winds.

A study was carried out to (1) monitor ozone and sulfur dioxide concentrations at various distances east of Bakersfield in mountain terrain; (2) measure sulfate content of surface soils and subsoils, and the sulfur content of pine needles and lichen thalli; (3) fumigate seedlings of native species (ponderosa pine, Jeffrey pine, and digger pine) with mixtures of ozone and sulfur dioxide; and (4) survey tree vegetation to determine if other than typical ozone symptoms were present (Taylor 1986).

Mountain air monitoring sites were located at 2400 and 6600 feet, with the highest one about 35 miles northeast of Bakersfield in the mixed conifer forest type. Ozone concentrations were within 1–2 parts per hundred million (pphm) of each other at Bakersfield and the 6600-foot site, but sulfur dioxide was 10 times lower at the 6600-foot site. Thus the opportunity for mixed gas fumigations was remote under natural conditions. Only ozone symptoms were seen.

The concentrations of extractable sulfate in surface soils (0–5 centimeter layer) and sulfur in pine needles and lichen thalli tended to decrease ($p < 0.01$) with increasing elevation and distance northeast or east. Sulfur concentrations in living tissues were considerably below toxic levels. Mixed-gas fumigations of ponderosa pine and Jeffrey pine seedlings with adult foliage resulted in more foliage injury from gas mixtures than from single pollutants when the ozone concentration was 20 pphm and sulfur dioxide concentrations were 5, 10, or 20 pphm. Symptoms resembled a necrotic mottle in ozone–sulfur dioxide mixtures. Recently germinated seedlings of ponderosa, Jeffrey, and digger pines showed reductions in root growth ($p < 0.05$) at 10 pphm sulfur dioxide and 20 pphm ozone after 64-day fumigations. These results suggest that future air quality would have to deteriorate enormously with respect to sulfur dioxide before additive injury from ozone could be expected.

A method of rating ozone injury to California black oak was tested as an additional bioindicator of the annual ozone dose in Sequoia and Kings Canyon national parks. Preliminary results suggested that black oak would provide better evidence of an extraordinarily high single season of exposure than pine needles, which show the accumulated effects of several years' exposure.

Experiments are in progress to investigate the possible effects of ozone on the growth and survival of newly emerged giant sequoia seedlings. Open-top chamber fumigations are in progress to test effects of carbon-filtered air, ambient air, and ambient air with an addition of 50 percent more ozone on the foliage appearance and growth of roots and tops of newly emerged giant sequoia seedlings.

In conclusion, monitoring trends of ozone injury to pines in the southern Sierra Nevada and continuing experiments to define the effects of possible future air quality scenarios have considerable significance for this region. The Central Valley of California has a high meteorological potential as a source for higher photochemical air pollution levels. Population increases in the valley and even in the distant San Francisco Bay area may eventually outstrip present means of emission control from automobiles and other hard-to-limit sources. The natural up-slope, up-canyon flow of polluted air from the valley will continue to threaten the forest resource, recreation value, and wilderness heritage of the national parks and national forests of the southern Sierra Nevada.

Surveys for Ozone Injury to Forests in Other Western States. The Air Quality Division of the National Park Service has supported surveys of pine forests in Saguaro National Monument near Tucson, Arizona, and in Rocky Mountain National Park near Denver, Colorado. Current evidence indicates that 20 percent of pines in plots in the Rincon Mountains of the Saguaro National Monument have chlorotic mottle symptoms attributable to ozone. No symptoms have been observed on ponderosa pines at plots in Rocky Mountain National Park (Stolte 1987).

Inventory of Ozone and Acidic Deposition in Selected Regions of the West

In this section some trend data will be reported for ozone and acidic deposition in certain areas of the West. In addition, the issue of global carbon dioxide increase will be introduced because of its primary effects on plant productivity and its possible secondary interactions with other abiotic stresses.

Effects on forest vegetation may be due to wet deposition of hydrogen ions, sulfates, nitrates, and perhaps hydrogen peroxide and to dry deposition of gases, including ozone, sulfur dioxide, nitrogen dioxide, nitric acid, and acidic particulates (mostly sulfates and nitrates). In the West, ozone and the dry and wet deposition of acidic compounds are emphasized.

Urban centers are currently the major source of these pollutants in the West; outside such areas there are occasional point sources of a specific pollutant that emit sufficient quantities to be detectable over wide regions. For example, slightly elevated sulfate levels were detectable over wide areas when copper smelters were in full operation in Arizona (Oppenheimer et al. 1985).

Ozone

The various centers of ozone pollution in the West are determined by a combination of population density and the general climate of each region (see figure 3.3). During summer the average daylight (0600 to 2000 Standard Time) concentrations are clustered mainly in the 0.04 to 0.06 ppm range in urban areas (USEPA 1986). California leads in exceeding 0.12 ppm because of its large population and a climate favorable to pollutant accumulation (Ludwig and Shelar 1980). Forest regions in California are often located within the pollutant trans-

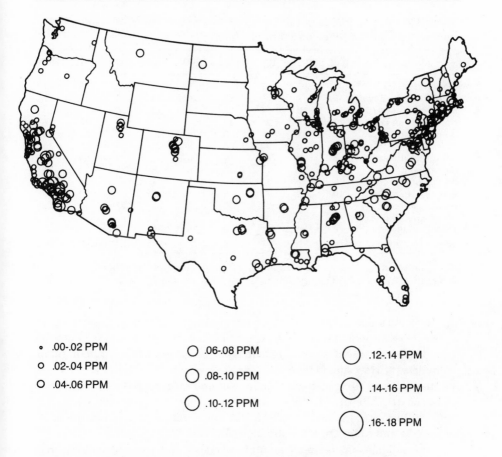

• .00-.02 PPM   ◯ .06-.08 PPM   ◯ .12-.14 PPM

o .02-.04 PPM   ◯ .08-.10 PPM

O .04-.06 PPM   ◯ .14-.16 PPM

                       ◯ .10-.12 PPM

                                               ◯ .16-.18 PPM

Figure 3.3. Average daylight (0600 to 2000 standard time) concentrations of ozone, April–September 1981.

port range of the major urban centers, including the Los Angeles and San Francisco Bay areas. Pollutant transport mechanisms operate effectively over long distances. For example, the pollution of the San Francisco Bay area has been traced 300 kilometers to Yosemite National Park (Carroll and Baskett 1979). The Los Angeles urban plume has been traced as far as 350 kilometers east, across the Mojave Desert to the Colorado River, near Needles, California. Daily maximum ozone concentrations were 0.07 to 0.08 ppm during one episode of high pollution. The long-term record for daily peak ozone concentrations ranged from 0.019 to 0.044 ppm at Needles, California (Hoffer et al. 1982).

The ozone air quality trends remained stable in California in all 14 designated air basins during 1977–85 (see table 3.7) except for the South Coast Air Basin (Los Angeles area), where the 3-year mean maximum hour ozone concentration declined from 0.147 to 0.127 ppm between 1978 and 1984 (California

Table 3.7. *Ozone trends in California's fourteen air basins, 1977–85 (maximum hourly ozone concentration [pphm]).*

Air basin	1985	1984	1983	1982	1981	1980	1979	1978	1977
North coast	7	7	7	8	9	8	10	9	—
San Francisco Bay	16	17	20	15	18	20	19	23	17
North central coast	11	10	11	11	14	14	10	11	14
South central coast	23	19	23	23	24	21	23	25	26
South coast	39	34	39	40	39	49	47	46	39
San Diego	22	28	28	23	29	23	26	39	25
Northeast plateau	8	9	7	7	6	—	9	—	—
Sacramento Valley	20	21	17	16	18	18	25	22	19
San Joaquin Valley	16	17	17	18	18	21	18	21	21
Great Basin valleys	10	9	9	9	8	10	8	7	—
Southeast desert	29	25	26	24	33	29	27	30	27
Mountain counties	—	—	—	—	—	—	—	—	—
Lake County	8	8	7	8	8	9	—	—	—
Lake Tahoe	10	8	8	9	10	9	8	10	10

Source: California Air Resources Board, *Effect of Ozone on Vegetation and Possible Alternative Ambient Air Quality Standards* (Sacramento, Calif.: 1987).

Air Resources Board 1987). Most air-monitoring stations in California are in urban areas; the mountain counties in particular have a poor ozone-monitoring record. Natural resource management agencies have established continuous monitoring at a few sites in the Sierra Nevada (see table 3.8). The 6-year record at Whittaker Forest (1654 meters), near Sequoia and Kings Canyon national parks, shows no detectable trend from 1976 to 1981 (Vogler 1982). After 1981, continuous monitoring records are available from Giant Forest, and from Ash Mountain in Sequoia and Kings Canyon national parks.

The relationship between monthly averages and the maximum 1-hour ozone concentrations for 1981 at Tucson, Denver, and two California cities

Table 3.8. *Average ozone levels during summer months at monitoring sites in the southern Sierra Nevada, 1976–81.*

Site name	1976	1977	1978	1979	1980	1981
(mean daytime hourly ozone at 0900–2000 hours, June–September)						
Whittaker Forest	.07[a]	.084	.078	.079	.075	.082
Mountain Home	—[b]	.073[c]	.068	.071	.057	.073
Greenhorn Summit	—	—	.085	.084	.082[a]	.086

Source: D. R. Vogler, *Forest Pest Management Report 82–17* (San Francisco, Calif.: Forest Service, U.S. Department of Agriculture, 1982).

Note: Each year includes 85 percent or more of possible hours except as footnoted.

[a]Only 60 percent of total possible hours.

[b]No data.

[c]Only 45 percent of total possible hours.

(Pomona and Lennox) have been compared (USEPA 1986) (see figure 3.4). Pomona is upwind from the San Bernardino National Forest, where pine damage is severe nearest the basin. The consistent inversion layer allows for pollutant buildup on many summer days, and on-shore breezes transport the polluted air eastward to the San Gabriel and San Bernardino mountains. The summer continental climate of Denver and Tucson does not result in frequent inversion layer formation, for example, fewer than 5 percent of summer days in Colorado and Arizona compared with 83 to 86 percent in coastal California and 30 percent in coastal Washington (Holzworth and Fischer 1979). Wind directions are also more variable in a continental climate. The greater climatic variability and re-

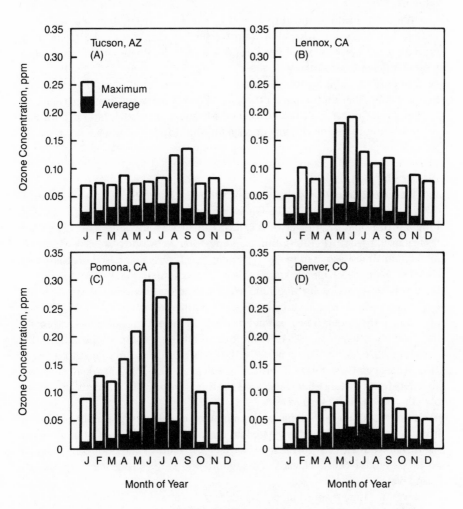

Figure 3.4. Seasonal variations in ozone concentrations in four cities, 1981 (monthly averages and 1-hour maximums).

sulting lower ozone concentrations (compared with Pasadena and Lennox) make it less probable that forests near Denver and Tucson will receive sufficient ozone exposure to cause significant foliage injury.

Ozone air quality data from the Pacific Northwest (Puget Sound) show concentrations similar to those recorded in the southern Sierra Nevada (Edmunds 1988). In addition, data from the region of Vancouver, British Columbia, also yield concentrations greater than 0.08 ppm from 1–3 percent of the time from 1978 to 1985, with concentrations above that level decreasing in recent years (Greater Vancouver Regional District 1985). The Pacific Coast summer climate is clearly much more conducive to ozone pollution than inland or Rocky Mountain climates.

Winter conditions in continental sites may be more conducive to pollutant accumulation, particularly in the Southwest. For example, the Lake Powell area, including southern Utah and northern Arizona, is dominated in winter by stagnant conditions associated with slow moving high-pressure systems. Pollutant accumulation is enhanced considerably under these conditions (Chang-Han and Pielke 1986). The emissions in the Lake Powell basin are mainly from coal combustion. Denver, Salt Lake City, and Missoula are also noted for pollutant accumulation in winter. Effects on vegetation during winter have not been reported.

Acidic Deposition

Wet Deposition. The pH of wet deposition in the western states ranged from 4.7 to 5.7 in 1986 (see figure 3.5). Only 3 of 30 National Atmospheric Deposition Program/National Trend Network stations in 11 western states had pH values less than 5.0 (NAPAP 1986). Sulfate deposition in western states is approximately ten times less than in eastern states (see figure 3.6), based on 1986 data (NAPAP 1986). A sulfur transport scale of 1000 kilometers was proposed for the intermountain and Rocky Mountain West, including Idaho, Wyoming, Nevada, Utah, Colorado, Arizona, and New Mexico (Oppenheimer et al. 1985). The implication of this report for emission limitation in the entire region was extremely important because increases of emissions in one area could supposedly influence the entire region. This report and the 1985 World Resources Institute report "The American West's Acid Rain Test" (Roth et al. 1985) have sparked considerable debate about data interpretation regarding transport of pollutants and possible effects on natural resources.

Reiter (1984) reported that under monsoon conditions, pollutant emissions from sources in Arizona, New Mexico, the Mojave Desert, the Colorado River Basin, and southern California may be transported northward aloft within moist air masses to the Sierra Nevada. This would add to the background wet deposition levels.

Overall a finer resolution look near urban centers in California shows the pH of rainfall to be lower than indicated by the National Atmospheric Deposition Program stations that are generally distant from sources. Urban sites along the

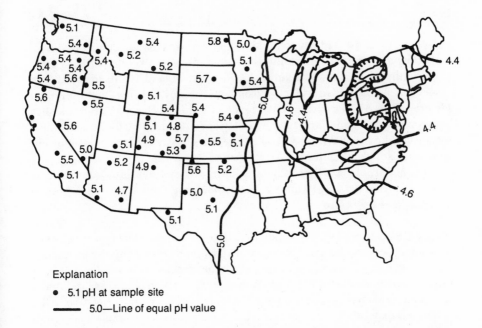

Figure 3.5. Precipitation-weighted annual average pH of wet deposition, 1986, based on NADP/NTN data.

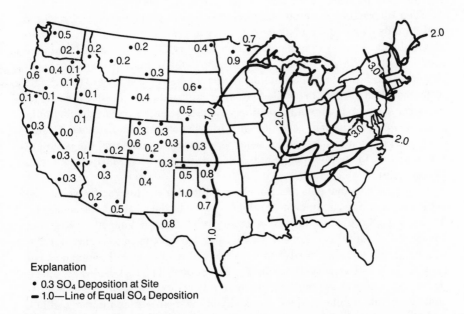

Figure 3.6. Annual sulfate iron deposition in the United States, 1986. Based on NADP/NTN data (g/m²). Source: National Acid Precipitation Assessment Program 1986.

coast have the most acidic precipitation due to a combination of high pollutant emissions and lower concentrations of ammonia and alkaline particles compared with inland areas (California Air Resources Board 1986), and acidity is associated mainly with nitric acid except where industries emitting sulfur dioxide are clustered. For example, the range of rainfall pH at 33 California sites between July 1984 and June 1986 was between 5.10 and 5.60 for 27 sites; the 6 remaining sites ranged from 4.71 to 5.02. All 6 sites were in urban locations in southern California (California Air Resources Board 1986).

The acidity of fog in the southern California coastal area throughout the year can be as low as pH 1.69 (Waldman et al. 1982), and the acidity of San Joaquin Valley winter fog is also higher than rain.

Dry Deposition. Dry deposition is 15 times more important than wet deposition in the Los Angeles area (Morgan and Liljestrand 1980). In this region the dry deposition of nitric acid is as much as 12 times greater than wet deposition of nitric acid (Russel et al. 1983). Substantial dry deposition of nitric acid, nitrate, and ammonium has been documented in an area of chaparral vegetation at the San Dimas Experimental Forest east of Los Angeles (Bytnerowicz et al. 1987) and at Giant Forest in Sequoia and Kings Canyon national parks (California Air Resources Board 1986). One of the hypotheses for the cause of certain tree declines invokes the untimely addition of nitrogen, which delays the development of winter hardiness in new foliage (Friedland et al. 1984).

The information base on temporal and spatial characteristics of regional pollution by ozone and on both wet and dry deposited nitrate, nitric acid, and sulfate in the West is dominated by reports from California, with its unique Mediterranean climate. It is difficult to extrapolate information on present conditions in California to the continental regions of the West because of the climate difference. It is prudent to conclude from existing data that forested areas must be in a consistent downwind direction from urban centers before visible injury from ozone or significant accretion of acidic deposition can be noted. Pacific coastal climates and topography seem to enhance the potential for pollutant accumulation in summer compared with continental climates (Holzworth and Fisher 1979).

It is fortunate that data sets from both California and the Pacific Northwest show that ozone has not increased, and may even be decreasing over the past 8–10 years (California Air Resources Board 1987, Greater Vancouver Regional District 1985). But this is not a cause for relaxation of concern because population increases and associated land development tend to spread photochemical air pollution and associated acidic deposition over wider areas. Traffic speed tends to decrease as regions become more intensely developed. Primary emissions from automobiles increase markedly at lower speeds and under gridlock conditions. The capacity of antipollution equipment to compensate as the automobile population increases may be outstripped in critically conjested areas.

Global Increase of Carbon Dioxide

Atmospheric carbon dioxide levels have increased by 25 percent from 1800 to 1985. The rate of increase accelerated after 1958 due to an unprecedented increase in the use of fossil fuels. A doubling of the atmospheric carbon dioxide level of the year 1800 may produce a warming of the global average temperature greater than during any period in the last 100,000 years (Solomon et al. 1985). The increase in temperature coupled with the increase in carbon dioxide may have effects on forest growth and community succession that could be profound, as discussed in the next section.

The Multiple Stress Hypothesis in Tree and Forest Decline

An earlier section considered the various biotic and abiotic factors that influence forest health simultaneously or sequentially. This section attempts to show that a diverse collection of diseases and insects along with extremes of the normal physical environment and well-known air pollutants are currently, or have been in the past, important agents leading to growth decline and mortality of western forest species. Usually there is no difficulty in identifying the agent or agents responsible if careful diagnostic work is done. Following the well-publicized descriptions of the forest decline in central Europe, for which the primary cause or causes remain under dispute, there has been a wave of reaction in the United States whereby unhealthy appearances of certain tree species are sometimes ascribed to the complex decline category before sufficient diagnostic investigation has been completed.

An example of this quick reaction to reports of tree decline occurred in Gothic, Colorado, in 1984. After a field trip two scientists, one from New England and one from West Germany, announced a discovery of Engelmann spruce trees suffering from "acid rain damage." This excursion to Gothic was held in conjunction with a symposium on acid rain at nearby Gunnison, Colorado. The situation escalated immediately when the press reported the incident, and it struck a sensitive nerve with environmental organizations and the state's significant tourism industry. Local forest pathologists quickly recognized that the trees in question were suffering from root disease caused by one or more fungi (*Ionotus circinatus* and *Armillaria mellea*) known to be common in the area. Nevertheless, a panel of scientists from outside the state was called in to assist in an additional investigation and to evaluate carefully all other possible causes. The investigation, sponsored by the Environmental Protection Agency, culminated in a report and news release (Bruck et al. 1985).

Conceptual Framework for Studying Multiple Stress Effects

As work on *Waldsterben* (forest death) progressed in West Germany, scientists in the United States acquired a new viewpoint regarding forest health. This viewpoint asks whether there are subtle changes in the forest environment on this continent that may shift the balance of physiological processes to different homeostatic states of growth and development, characterized as a slow loss of tree vigor. No significant response may be seen until a chain of relatively minor incidents triggers an acceleration of morbidity and mortality. In the context of studying and explaining such a phenomenon, a process-driven modeling approach was suggested (Southern Commercial Forest Research Cooperative 1986).

At least six hypotheses have been proposed for the cause of the European forest decline that could be investigated with the process model approach. They involve the following factors: acidification-aluminum toxicity, ozone, magnesium deficiency, general stress, excess nitrogen, and an airborne organic chemical (Schutt and Cowling 1985). The elements of all hypotheses could conceivably interact to cause Waldsterben, or the "new type of forest decline." There is evidence that the characteristics of decline differ sufficiently from region to region in central Europe so that it is not necessary to be concerned about all the hypotheses in a given area; a single hypothesis may be appropriate in a particular region. An example is the deficiency of soil magnesium in parts of southwest Germany (Zoettl and Huettl 1986).

A model for understanding the role of multiple stresses suggested by Sharpe and Scheld (1986) includes the interplay of cause(s), interaction(s), mechanism(s), and effect(s) (see table 3.9). Modeling therefore is essential for relating causes that act at several time scales to changes in mechanisms and processes that influence tree or stand health. The goal is to provide and test predictive capability, which will lead to actions that can avoid or remedy the effects of multiple stresses. In the West, one example of experimentation with multiple stresses—centered on ponderosa pine seedlings, ozone, acid rain, and

Table 3.9. *Suggested components of a multiple stress hypothesis.*

Causes	Interactions	Mechanisms	Effects
Ozone	Cumulative	Increased membrane permeability	Leaf senescence
Acidic deposition	Catastrophic	Reduced photosynthesis	Canopy dieback
		Increased respiration	Growth
Drought		Reduced translocation	Susceptibility to insects
		Root biomass reduction	Increased drought stress
		Reduced defensive chemical synthesis	Increased tree mortality
		Decreased water uptake	

Source: P. J. H. Sharpe and H. W. Scheld, Role of mechanistic modeling in estimating long-term pollution effects upon natural and man-influenced ecosystems, *Workshop on Controlled Exposure Techniques and Evaluation of Tree Response to Airborne Chemicals,* NCAS, Technical Bulletin No. 500 (1986).

drought—is the study sponsored by the Electric Power Research Institute at Whitaker Forest (4 miles west of the Sequoia and Kings Canyon national parks boundary near Grant Grove) (Hakkarinen 1987). In addition, the Synthesis and Integration Project of the Forest Response Program will use the process modeling approach in conjunction with experiments under its sponsorship and with archival information from previous work (Kiester 1987). The predictive capability that will emerge from this work is limited, as always, by our capacity to clearly distinguish the effects of individual stresses in the complex forest environment.

Present State of Methodology

Tree age, interspecific competition, and site characteristics (light, nutrients, and water) are among the most important variables that modify tree and stand responses to stresses. These same factors are very influential in regulating responses to factors that favor improved vigor and growth. These variables form the background against which both diagnosis and prognosis are attempted.

Tree growth can be divided into three phases with respect to tree age. In total it resembles a normal curve. In the first phase, rapid growth is proportional to tree age; the main factors limiting growth are water, nutrients, and light. Competition induces the mortality of the least competitive individuals. The second phase is represented by a span of years when growth rate remains fairly constant. Trees of this age reflect sensitivity to those factors that limit carbohydrate production, such as moisture and temperature. The third phase is a decline in growth because of a natural reduction in leaf area and a higher respiration demand.

Unfavorable climate usually increases the effects of diseases and insects; organisms that are weak pathogens may become more agressive in older trees. As indicated earlier, 59 percent of the nation's softwood reserves are located in the western states, particularly the Northwest, where there are large volumes of timber in older-age stands. This statistic underscores the vulnerability of this part of our timber reserve to the effects of environmental stresses. It is necessary to conserve the supply and to avoid crisis-solving practices associated with stand management, such as salvage logging. Therefore older forest stands must be protected from preventable stresses (air pollution). Therefore, we need sensitive methods to detect early signs of stress on tree health.

Tree ring analysis has been applied in western forests for several purposes, including retrospective examination of the effects of tussock moth and western spruce budworm on the radial growth of grand fir and Douglas fir (Brubaker and Greene 1979) and the recovery of larch formerly affected by sulfur dioxide from the smelter at Trail, British Columbia (Fox et al. 1986). In each of these cases the interpretation of tree ring analysis was aided by the availability of records that conclusively dated the years during which insects or air pollution was present. There was strong support for eliminating or deemphasizing other possible causes of the observed growth declines.

In situations without sufficient historical documentation or other corroborative evidence, it is much more difficult to determine with certainty whether changing variables in the environment are responsible for tree growth changes. For example, LaMarche et al. (1984) reported increased growth rates of bristlecone pine at high-altitude sites in Colorado, New Mexico, and California and, in particular, of limber pine in Nevada in recent times. Growth increases since the mid-nineteenth century exceed those expected from climatic trends but are consistent in magnitude with global trends in carbon dioxide increase, especially in recent decades. An alternate hypothesis, however, discussed by Solomon et al. (1985), suggests that precipitation increases in August and September may also be the cause of growth increases. Both interpretations are vulnerable to rejection until additional independent evidence is available.

A study of the growth of Jeffrey pine in Sequoia and Kings Canyon national parks provides further illustration of the difficulty of establishing a firm cause-and-effect relationship when recent growth declines are discovered (Peterson et al. 1987). This study included samples from trees growing in the western parts of the parks, where ozone exposure can be verified by foliar symptoms (Wallner and Fong 1982), and in the eastern parts, where distance and high terrain reduce ozone transport drastically. All age classes of trees were included in the sample but only older trees (more than 40 centimeters in diameter and over 100 years old) showed a significant (11 percent) decline in growth since 1965. This growth decline was derived by comparing populations in the "exposed" and "control" areas. After the variation in growth due to other factors has been removed by appropriate techniques, there is still need for further evidence to corroborate the proposal that ozone air pollution is the cause of growth decline in recent years. In this situation there is reasonable historical evidence for the claim that 1965 was approximately the beginning of significant ozone pollution in the nearby San Joaquin Valley.

It is not unexpected that older trees were those that indicated the most growth reduction because they may already have been in a natural decline of growth. Another important constant in this study was that only open-grown trees on shallow, rocky soils were sampled. Such trees are considered to be "sensitive" in the nomenclature of tree ring specialists because they respond more dramatically to high and low precipitation years. These results cannot be extrapolated to more favorable sites where there is more intertree competition.

Finally, tree ring analysis can be very useful for detecting the early evidence of chronic air pollution stress if certain requirements are met concerning sampling design, data analysis, and the acquisition of additional historical evidence about air pollution in the region being studied. Without the latter evidence, the results will remain unconvincing.

The interpretation of foliar symptoms as a means of diagnosing the cause of injury to conifers will continue to be the most practical method. Keys to the identification of symptoms caused by both biotic and abiotic agents are helpful (Bega 1978). Histological examination provides reasonable specificity for identifying the cause of different symptoms on the same needle (Miller and Evans

1974). Biochemical assays are useful. For example, the measurement of peroxidase activity in plant tissue is a very sensitive indicator of stress, but many stresses cause the same reaction.

Perhaps combinations of observations and measurements involving several processes at different levels of biological organization (leaf, tree, stand) will be the most useful for identifying the early evidence of tree injury and pinpointing the specific cause.

Scientific Questions on Effects of Combined Stresses

The statistics regarding composition and age of forests in the West (USDA 1982) emphasize the prominence of uneven-aged stands in which old-growth trees represent a very important component of the nation's softwood reserves. Mortality in old-growth stands offsets much of the total annual growth. Net annual growth of growing stock in the West was 5.2 billion cubic feet—less than one-quarter of the national total (USDA 1982). This information strengthens concern about managing these stands properly so that conditions do not develop that lead to greater morbidity and mortality. It is necessary to improve the fundamental understanding of the relationship between biotic and abiotic stress agents (air pollutants) and the basic physiological processes governing tree growth and stand development, as represented by Kercher and King (1985) (see figure 3.7).

An improvement in the predictive capability of mechanistic models of photosynthesis, carbon allocation, nutrient flux, and water flux can provide the foundation of knowledge needed to identify whether imbalances are induced as

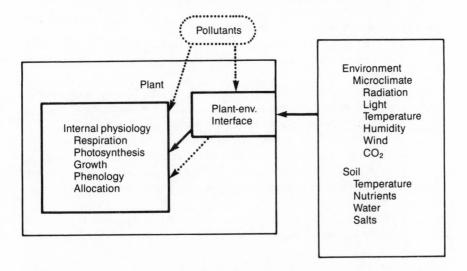

Figure 3.7. Conceptualization of plant processes as influenced by common environmental factors and air pollutants.

a result of particular combinations of stresses. For example, are there sequential and simultaneous combinations of conditions that together create subtle imbalances that decrease tree health to a phase of rapid morbidity and mortality after the action of one or more triggering variables? A major aspect of this investigation is to determine how low concentrations of ozone and acidic compounds may interact with other variables to alter basic physiological processes. Most of these questions can be modified to inquire about response as a function of tree age and/or competition class. Suggested research categories and some specific questions are given in this section.

A. Photosynthesis and Allocation of Carbon
1. Does chronic ozone injury result in an unfavorable allocation pattern of photosynthate among leaves, stems, and roots of conifer species?
2. Can the causes of shifts of photosynthesis within and between seasons be confidently partitioned among possible causal factors (ozone injury, nutrient imbalance, water deficits, and pathogenesis by infectious organisms)?
3. Will patterns of enzyme response to free radicals in injured leaf tissue be used to predict changes in photosynthetic rate?
4. How does the increase of atmospheric carbon dioxide influence basic physiological processes of conifers compared with deciduous hardwoods in regard to the mediation of other stresses?
5. Can tree ring analysis techniques and within-season observations of wood cell production be refined to better identify the effect of stresses occurring at a particular time or in a particular sequence?
B. Nutrient Flux
1. Does the cumulative atmospheric deposition of protons and nutrients, with or without ozone injury, produce changes in nutrient balance or allocation in trees or stands as reflected by present pool sizes or pool nutrient ratios?
2. How do edaphic factors (soil parent material, depth, and water-holding capacity) influence nutrient balance or uptake of nonnutrient ions following cumulative exposure to ozone and/or acid deposition?
3. Does nitrate and proton deposition to trees and stands from fogs affect the foliar nutrient pool under spring and autumn conditions?
4. Does nitrogen enrichment from atmospheric deposition alter tree (foliage) sensitivity to water stress or freezing temperatures?
C. Water Flux
1. Do different distributions of soil moisture availability during the growing season exercise a major control over ozone uptake and development of leaf injury?
2. Does ozone injury to foliage or changes in leaf tissue from the deposition of acidic compounds alter tolerance to leaf dessication during the winter?
D. Morbidity and Mortality Processes

1. Do reductions of foliage surface area and changes in carbon allocation due to ozone injury influence the pathogenesis of single diseases or complexes including dwarf mistletoes and root decay fungi?
2. Do reductions of foliage surface area and changes in carbon allocation due to ozone injury reduce the production of chemicals that aid in defense against insects and diseases?
E. Pollutant Deposition Processes
1. What level of influence do topography and stand density have on mass transport and deposition of pollutants to foliage?
2. Do nocturnal ozone concentrations, which generally remain high at mountain sites, constitute a threat to sensitive tree species greater than lowland agricultural species would receive?
3. How do different distributions of ozone peaks or respite days during the growing season influence tree injury as measured by visible symptoms and leaf or needle senescence?

Conclusions

For most of the forested areas of the West, including Alaska, conventional knowledge about the effects of diseases, insect pests, and abiotic stresses is sufficient to explain the trends in forest health. At the same time, there is certain evidence that some forested areas in California, and less certain evidence that particular areas in Arizona and Washington, are repeatedly subject to ozone and acidic deposition, and that forest health has been reduced from the combined effects of traditional pests and stresses with increased levels of pollutants. Conventional knowledge is not adequate to predict how pollutant stress will interact with the usual set of biotic and abiotic stresses.

The predictions of future needs for timber products indicate that demand may exceed supply in the next century at current pricing levels. Public demands for other amenities offered by western forests can only increase. Research programs should therefore concentrate on the task of learning how fundamental physiological processes may be affected by interactions of pollutants with both biotic and abiotic stresses. Emphasizing the development of process-driven models will make predictive capability transferable to more than one problem area.

References

Adams, D. F., C. G. Shaw, and W. D. Yerkes, Jr. 1956. *Phytopathology* 46:587.

Arkley, R. J. and R. Glauser. 1980. Symposium on Effects of Air Pollutants on Mediterranean and Temperate Forest Ecosystems, Riverside, Calif. *U.S.D.A. Gen. Tech. Bul.* PSW–43, p. 225.

Bega, R. V. 1978. Diseases of Pacific Coast Conifers. USDA, Forest Service, *Agricultural Handbook No. 521.* 206 pp.

Brubaker, L. B., and S. K. Greene. 1979. *Canadian Journal of Forest Research* 9:95.

Bruck, R. I., P. R. Miller, J. Laut, W. Jacobi, and D. Johnson. 1985. USEPA Region VIII. Investigation into the health of forests in the vicinity of Gothic, Colo. EPA–908/9–85–001. 69 pp.

Bruhn, J. N. 1980. Ph.D. Dissertation. University of California, Berkeley. Effect of oxidant air pollution on ponderosa and Jeffrey pine foliage decomposition. 273 pp.

Bytnerowicz, A., P. R. Miller, D. M. Olszyk, P. J. Dawson, and C. A. Fox. 1987. *Atmos. Environ.* 21:1805.

California Air Resources Board. 1986. *Fourth Annual Rept.* Acid Dep. Res. Mon. Prog. State of California, Sacramento, Calif.

California Air Resources Board. 1987. Effect of ozone on vegetation and possible alternative ambient air quality standards. Tech. Serv. Div. Sacramento, Calif.

Carlson, C. E., and J. E. Dewey. 1971. *Environmental Pollution by Fluorides in Flathead National Forest and Glacier National Park.* USDA, Forest Service, State and Private Forestry, Missoula, Mont. 57 pp.

Carroll, J. J., and R. L. Baskett. 1979. *Journal of Applied Meteorology* 18:474.

Chang-Han, Yu, and R. A. Pielke. 1986 *Atmos. Environ.* 20:1751.

Coffin, H. 1959. USDA, Forest Service, Pacific Southwest Forest and Range Experiment Station Technical Paper 39. 30 pp.

Compton, O. C., L. F. Remmert, and J. A. Rudinsky. 1961. Oregon Agricultural Experiment Station Misccellaneous Paper 120.

Dahlsten, D. L., and D. L. Rowney. 1980. Symposium on Effects of Air Pollutants on Mediterranean and Temperate Forest Ecosystems, Riverside, Calif. pp. 125–130. *U.S.D.A. General Technical Bulletin* PSW–43.

Davidson, A. G., and R. M. Prentice. 1967. *Important Forest Insects and Diseases of Mutual Concern to Canada, the United States and Mexico.* Queen's Printer, Ottawa. No. Fo 47–1180. 248 pp.

Duckworth, S., and D. Crowe. 1979. Sulfur dioxide and sulfate survey, Bakersfield 1977–78. Calif. Air Resources Bd., Tech. Serv. Div. Sacramento, Calif. 34 pp.

Edmunds, R. L. 1988. Ann. Meeting, Forest Response Program. Corpus Christi, Tex. Vol. II. Project Status Reports. pp. 289–293. North Carolina State University.

Errington, J. C., and J. V. Thirgood. 1971. Northern Miner, Ann. Rev., p. 72.

Eyre, F. H., ed. 1980. Society Amer. Foresters. Washington, D.C.

Fox, C. A., W. B. Kincaid, T. H. Nash, D. L. Young, and H. C. Fritts. 1986. *Canad. J. For. Res.* 16:283.

Friedland, A. J., R. A. Gregory, L. Karenlampi, and A. H. Johnson. 1984. *Canad. J. For. Res.* 14:963.

Gemmill, B., J. R. McBride, and R. D. Laven. 1982. *Tree Ring Bulletin.* 42:23.

Greater Vancouver Regional District. 1985. Ambient air quality. Annual report, Appendix A. Vancouver, British Columbia.

Hakkarinen, C. 1987. Electric Power Research Institute. Forest Health and Ozone. Spec. Rpt.

Haywood, J. K. 1905. U.S.D.A. Bur. Chem. Bul. 89:23.

Hiratsuka, Y. 1987. Forest Tree Diseases of the Prairie Provinces. Info. Report NOR-X–286. N. Forest. Cen. Edmonton, Alberta. 142 pp.

Hoffer, T. E., R. J. Farber, and E. C. Ellis. 1982. Sci. Total Environ. 23:17.

Holzworth, G. C., and R. W. Fisher. 1979. USEPA Report No. EPA 600/4–79–206. Research Triangle Park, N.C.

Houston, D. B. 1981. USDA, Forest Service. NE-INF–41–81.

James, R. L., F. W. Cobb, P. R. Miller, and J. R. Parmeter, Jr. 1980. *Phytopathology* 70:560.

Kercher, J. R., and D. A. King. 1985. In *Sulfur Dioxide and Vegetation,* ed. W. E. Winner, H. A. Mooney, and R. A. Goldstein, pp. 357–372. Stanford University Press. Stanford, Calif. 593 pp.

Kickert, R. N., and B. Gemmill. 1980. Symposium on Effects of Air Pollutants on Mediterranean and Temperate Forest Ecosystems, Riverside, Calif. pp.181–186. U.S.D.A Gen. Tech. Bul. PSW–43.

Kiester, R. 1987. USEPA, NAPAP. Synthesis and Integration Project Research Plan, Forest Response Program, EPA. Corvallis, Oreg.

LaMarche, Jr., V. C., D. A. Graybill, H. C. Fritts, and M. R. Rose. 1984. *Science.* 225:1019.

Leapheart, C. D., and E. F. Wicker. 1966 *Canadian Journal of Botany.* 44:121.

Luck, R. F. 1980. Symposium on Effects of Air Pollutants on Mediterranean and Temperate Forest Ecosystems, Riverside, Calif. USDA Gen. Tech. Bul. PSW–43. 240 pp.

Ludwig, F. L., and E. Shelar, Jr. 1980 *J. Air Pollution Control Association.* 30:894.

Manion, P. D. 1981 *Tree Disease Concepts.* Englewood Cliffs, N.J.: Prentice Hall. 399 pp.

McBride, J. R., P. R. Miller, and R. D. Laven. 1985. Air Pollutants Effects on Forest Ecosystems, St. Paul, Minn. pp. 157–167. Acid Rain Foundation.

McMinn, R. G. 1956. Canada. Dep. Agr., Forest Biol. Div. Sci. Serv. *Bimonthly Progress Report.* 12:3.

Middleton, J. T., J. B. Kendrick, Jr., and H. W. Schwalm. 1950. *Plant Disease Reporter* 34:245.

Miller, P. R. 1983. Symposium on Air Pollution and Productivity of the Forest, Washington, D.C. pp. 161–197. Izaak Walton League and Penn State University.

Miller, P. R., and M. J. Elderman, eds. 1977. USEPA Report No. EPA 600/3–77–104.

Miller, P. R., and L. S. Evans. 1974. *Phytopathology* 64:801.

Miller, P. R., M. H. McCutchan, and H. P. Milligan. 1972. *Atmos. Environ.* 6:623.

Miller, P. R., and A. A. Millecan. 1971. *Plant Disease Reporter.* 55:555.

Miller, P. R., J. R. Parmeter, Jr., O. C. Taylor, and E. A. Cardiff. 1963. *Phytopathology.* 53:1072.

Miller, P. R., et al. 1980. Symposium on Ecology, Management, and Utilization of California Oaks, Claremont, Calif. USDA Gen. Tech. Bul. PSW–44, pp. 220–229.

Morgan, J. J., and H. M. Liljestrand. 1980. Final Report, Measurement and interpretation of acid rainfall in the Los Angeles Basin. California Air Resources Board, Sacramento, Calif.

Mueller-Dombois, D., J. E. Canfield, R. A. Holt, and G. P. Buelow. 1983. *Phytocoenologia* 11:117.

National Acid Precipitation Assesment Program. 1986. Annual Report. 163 pp.

National Park Service, Air Quality Division. 1986. Draft Report. Summary of visible ozone injury on pines and oaks in Sequoia and Yosemite National Parks.

Oppenheimer, M., C. B. Epstein, and R. E. Yuhnke. 1985. *Science* 229:859.

Parmeter, Jr., J. R., and R. V. Bega, and T. Neff. 1962. *Plant Disease Reporter.* 46:269.

Perry, T. O., and G. W. Baldwin. 1966. *Forest Science.* 12:298.

Peterson, D. L., M. J. Arbaugh, V. A. Wakefield, and P. R. Miller, 1987. *Journal of Air Pollution Control Association.* 37:906.

Pronos, J., and D. R. Vogler. 1981. USDA, Forest Service, PSW, FPM Rept. 81–20. San Francisco, Calif.

Reiter, E. R. 1984. Workshop on Atmospheric Tracers. Sante Fe, N.M.

Richards, Sr., B. L., O. C. Taylor, and G. F. Edmunds, Jr. 1968. *Journal of the Air Pollution Control Association.* 18:73.

Roth, P., C. Blanchard, J. Harte, H. Michaels, and M. T. El-Ashry. 1985. The American West's Acid Rain Test, Research Rept. No. 1. Washington, D.C.: World Resources Institute, 50 pp.

Russel, A. G., G. J. McRae, and G. R. Cass. 1983. Atmos. Environ. 17:949.

Scheffer, T. C., and G. G. Hedgecock. 1955. USDA Forest Service. Tech. Bul. 1117. 49 pp.

Schutt, P., and E. B. Cowling. 1985. *Plant Disease* 69:548.

Sharpe, P. J. H., and H. W. Scheld. 1986. Role of mechanistic modeling in estimating long-term pollution effects upon natural and man-influenced ecosystems, pp.76–82. Workshop on controlled exposure techniques and evaluation of tree response to airborne chemicals. NCASI Tech. Bull. No. 500. 82 pp.

Shaw, C. G., G. W. Fischer, D. F. Adams, M. F. Adams, and D. W. Lynch. 1951. *Northwest Science.* 15:156.

Sigal, L. L., and T. H. Nash. 1983. *Ecology* 64:1343.

Solomon, A. M., J. R. Trabalka, D. E. Reichle, and L. D. Voorhees. 1985. USDOE Atmospheric Carbon Dioxide and the Global Carbon Cycle, DOE/ER–0239, pp. 1–13.

Southern Commercial Forest Research Cooperative. 1986. Proc. Mature Tree Response Workshop, St. Louis, Mo.

Stolte, K. W. 1987. Personal communication.

Taylor, O. C. 1986. Effects of ozone and sulfur dioxide mixtures on forest vegetation of the southern Sierra Nevada. Final Rept. California Air Resources Board, Sacramento, Calif.

Treshow, M., F. K. Anderson, and F. Harner. 1967. *Forest Science.* 13:114.

USDA. 1973. Forest Service, Ag. Handbk. No. 445.

USDA. 1982. Forest Service, Forest Resource Report No. 23.

USEPA. 1983. Air Quality Criteria for Ozone and Other Photochemical Oxidants. Vol. II. EPA/800/8–84/020bf.

Vogler, D. R. 1982. USDA, Forest Service, PSW, FPM Rept. 82–17. San Francisco, Calif.

Waldman, J. M., J. W. Munger, D. J. Jacob, R. C. Flagan, J. J. Morgan, and M. R. Hoffman. 1982. *Science* 218:677.

Wallner, D. W., and M. Fong. 1982. Survey Report, National Park Service, Three Rivers, Calif.

Wert, S. L. 1969. Sixth International Symposium on Remote Sensing of Environment, Ann Arbor, Mich. pp. 1169–1177.

Williams, W. T., M. Brady, S. C. Willison. 1977. *Journal of the Air Pollution Control Association.* 27:230.

Williams, W. T., and J. A. Williams. 1986. *Environ. Conserv.* 13:229.

Zoettl, H. W., and R. F. Huettl. 1986. *Journal of Water, Air, Soil, Pollution.* 31:449.

4

Forest Decline Syndromes in the Southeastern United States

ROBERT IAN BRUCK

Evidence of Forest Declines in the Eastern United States

The evidence of forest declines in the eastern United States has accumulated principally over the past ten years. Several researchers have reported significant decreases in annual ring width growth of several tree species in nonurbanized parts of the eastern United States (Bruck 1984, Bruck and Robarge 1988, Johnson et al. 1984). The identification of existing or abnormal growth trends is difficult because long-term data on natural variations in these ecosystems are lacking. Influences of previous natural and human disturbances are often ignored or difficult to determine. Tree ring data collected at any point in time represent a biased account of forest dynamics because only the survivors are present, and therefore past episodes of lethal stress are inherently underrepresented.

The Pine Barrens

In the Pine Barrens of New Jersey, a strong statistical relationship between annual ring size variations and extreme soil pH led to the hypothesis that acid deposition may have been a growth-limiting factor for the past two decades. Radial ring width growth of pitch, short-leaf, and loblolly pine has declined over the past 26 years (A. H. Johnson et al. 1981). Of the trees examined, approximately one-third showed normal annual ring size growth, another third showed noticeable abnormal compression of ring sizes during the past 20–25 years, and

This chapter is the result of the dedicated efforts of numerous scientists, graduate students, and technicians at North Carolina State University. They include the following: principal investigators: W. Robarge, R. Bradow, E. Cowling, S. Khorram, J. Brockhaus, and W. Cure; students: S. Meier, A. McDaniel, and P. Smithson; and Technicians: K. Reynolds, J. Pye, M. Campbell, M. Huster, J. Bograd, T. Anderson, J. Power, L. Pohl, and D. Reilly.

Finally, many thanks are offered to T. Nowaczyk and P. Wissinger for the preparation of this manuscript.

the remainder showed dramatic reduction of annual ring width over this time interval. The effects were evident in trees of different species and different sites and occurred regardless of age or whether the trees were planted or native. This suggests involvement of a broad-scale environmental factor such as atmospheric deposition, ozone, or climate change.

High-Elevation Spruce-Fir Forests

A systematic decline of red spruce has been documented quantitatively in the Green Mountains of Vermont (Siccama et al. 1982) and observed in New York, New Hampshire (A. H. Johnson et al. 1983), Virginia, North Carolina, and Tennessee (Bruck 1984, Bruck and Robarge 1988). Quantitative vegetation surveys conducted in the mid–1960s in six forests in New York, Vermont, and New Hampshire were repeated 15–18 years later (Siccama et al. 1982). Basal area reduction in live spruce averaged approximately 45 percent during that interval, with the basal area decreasing an average of 66 percent for trees with a diameter breast height (DBH) of 2–10 centimeters, and spruce seedling density decreasing by 50 percent.

Additional quantitative evidence of spruce dieback and decline is based on a survey of 24 forest stands in the Catskill, Adirondack, Green, and White mountains. Standing dead spruce trees account for approximately 22 percent of the spruce present, with an additional 40 percent of the spruce showing symptoms of dieback (death at the top). Mortality is correlated directly with elevation, with mortality in spruce ranging from 10 percent in the hardwood forest at a lower elevation to 30 percent at an altitude of 1175 meters. Trees in all size classes are affected. The primary cause is unknown but does not appear to be successional dynamics, insect damage, or primary pathogens (Hadfield 1968, Roman and Raynal 1980, Siccama et al. 1982). However, recent data concerning potential climatic anomalies indicate that climatic and physiological drought may have played a role in these observations (A. H. Johnson et al. 1984). Studies of declining red spruce indicate the presence of many secondary pathogens that may or may not be influencing stand dynamics.

Surveys begun during 1983 on Mount Mitchell (North Carolina) and six other southern Appalachian peaks show that red spruce and, to a lesser degree, Fraser fir exhibited marked growth reduction beginning in the early 1960s (Bruck 1984). This observation was made for those trees growing above 1920 meters (6300 feet) elevation, regardless of present vigor. Surveys also indicate that visual forest decline is occurring in varying degrees throughout the southern Appalachian Mountains. West-facing slopes appear to have greater decline and dieback incidence along with greater annual ring width suppression. This was observed in an average of 82 percent of all sampled red spruce dominant and codominant trees. Preliminary soil analyses suggest higher loading of lead and other heavy metals on west-facing slopes and at higher elevations.

A rapid deterioration in the physical appearance of Appalachian spruce-fir ecosystems seems to be in progress. No cause-and-effect mechanism has yet been firmly established to explain this forest decline. Within the framework of Manion's decline factors (1981), the spruce decline has the characteristics of a complex biotic-abiotic disease related primarily to environmental stress. With the limited conclusive information on the mechanisms of this decline, it is difficult to accept or reject any of the potential pathways.

Southern Commercial Forests

In November 1985, the U.S. Forest Service published findings of a comprehensive survey of timber in the Southeast (see figure 4.1) showing that net annual growth of softwood timber had peaked and turned downward after a long upward trend (Sheffield et al. 1985). The most pronounced reductions were measured in the growth of yellow pines on nonindustrial private forest land, which accounts for about 69 percent of the timberland in this five-state region (Florida, Georgia, North Carolina, South Carolina, and Virginia). Comparisons of radial growth rates over the same time periods for the coastal plains of Georgia, South Carolina, and North Carolina revealed a reduction of similar magnitude. These reductions are important because the trend in net volume growth strongly influences amounts of timber available for future harvest.

Recent analyses of U.S. Forest Service Forest Inventory and Assessment (FIA) data indicate a 16 to 20 percent reduction of radial growth of some pine species in Piedmont Survey Units in Georgia and South Carolina. This trend was discovered following the fifth survey of Georgia's Piedmont region (1982), when a loss in volume was exhibited in comparison with the fourth survey (1972). Follow-up comparisons of surveys in South Carolina indicated similar average growth reductions in their Piedmont units. Growth comparisons on Coastal Plains Survey Units indicated slight or no growth losses between the fourth and fifth survey periods.

The apparent reductions implied in these preliminary reports were regarded with some skepticism by FIA personnel and most southern forest biometricians. The FIA survey system was designed to provide an indication of the standing timber inventories in each state rather than average growth rates. Many foresters familiar with the inventory systems have expressed doubts that reliable growth estimates can be made from these data. They suggest that losses or gains of 25 percent could be within the range of uncertainty inherent in FIA survey techniques.

Most analyses of the growth reductions were made using radial increments. Normally this parameter is considered to be a poor measure of forest productivity since reductions in radial increment occur naturally over time in individual trees even though their volume increment may be increasing. The traditional growth measurement is net volume change per acre. Net volume growth is determined from (1) in-growth or volume of trees that grew into a measurable size since the last inventory, (2) volume loss to mortality and havesting since the last inventory, and (3) the change in volume of the surviving trees between inventories. In the situation for the Piedmont units, in-growth volumes had decreased.

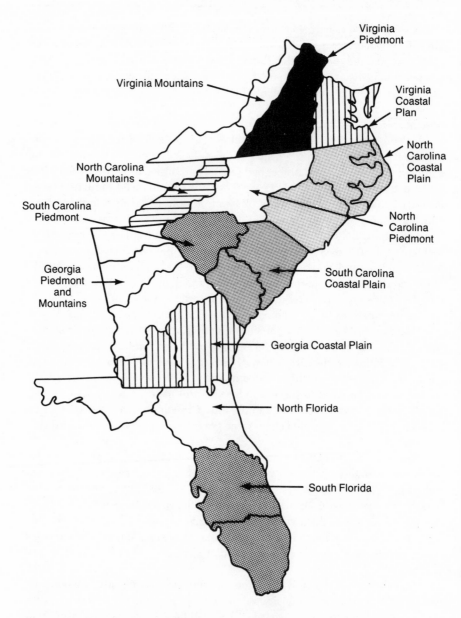

Figure 4.1. Forest survey delinations of growth-loss study areas in the Southeast.

These changes were uniformly distributed across all units. Radial growth of surviving trees was used because it was more accurate and reliable than data on in-growth and tree mortality.

The observed average annual radial increments for natural stands of loblolly and shortleaf pine from the fourth (1972) and fifth (1982) surveys of the Georgia Piedmont Survey Units are shown in figures 4.2 and 4.3. These data reveal that

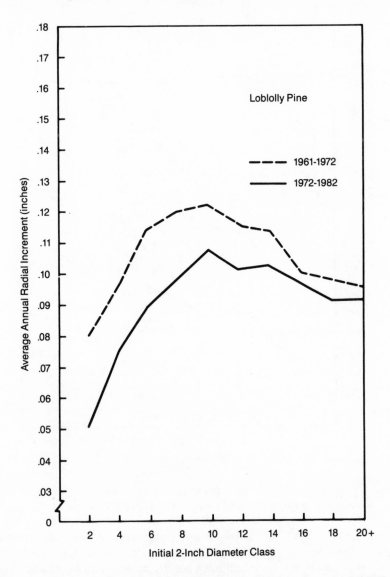

Figure 4.2. Average annual radial increment, by diameter class, for loblolly pine in the Georgia Piedmont and mountains. Source: Sheffield et al. 1985.

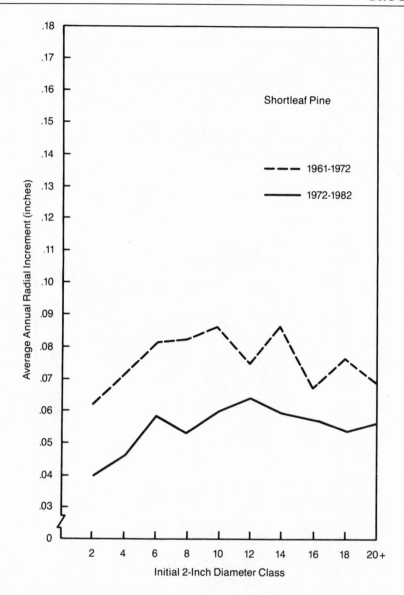

Figure 4.3. Average annual radial increment by diameter class, for shortleaf
pine, in the Georgia Piedmont and mountains. Source: Sheffield et al. 1985.

diameter growth of all but the largest pines decreased in these units. Similar
growth differences occurred in South Carolina (figures 4.4 and 4.5). Comparison
of coastal plain units did not indicate a similar decline between the 1972 and
1982 measurements.

Additional comparisons have been made using data from Georgia's third

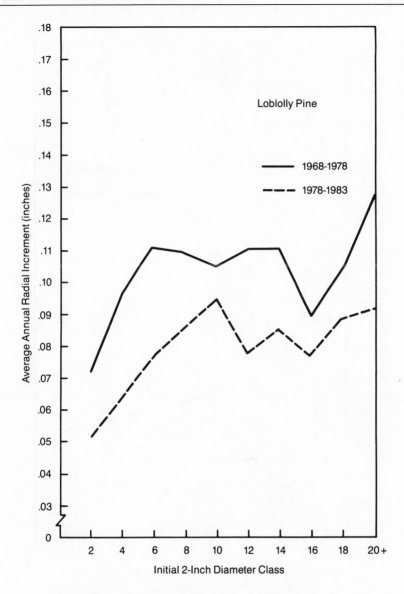

Figure 4.4. Average annual radial increment, by diameter class, for loblolly pine in the South Carolina Piedmont. Source: Sheffield et al. 1985.

survey for the period 1956–61. Initial impressions are that radial increments of pine in natural stands in the coastal plains may have significantly decreased between the third and fourth surveys. Comparisons with Georgia's Piedmont data suggest that these growth rates also may have declined between the third and fourth surveys.

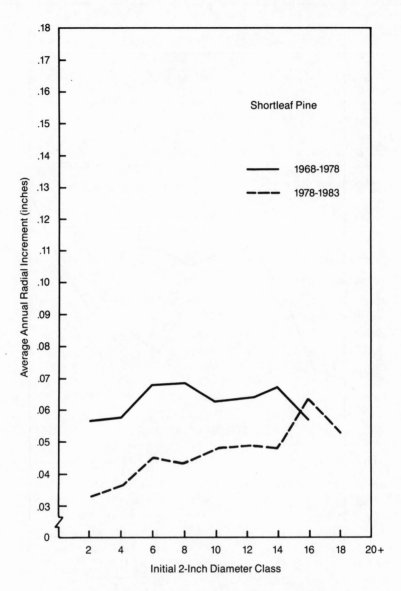

Figure 4.5. Average annual radial increment, by diameter class, for shortleaf pine in the South Carolina Piedmont. Source: Sheffield et al. 1985.

Understanding Forest Declines

After more than a quarter century of research, forest declines are still not well understood. They are commonly the result of multiple primary and secondary stress agents that eventually cause the deterioration and death of individual trees, and on rare occasions entire ecosystems (Manion 1981). Environmental stresses may include major climatic shifts, severe droughts, and abnormally cold or warm temperatures.

The potential ramifications of environmental stresses for modern declines are great because the ecosystems have been present for centuries, and the tree species (for example, red and white spruce) are genetically relatively uniform. The forest tree species involved in declines observed in central Europe and in boreal montane ecosystems of the eastern United States have lifespans of 350–500 years (A. H. Johnson et al. 1983).

Because of the potential multiplicity of causes, forest declines are often termed forest decline syndromes. A forest decline syndrome involves a series of specific stress factors (Manion 1981):

- *Predisposing factors* are inherent to the tree and site and include climate, air quality, soil type, site quality, and genetic strain and age of the tree. These predisposing factors determine the vigor and susceptibility of the tree.
- *Inciting factors* are short in duration and may be physical or biological in nature—for example, insect defoliators, late spring frosts, drought, or air pollution. Trees predisposed to injury may have difficulty recovering from the influences of inciting factors.
- *Contributing factors* lead directly to mortality of the tree and are usually biotic factors such as bark beetles, canker fungi, root and sap rot fungi, and virus-like organisms. Contributing factors will accelerate specific symptoms on weakened hosts and are the best understood of all the indicators of decline.

Eventually a plant dies from the "spiral" or "domino" effect of long-term predisposing factors, short-term inciting factors, and the final stress of biotic contributing factors in the forest decline syndrome.

Because of the complex relationships between biotic and abiotic factors, the exact causes of very few decline syndromes have been demonstrated satisfactorily. Forest tree declines during the past 50 years have been caused by combinations of predisposing, inciting, and contributing factors. Cases of recent explainable declines that illustrate this process include birch dieback, ash dieback, maple decline, and oak decline.

Several hypotheses relating to air pollution have emerged in the United States and West Germany to explain one portion or another of recent forest decline syndromes (Schutt and Cowling 1985):

- photochemical oxidants (for example, ozone) causing direct anatomical and/or physiological perturbation to tree foliage;

- hydrogen ion deposition leading to cation leaching and depletion of the soil nutrients;
- excess nitrogen deposition resulting in suppression of beneficial root fungi development;
- soil acidification and aluminum mobilization in the soil leading to toxicity and/or physiological drought effects on tree roots;
- soil acidification and metal mobilization in soil leading to toxicity of soil microorganisms, resulting in decreased nutrient availability;
- altered susceptibility of trees to pests and pathogens from nitrogen- and sulfur-derived pollutants;
- altered reproductive or regenerative processes from nitrogen- and sulfur-derived pollutants;
- sulfur and/or nitrogen oxides, alone or in combination with oxidants, leading to nutrient stress because of leaching from foliage; and
- altered photosynthesis, respiration, translocation, and carbon allocation processes from exposure to sulfur and nitrogen oxides leading to water and/or nutrient stress.

Forest Decline Syndromes as Part of Natural Processes

Forest declines may occur as part of normal ecological processes. Tree species expand their range to the limit of their tolerance of site conditions. As a rule, trees at the edge of their site tolerance are more stressed and most susceptible to declines. (Note, however, that recent declines have occurred well within the natural geographic range of many species, not only on the periphery of their range.) Pathologists are often asked if we are really observing natural ecological succession when we see a forest decline. Natural factors that can influence forest declines are discussed in this section. From this ecological point of view, currently observed declines need much more study to clarify this cause-and-effect mechanism.

Succession

Forest succession is the dynamic process in which a newly disturbed site goes through a series of biological changes leading up to a stable "climax" forest ecosystem. Succession works in conjunction with other natural environmental trends such as climate, and it can involve temperature and precipitation, availability and genetics of herbaceous and forest tree species, and the presence of insect and disease pests. Although successional dynamics have been studied by many forest ecologists over the past century, little is known about the specific processes leading from completely disturbed ecosystems, through transitional communities, into climax forest conditions in the spruce-fir ecosystems of the high-altitude Appalachian crest.

The significance of forest succession cannot be ignored when studying a forest decline syndrome. The forest declines observed in central Europe and in

high-elevation eastern North America could conceivably be the result of previously unknown natural successional processes in which, for example, every
5000 to 10,000 years these ecosystems basically self-destruct (Manion 1981).
Although such occurrences are unknown at this time, the possibility should be
recognized. Investigating this hypothesis within a reasonable time frame will be
difficult, partly because many of the areas with symptoms of forest decline are
classified as Class 1 Wilderness Areas and are protected by law from perturbations by humans. This makes it difficult to initiate long-term destructive plots for
the purpose of addressing forest successional dynamics.

Research modeled after the Hubbard Brook Ecosystem Study in New
Hampshire should be initiated in certain areas where forest decline is taking
place, and perhaps more important in areas that are "susceptible" to a forest
decline syndrome. The major drawbacks of this type of ecosystem approach are
the cost and length of study.

Competition

Competition both between different forest tree species and within a tree
species is a subject of great interest to forest ecologists. For example, encroachment of hardwoods such as birch, beech, and maple into coniferous spruce-fir
areas has been noted and attributed to various processes such as disturbance,
insect and disease infestations causing widespread mortality of spruce and fir,
and climatic change. Because spruce-fir ecosystems have not been studied extensively, the actual dynamics of hardwood encroachment into areas of spruce
and fir are not well understood.

The adult trees in a forest are the surviving individuals. Trees that could not
survive the competition for nutrients, water, and light within the forest ecosystem are no longer part of the present ecosystem. The absence of these trees
and the information they contained makes it difficult to reconstruct historically
the competition dynamics of 30–100 years ago.

The significance of competition may be important in understanding why
spruce and fir trees currently are being stressed. The type of intensive research
that needs to be conducted will be costly and lengthy. The research needs for
succession studies would apply to competition dynamics. Long-term permanent
plot ecosystem studies must be established and funded over a period of 20 years
or more in order to detect trends in which climatic and stand successional
dynamic features can be correlated with the ingress of hardwoods and the competition between and among spruce and fir trees.

Climate Changes

Little is known about long-term climatic changes in areas where forest declines are occurring. Some evidence points toward a long-term warming trend
while other evidence indicates the climate has been growing cooler during the
present century. The subject is a topic of debate for bioclimatologists and mete-
orologists.

Early investigators suggested the involvement of significant climatic changes in forest declines. Plant development and deterioration are linked to weather conditions, but weather is simply a short-term reflection of climate. The role of continental climatic change in decline causality is difficult to interpret.

Accurate weather records are not available from before about 1930 for the northern and southern Appalachians. The lack of high-quality data to address climatic trends in high-altitude ecosystems makes the detection of long-term climatic change difficult. However, climatic data collected since the 1930s in the southern Appalachian Mountains at three high-altitude meteorological stations (Grandfather Mountain, Mount Mitchell, and Clingman's Dome) show little variation in terms of minimum, maximum, and average temperatures and average weekly, monthly, and yearly precipitation. Analysis indicates little significant variation among or between years of data (Bruck 1984).

Temperature and precipitation trends are critical factors in understanding forest decline syndromes. The phenomenon of the greenhouse effect has been documented to result from the buildup of carbon dioxide (CO_2) and other anthropogenic trace gases in the atmosphere. Some scientists propose that increasing atmospheric concentrations of CO_2 and other gases have raised the average surface temperature and that higher overall temperatures may have an effect on particularly long-lived forest ecosystems. Hepting (1961) of the U.S. Forest Service presented evidence of a climatic warming trend during the middle part of this century. He postulated that a 2-degree rise in the mean annual temperature would favor forest pathogenic organisms (that is, very rapid genetic selection) over the successionally stable long-term forest ecosystem of the boreal montane zones throughout the Appalachian crest. The trees present during the 2-degree shift would not be able to respond effectively because of their slow reproductive capacity and genetic recombination. Hence, a temperature shift could favor a disease or insect infestation syndrome where none was observed previously.

Under research sponsored by the National Acid Precipitation Assessment Program (NAPAP), many existing uncertainties are being addressed by the Mountain Cloud Chemistry Program established recently by the Environmental Protection Agency (EPA) along the Appalachian Mountains. Micrometeorological data as well as meteorological trends and atmospheric chemistry data are being collected at these sites to determine long-term chemical and climatic trends. It will take many years, perhaps decades, of data collection and analysis to discern the atmospheric and climatic trends in the areas where forest decline syndromes are observed.

Pathogens and Pests

Although compendia of the incidence of pathogens and pests have been produced (Hepting 1971), often little is known about the epidemiological significance of these pests and pathogens in causing tree disease or mortality. When an ecosystem is under stress, indigenous root fungi, bacteria, and viruses often play an important role as contributing factors in damaging weakened trees and causing tree mortality.

The balsam wooly adelgid (BWA) has contributed greatly to Fraser fir mortality in the southern Appalachians. But it is unknown whether the aphid has acted as a primary and inciting pest or if it is actually a secondary manifestation of stress. No pathogens of significance have been detected in the crowns, bole, or root of red spruce trees in the southern Appalachian Mountains. The fungal pathogen *Armillaria mellea* is believed to be of secondary importance as a stress pathogen in the northern Appalachian Mountain red spruce zone.

Research should be initiated to collect baseline data on the incidence and magnitude of pests and pathogens throughout areas where forest decline is observed. Tree damage and mortality cannot be attributed to secondary extrinsic factors until comprehensive surveys and experimentation determine the significance of the pathogen and pest populations. Research has been initiated under the Southern Appalachian Resource Research Management Cooperative program sponsored by the U.S. Forest Service to address comprehensively the status of pests and pathogens in the spruce-fir zone of the southern Appalachian Mountains.

Contribution of Human Activities to Forest Declines

A number of human activities can potentially contribute to forest decline. These include emissions of gaseous pollutants such as ozone (O_3) and sulfur dioxide (SO_2), acid-deposition-related effects, and land management practices, as well as the synergism of natural and human influences. The evidence that such activities can contribute to forest declines is considered in this section.

Gaseous Pollutants

In recent decades, emissions of acid deposition precursors (SO_x and NO_x) and other gaseous pollutants have increased the exposure of extensive forest areas in both central Europe and North America to these substances. Although no conclusive proof exists that air pollutants are currently limiting the growth of forests in either Europe or the United States, they are suspected to be involved because of the unique nature of the symptoms, the location of the affected forests in relatively high pollution regions, and the simultaneous appearance of symptoms across wide areas.

Reports of decreased growth and increased mortality of forest trees in areas receiving high rates of atmospheric deposition have increased scientific efforts to quantify changes in forest productivity and to identify the causes. The potential synergistic nature of combined pollutant exposure and the range of direct and indirect effects make quantification of observed effects particularly challenging. The complexity of forest growth processes and the sensitivity of

forest trees to natural environmental stresses add to the difficulties of quantifying the effects of pollutants on forests.

Detecting responses in mature forest trees is made difficult by several factors: the complexity of competition, climate, and site-specific factors; the potential interactions among acid deposition, gaseous pollutants, and trace metals;
and the lack of unaffected (control) sites. Although the task of assessing possible
impacts on forest productivity is difficult, the potential economic and ecological
consequences of even subtle changes of forest growth over large regions suggest
that assessments should be conducted rapidly and efficiently.

Ozone. The direct acute and long-term chronic effects of photochemical
oxidants, especially ozone, on trees may be significant factors in forest decline
syndromes. As documented by Huettl in chapter 2 in this volume, considerable
research in West Germany has focused on ozone, alone or in combination with
other factors, as an important causal agent in *Waldsterben* (see also Prinz et al.
1982). Ozone is the product of the photooxidation of nitrogen compounds and
hydrocarbons in the atmosphere. The role of nitrogen in ozone production is of
particular interest in the southeastern United States, a known epicenter of air
stagnation and acute ozone events (U.S. Environmental Protection Agency, R.
Bradow pers. comm. 1988).

Hepting (1961) characterized the direct effects of ozone on eastern white
pine in the southern Appalachian Mountains and demonstrated the broad genetic susceptibility and low tolerance of this forest tree species to ozone. In the San
Bernardino Mountains of the eastern Los Angeles Basin, damage of ponderosa
pine has been attributed to ozone and other photochemical oxidants, as Miller
discusses in chapter 3.

Various German experiments conducted by Prinz et al. (1982) sought to
characterize the potential effects of ozone, acid fog, SO_2, and NO_x on decline
symptomatology. In central Europe, ambient ozone concentrations of 150 parts
per billion (ppb) are exceeded frequently and concentrations in excess of 240
ppb occur occasionally. In controlled experiments, ozone fumigations for 6–8
weeks, 24 hours per day, using 150–250 ppb ozone caused a degree of yellowing
(chlorosis) and mottling similar to that observed under actual conditions in the
field. Not all the symptoms observed in the field, however, could be reproduced
in fumigation chambers. Combinations of acute ozone exposure and acid fog
treatments enhanced the loss of magnesium from the foliage. Analysis of concentrations and exposure time calculated from field data was consistent with
accumulated doses that caused yellowing in the controlled experiments.

In the United States, a comprehensive research program is in the planning
stages to address the potential involvement of ozone and other photochemical
oxidants in areas of forest decline. Experiments dealing with the combined
effects of drought, ozone, and acid deposition under simulation conditions may
yield important experimental data on the potential of these interacting factors to
contribute to forest decline symptoms. As part of the Mountain Cloud Chemistry

Program and other research cooperatives, more data are being collected to determine the levels and effects of nitrogen in high-altitude forest ecosystems.

Sulfur Dioxide. Fumigation with high levels of gaseous sulfur dioxide can have direct adverse effects on forest tree species. Classic symptoms of acute SO_2 damage are lacking in the German forest, but some scientists suspect that the levels observed in West Germany may contribute to cell membrane damage and alter basic physiological processes of forest trees (Tomlinson 1983). In the forested areas of Germany where decline and dieback have been observed, ambient levels of SO_2 range from an average annual value of about 7 to 21 $\mu g/m^3$ (Tomlinson 1983).

In the United States, the available data do not suggest a major role for direct damage from gaseous SO_2. Typical observations made in the northeastern United States indicate that SO_2 concentrations are approximately one-tenth those observed in central Europe (Friedland pers. comm. 1988). There are no data from declining forests suggesting alteration in structure or functions caused by chronic SO_2 levels. However, this possibility has not been thoroughly investigated.

Effects of Acid Deposition on Forest Processes

The effects of acid deposition on forest growth depend on several site-specific factors such as the soil nutrient status and composition and level of atmospheric inputs. Ions such as sulfate (SO_4) and nitrate (NO_3) are already in the ecosystem, and the hydrogen ion (H^+) is generated naturally by the plant community (Ulrich 1982).

Acids are produced naturally within soils (Reuss 1977, Rosenqvist 1977). Atmospheric acid inputs must be viewed as an addition to continual natural acidification and leaching processes from carbonic acid formation, organic acid formation, vegetative cation uptake, and a variety of forest management practices (Anderson et al. 1980, D. W. Johnson and Cole 1977, Reuss 1977, Sollins et al. 1980).

A key question is whether atmospheric H^+ inputs significantly add to or exceed natural H^+ production within the soil. The spruce-fir ecosystems where symptoms of forest decline are occurring are highly acidic naturally; thus inputs of atmospheric hydrogen ions may be expected to have a small effect on soil acidification processes. However, the nutrient supply may be influenced by effects of acid deposition on cation leaching or by pH-induced changes in mineral solubility, microbial processes, and weathering rates, in addition to the direct effects of nitrogen and sulfur inputs. Leaching of cations such as calcium (Ca^{2+}), magnesium (Mg^{2+}), potassium (K^+), and sodium (Na^+) from upper soil layers may lead to loss of available plant nutrients.

Acid deposition can cause increases as well as decreases in forest productivity (Abrahamsen 1980, Cowling and Dochinger 1980). Deficiencies of sulfur have been indicated in forests remote from pollutant inputs in eastern Australia

(Humphreys et al. 1975) and in the northwestern United States (Will and Young-berg 1978, Youngberg and Dyrness 1965). Atmospheric sulfur inputs may benefit those forest soils with little sulfate absorption capacity as a source of essential nutrients.

Nitrogen deficiencies are common in forests throughout the world. Inputs of NO_3 as well as ammonium (NH_4^+) and other forms of nitrogen are likely to improve forest nutrient status and productivity in many cases. Nitrate, in con-trast to sulfate, is adsorbed very poorly in most forest soils (D. W. Johnson and Cole 1977, Vitousek et al. 1979), and additional inputs of nitrogen are rapidly immobilized through biological processes. Because forest nitrogen requirements are relatively high compared with sulfur requirements, additional inputs seldom result in excess nitrogen. An exception is the Solling site in West Germany, where atmospheric inputs of N, S, and H^+ probably exceed the requirements of the forest. High-altitude spruce-fir ecosystems in the eastern United States, which are estimated to receive 50–75 kilograms per hectare a year (kg/[ha-yr]) of nitrogen, also may be nitrogen saturated.

If atmospheric nitrogen inputs exceed forest requirements and excess nitro-gen were available in soils, nitrification might be stimulated. Nitrification pulses are thought to be responsible for a large percentage of the nutrient leaching at the heavily affected Solling site in West Germany (Ulrich et al. 1980). Thus, nitrogen saturation of forest ecosystems could result in significantly increased cation leaching and, under extreme circumstances, further soil acidification. Such sat-uration would occur most readily in forests with low nitrogen requirements, that is, boreal coniferous ecosystems (Cole and Rapp 1981), or in forests with ade-quate or excessive nitrogen supplied by nitrogen-fixing species. However, evi-dence of widespread imminent nitrogen saturation of forests does not exist.

Most soils are not likely to be acidified significantly by atmospheric depos-ition at current levels in the United States (Reuss 1977). In general, only acid soils highly leached of nutrients—sandy soils or those with low cation exchange capacity—are likely to be sufficiently low in calcium to allow deposition inputs to affect growth of higher plants (Reuss 1977). Elevated concentrations of alumi-num (Al) in acidified soils may be a primary factor in limiting plant root develop-ment (Foy 1981). If aluminum is not present in excess, most acid soils will have adequate calcium for growth of most plant species including forest trees. Many if not all of the calcium deficiencies reported on acid soils in field studies derive from Al-Ca antagonisms rather than simply low calcium (Adams 1984). Sim-ilarly, magnesium (Mg) deficiencies observed in plants grown on acid soils are often a result of Ca-Mg antagonisms rather than low total soil magnesium levels.

Although the data by Reuss (1977) are probably adequate to estimate current atmospheric deposition rates at low elevations in the United States, recent data from the EPA Mountain Cloud Chemistry Program (Mohnen pers. comm. 1988) indicate that sulfur and nitrogen deposition at high elevations occurs primarily through cloud moisture deposition. Acidic cloud deposition to surface soils in boreal montane ecosystems is so high that soil acidification over time may in fact take place. Recent data yield annual deposition rate estimates of 122 kilograms per hectare (kg/ha) of sulfate, 65 kg/ha of nitrate, 15 kg/ha of ammonium, offset by only 5 kg/ha of base cations. The average pH of cloud moisture ranges from 2.6 to 3.4.

Deposited metals can be toxic if they are soluble and in sufficient quantities. Evidence of high deposition rates has been found only in the vicinity of smelters (Friedland et al. 1983a). In near neutral soils, heavy metals occur in inorganic compounds or bound with organic matter, clays, or hydrous oxides of iron (Fe), manganese (Mn), and aluminum. If soils become more acidic, the solubility increases and can result in metal concentrations that are toxic to many forms of vegetation. For example, zinc, copper, and nickel toxicities have occurred frequently in a variety of acid soils. Iron toxicity occurs only under flooded conditions where Fe occurs as the reduced soluble Fe^{2+} form (Foy et al. 1978). Toxicity to plants from lead, cobalt, beryllium, arsenic, or cadmium occurs only under very unusual conditions.

Soil Acidification and Nutrient Availability. Low alkalinity soils theoretically are susceptible to acidification and loss of nutrients from acid inputs via atmospheric deposition. However, few studies have been conducted to establish a direct link between acid deposition and significant changes in soil pH. The effects, if any, of acid deposition on soil nutrient cycling in ecosystems are of a chronic long-term nature.

Only long-term, quantitative studies of atmospheric input and soil nutrient cycling changes can establish a cause-and-effect relationship between acid deposition and direct or indirect soil-mediated effects. If a link between atmospheric deposition and acidification of ecosystems were established, biotic effects still might not be realized for many decades. However, these effects have potential tree-growth ramifications, and ecosystem changes because of pollution from human activities should be investigated.

Aluminum toxicity is the most prominent growth-limiting factor in many if not all acid soils (Foy 1973, 1974, 1981; Tanaka and Hayakawa 1974) and is believed to be a primary factor in limiting plant root development (Foy 1981). Aluminum leaching has long been implicated in agricultural crop damage. Kokorina (1977) noted that acid soil toxicity was more harmful during dry years. This dry-season phenomenon in concert with the acid deposition also may be a factor in the forest growth reduction among spruce and fir in West Germany reported by Ulrich et al. (1980).

From long-term intensive ecosystem level studies in West Germany, Ulrich et al. (1980) and Ulrich (1982) suggested that acid deposition has contributed to changes in H^+ generation and consumption, causing soil acidification, mobilization of aluminum, mortality of roots, interference in water and nutrient uptake, and ultimately the dieback of spruce, fir, and European beech. This hypothesis is based on observed changes in soil solution chemistry and a nearly parallel decrease in fine root biomass and increase in soil solution aluminum concentrations during the growing season. Nutrient solution studies indicated that the ratio of uncomplexed aluminum (Al^{3+}) to calcium found in the soil solution was sufficient to cause abnormal root growth and development. Although these findings suggest the possibility of aluminum toxicity, they cannot be considered definitive at this time because correlative data are lacking.

Recent, preliminary data indicate that the aluminum concentrations in soil water (measured using tension lysimeters) due to acid input may be affecting root systems of spruce and fir trees, particularly at high elevations. Now that the deposition rates of hydrogen ion, sulfate, and nitrate are better understood, the aluminum toxicity issue should be further explored in both laboratory and field experiments to determine whether acid-mobilized aluminum is contributing to the mortality of fine roots.

Work by scientists from the University of Göttingen in West Germany revealed that soils in the Solling forest are being severely affected by ecosystem acidification. Aluminum concentrations in soil solutions at the Solling site have increased twofold beneath the beech stands and tenfold beneath spruce stands over the last decade (Matzner and Ulrich 1981).

Although the Solling site appears to be susceptible to the effects of acidification on aluminum leaching, many other sites with the same tree symptoms have little or no susceptibility to this acidification process. In the calcareous Alps south of Munich, similar symptoms of high-altitude yellowing and browning are occurring, with soil pH ranging from 6.5 to 8.0. Preliminary data indicate that little or no aluminum leaches into soil solutions on these calcareous sites.

Studies of basic cation leaching from acid inputs often give inconsistent results. Under ambient conditions Mayer and Ulrich (1977) noted a net loss of calcium, magnesium, potassium, and sodium from the soils under a European beech forest. Except for sodium, however, the loss was equal to or less than nutrient accumulation in the trees. Roberts et al. (1980) reported that acid precipitation on Delamere (pine) forest of central England may produce small changes in litter decomposition but Ca, Mg, K, or Na leaching rates were not affected. Cole and Johnson (1977) found no detectable effects of acid precipitation on the soil solution of a Douglas fir ecosystem in the United States. Conversely, Anderson et al. (1980) noted a net output of calcium from both a pine forest soil in Sweden and a beech forest soil in West Germany; both soils accumulated nitrogen but not sulfate. Cronan (1980) reported net losses of Ca, Mg, K, and Na from subalpine spruce fir soils in New Hampshire and attributed the losses to acid deposition. Studies by Mollitor and Raynal (1982) suggest that leaching of potassium may be the most serious problem of cation leaching in the Adirondack forest soils.

Bauch (1983) determined that the roots of declining spruce and fir were calcium deficient but had the same levels of aluminum as healthy spruce and fir. Rehfuess (1981) has observed declining fir on calcareous soils, which seems to preclude aluminum toxicity or calcium deficiency. More recently, however,

Rehfuess et al. (1982) noted foliar magnesium and possibly calcium deficiencies even in base rich soil. They speculate that accelerated foliar leaching may be responsible. Rehfuess points out that the parallel change in soil solution aluminum and fine root biomass noted by Ulrich was not synchronized and that marked decreases in fine root biomass preceded the increase in soil solution in aluminum.

The aluminum toxicity hypothesis proposed by Ulrich is not supported by the data collected for the northeastern and southeastern United States, where dieback and decline are most prominent in high elevations and where soils have 80 percent organic matter by weight (Friedland et al. 1983b). Aluminum toxicity will likely be masked by the complexation of aluminum with organic matter. Data on spruce root chemistry from Camels Hump Mountain in Vermont indicate that the calcium to aluminum ratio increases with increasing elevation. As mortality increases with elevation, it is not likely that imbalances of Al and Ca in root tissue are a major cause of spruce decline (Johnson et al. 1983, Lord 1982).

The potential role of acid deposition on leaching of aluminum in diverse ecosystems has not been established. Comprehensive nutrient cycling studies, using lysimetry, will be needed to determine acid deposition nutrient leaching and cycling effects. Intensive research is recommended of the caliber and quality of the Hubbard Brook Ecosystem Study in New Hampshire that spanned decades. Studies have been initiated by the Electric Power Research Institute and EPA to establish sites for continued research.

Plant Nutrient Deficiencies. Leaching of nutrients from coniferous foliage and nutrient deficiencies as the result of acid deposition are still points of uncertainty and debate. Few data are available on stand dynamics and soil mineral availability in stands where nutrients have been leached.

Scientists in West Germany (Zoettl pers. comm. 1987) have noted that where spruce and fir foliage chlorosis (yellowing), necrosis (browning), and mortality are occurring, a marked mineral nutrient deficiency also occurs involving magnesium and calcium. Through a series of fertilizer trials they have been able to cause the slightly chlorotic or yellowing trees to come back to health and retain their vigor over a 3-year period.

The role of acid deposition and its contribution to the observed chlorosis are not known at this time. If hydrogen ion deposition to coniferous tree canopies (the overarching foliage of the ecosystem) is indeed leaching essential plant nutrients, then acid deposition would have an indirect effect on tree decline. However, in-depth studies under controlled conditions are needed to establish this linkage. The strong correlation between yellowing and dying tissue in coniferous trees and nutrient depletion should focus research on the potential of extreme hydrogen ion deposition (below pH 3.8) to cause these effects. After controlled-environment experiments are conducted, field trials should be initiated to establish the linkage under natural conditions.

Nitrogen Compounds. Limited high-altitude cloud monitoring data show large loadings of nitrogen in areas where forest decline is occurring. Deposition estimates of 37 to 44 kg/(ha-yr) of nitrogen have been made for subalpine eco-sytems in the eastern United States (Lovett et al. 1982). Deposition of nitrogen is higher (between 50 and 70 kg/(ha-yr) in polluted areas in central Europe (Ulrich pers. comm. 1988). However, the significance of these nitrogen levels for the present forest decline syndrome has not been determined.

Nitrogen compounds, unlike sulfur compounds, are highly active in the forest ecosystem. Nitrogen is the primary essential nutrient for growth and re-productive processes of forest trees. Although generally limiting in a forest eco-system, high-altitude spruce-fir ecosystems are nitrogen rich (Bruck and Robarge 1986, 1988). It is hypothesized that atmospherically deposited nitrogen may upset the balance of nitrogen in this alpine environment and create a toxic nitrogen excess. Many of the observed decline symptoms could be attributed to excess nitrogen input.

For these high-altitude trees already exposed to severe climatic conditions, additions of nitrogen compounds could result in a "stress syndrome" predispos-ing an entire ecosystem to other disease-causing agents and/or the effects of climate. Effects include reduction in frost hardiness, cell wall thinning, and radical shifts in soil microorganism populations. Although many potential path-ways of the effects of nitrogen in forest decline have been identified, none has been demonstrated experimentally.

Work in the northeastern United States indicates that nitrate concentrations in cloud moisture average between 185 and 200 microequivalents per liter (μeq/liter). Whether or not such levels are detrimental to root fungi relationships of red spruce is not known. Recent studies reveal that simulated acid rain treat-ments consisting of sulfuric and/or nitric acids have deleterious effects on the incidence and vigor of beneficial root fungi on the short roots of loblolly pine (Shafer 1984). Preliminary experiments indicated that the nitrate fraction con-taining concentrations of 100 μeq/liter has the most deleterious effect on both incidence and vigor of the short roots.

Excess nitrogen may reduce frost hardiness (Friedland et al. 1984b). Obser-vations of frost damage to the previous years' needles of red spruce over the past 5 years in the northeastern United States strongly suggest that repeated winter damage to foliage is a component of the spruce decline syndromes in both the Appalachian Mountains and central Europe. Three plausible factors, alone or in combination, could predispose red spruce to winter damage:

- natural susceptibility to frost damage,
- decline of vigor caused by stresses in the decline syndrome, and
- lack of frost hardiness due to cell wall thinning, induced by excess nitrogen via atmospheric deposition.

Speculation exists that the nitrogen component of acid deposition may be of greater significance than sulfur in terrestrial ecosystems, but just as with sulfur, the association between nitrogen deposition and tree damage on high-elevation sites is purely correlative. No empirical research has been conducted to establish a cause-and-effect relationship between nitrogen and tree injury, and trend data on nitrogen deposition are lacking.

Intensive research projects have been initiated in West Germany and, to a lesser extent, in the eastern United States to address the potential role of nitrogen deposition on spruce and fir trees. The interaction of nitrogen emissions leading to anthropogenic ozone production is of additional interest.

Deposition of Metals

Elevated concentrations of elements such as lead in soil and vegetation have been attributed to the combustion of leaded gasoline, coal burning and metal smelting activities, and the use of lead arsenate pesticides. Some studies link lead levels to mortality in microorganisms and plants (Babich and Stotsky 1979, Bauch 1983).

In addition to the direct effects on metabolic activities in plants, animals, and microorganisms, heavy metals can affect many subtle microbial activities and interactions. For example, a concentration of 1040 μg of lead per gram of soil in the form of lead arsenate significantly inhibits nitrogen mineralization. Numerous investigators have shown that indigenous soil fungi can be greatly inhibited in pure culture with 50–100 μg/ml of lead in the culture media.

Recent observations in central Europe and in the northeastern and southeastern United States have noted relatively high concentrations of lead in forest floor samples collected in remote areas of the Appalachian Mountains and throughout central Europe where forest decline symptoms are appearing (A. H. Johnson et al. 1982). Analysis indicates that the lead levels often exceed 2–3 g/m^2 in the forest floor of some high-elevation forests. These levels are less than the levels shown to be toxic in laboratory studies. The lead concentration values in the high-elevation stands with most of the observed forest decline symptomatology are approximately 4 to 10 times higher than those observed in low-elevations stands that do not exhibit decline symptoms. The current levels of heavy metals are high compared with presettlement background levels, approximately a 10- to 15-fold increase from certain ambient observations.

Whether or not heavy metal accumulations in the boreal montane soils play a role in observed forest declines is unknown. Additional research is needed to understand the effects of lead and other metals, alone and in combination with other factors.

Management Practices

A major unresolved question is whether management practices over periods of hundreds of years in central Europe have caused a depletion of soil nutrients and, hence, the chlorosis or yellowing noted on spruce and fir foliage. Few data exist on the potential effects of long-term intensive management practices, as employed in West Germany, on site productivity, on the susceptibility of trees to air pollution, and on the role in forest decline syndromes. Data from agricultural literature showing the long-term detrimental effects of intense site preparation and monoculture can be extrapolated only roughly to the forestry situation. In many areas where forest decline is exhibited at high altitudes in the eastern United States, the ecosystems are in virgin forests and/or Class 1 Wilderness Areas; that is, the forests are in their natural state or were cut only once perhaps 100 years ago.

If forest decline syndromes were being observed only in areas of intense management, the question of management practice would be more important. If management practices directly trigger, induce, sustain, or mitigate forest decline syndromes, comprehensive understanding of these effects will be critical. Long-term site modification practices should be initiated in controlled ecosystems in order to address the contributory or mitigative effect of management practices on forest decline symptomatology and epidemiology. This question is of particular interest in central Europe, where many stands have been managed heavily for over 500 years, and in the southeastern United States, where large areas of forests are managed to some degree.

Specific Evidence on the Declines in the Southeast

High-Elevation Spruce-Fir Decline

This section reviews studies recently completed or actively under way in the Black Mountain range of North Carolina, including those on Mount Mitchell (2037 meters), the highest peak in eastern North America.

Approximately 10,000 years ago, as the glaciers retreated north, a limited population of red spruce and Fraser fir was left occupying the highest peaks and ridges of the southern Appalachian Mountains. The Black Mountain range hosts the second largest remaining population of these boreal species. High-elevation peaks and ridges (above 1400 meters) provide a suitable habitat for the spruce-fir forests. Air temperature is from 5 to 10° C lower than at mountain base during the growing season, and precipitation averages greater than 170 centimeters annually. In addition, the high incidence of stratus and orographic cloud interception by these forests provides approximately twice the moisture input as can be accounted for by precipitation alone.

The introduction of the balsam wooly adelgid (ca. 1935–55) caused significant deterioration in certain areas of the Fraser fir range. This phenomenon continues to ravage Fraser fir throughout the southern Appalachians even today. Recent surveys have determined, however, that no significant pest or pathogens of red spruce have been either reported or detected.

Air pollutants have been suspected to be involved in the spruce-fir decline in the Southeast because of the unique and unprecedented nature of the observed symptoms, the location of the affected forests in relatively high pollutant deposition regions, and the simultaneous appearance of symptoms across wide areas. There is evidence of potential acid deposition effects on high-elevation forest ecosystems along the Appalachian mountain crest from eastern Canada through North Carolina. Although measurements are difficult and controversial, the vegetation appears to be subjected to very acidic cloud moisture for a considerable portion of the year (A. H. Johnson et al. 1983). Concern has increased because of the quantitative documentation of red spruce decline in the northern and southern Appalachian Mountains, but the causes are obscure at the present time (Bruck 1984, Bruck and Robarge 1988, Siccama et al. 1982).

Although the mountain summits are remote in relation to direct effects from large point sources of sulfur and nitrogen oxide emissions, they do receive high rates of hydrogen ion, sulfur, nitrogen, and heavy metal deposition (Friedland et al. 1983, Lovett et al. 1982). Cloud moisture pH is approximately 0.5 to 1.0 pH units lower than the ambient pH of rain or snow. Montane boreal forest vegetation is exposed to cloud moisture in the pH 3.0 to 4.0 range for about 30 to 80 days per year in the northern Appalachians and from 200 to 280 days per year in the southern Appalachian Mountains. Several studies have been initiated recently, including the EPA Mountain Cloud Chemistry Program, but substantially more data are needed to characterize the biological impacts of cloud acidity.

During the past decade a rapid and alarming deterioration of the boreal montane ecosystem has been observed throughout the Black Mountains. Some of the recent mortality can probably be attributed to the combined effects of balsam wooly adelgid, a prolonged climatic drought during the summer of 1986, and severe damage caused by rime ice on December 1, 1986. My hypothesis is that the sudden deterioration in stand conditions probably would not have occurred if the spruce-fir ecosystem were not already stressed and predisposed to decline. In this section evidence is presented from a number of studies that were initiated as early as 1983 to consider mechanisms other than climate and biotic pests and pathogens as contributing to the observed tree mortality and decline syndrome in the Black Mountains.

Evidence of Severe Decline on Mount Mitchell. A systematic survey of fir-spruce stands has been carried out on Mount Mitchell since 1983. The data collected over this period indicate that a severe, rapid, and unexplained decline is taking place.

A total of sixteen 25- to 100-year-old stands were selected for sampling. The primary selection criterion was that the vegetation composition of the stand be representative of the community on a given aspect and elevation as determined by visual inspection. The secondary criterion was that the tree canopy within the stand should be as intact as possible in order to obtain a sufficient number of live trees for statistical analyses. As a consequence of the second criterion, selected stands typically exhibited the least BWA (on fir) for a given community. Contrasting stands were sampled at both high and low elevations on a given peak when an elevational difference within the spruce-fir zone of at least 150 meters between high and low stands could be obtained.

Stands selected for sampling were 1 hectare in circular horizontal projection. Decline and dieback assessments in each stand were made within three randomly located 30 × 4 meters rectangular permanent plots. A general description of each stand was recorded, including a list of the species in the tree and shrub layers and dominant species in the herb layer. In addition, the general slope and aspect of the stand, height of the tree canopy, indications of logging or fire, and any striking topographic and landform features were noted.

Species, diameter at breast height, crown class, and dieback and decline ratings were also recorded for each tree. Dieback and decline were assessed only on red spruce and Fraser fir. Dieback was measured as a percentage loss of live crown length from the top and outside of the crown and was set to a 10-point scale (1 = 0–10 percent, 2 = 11–20 percent, and so on). Decline was assessed as percentage of foliage loss upward from the lower, inner crown, and was set to a 4 point scale (1 = 0–10 percent, 2 = 11–50 percent, 3 = 51–90 percent, and 4 = 91–99 percent). Percentage of foliage loss was determined by first estimating the proportion of crown length affected and then determining the number of years of needles retained with the affected portion of the crown. Normal needle retention time for red spruce was assumed to be at least 6 years while that for Fraser fir was assumed to be 5 years. Throughout the southern Appalachian Mountains, needle loss has consistently been observed to take place from the inside outward, leaving a tufted appearance on the outer edge of the tree.

The presence of fungal pathogens, insects, or mechanical breakage was noted for each tree. In the evaluation of fir trees, any tree that received a decline or dieback rating greater than 1 was checked for signs of the BWA by examining the bole for the presence of wool and branches for the presence of gouting.

Two increment cores were taken from each of the five dominant and largest codominant trees on each plot. Cores were taken just below breast height on opposite sides of the tree along the contour.

All dead trees on a plot were recorded by species, if possible, and their diameters measured. Saplings were recorded by species and an estimate of height was made to the nearest meter. For both red spruce and Fraser fir saplings, decline and dieback assessments were made using the same rating schemes as for trees. The Black Mountain range exhibits a distinct north-south orientation. Hence almost the entire spruce-fir ecosystem falls away from the ridge tops on an east or west aspect exposure. Our permanent plots were specifically selected to represent east and west aspects at both high (above 1800 meters) and low (below 1800 meters) elevations. In addition, it is important to note that these plots were NOT selected at random but rather stratified as discussed above and selected for their relative "health."

Once an area was located that satisfied the criteria of elevation and aspect, a plot center tree was located that was surrounded by mature spruce and/or fir trees that were both representative of the area and had no dead individuals or heavy impact by the BWA (in the case of fir). The rationale behind plot selection was to maximize the longevity of site visitation. (Certain high-elevation areas that exhibited mortality in 1984 are now completely devoid of any living spruce or fir trees above 10 cm DBH.)

The 272 red spruce observed in 1984 exhibited excellent physical appearance, with 79 percent of all trees classified as crown class 1. Nineteen percent were rated in class 2 (slight but recognizable foliar loss), 2 percent in class 3 (serious and obvious crown thinning), and no individual trees were dead (class 4). Although all the Fraser fir observed (213) were alive, their crown class ratings exhibited greater deterioration, with 60 percent in class 1, 15 percent in class 2, and 25 percent in class 3. The higher crown class rating for many individuals was probably attributable to BWA infestation (17 percent of all fir were classified as being positive for BWA).

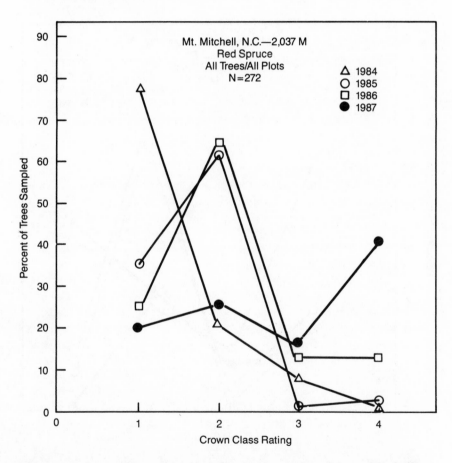

Figure 4.6. Crown class ratings for red spruce growing on Mount Mitchell, N.C. Surveys were conducted spring and summer of years 1984–87. Class 1 = 0–10% foliar loss, class 2 = 11–50% foliar loss, class 3 = 51–99% foliar loss, and class 4 = standing dead.

During subsequent evaluations of the permanently tagged trees (by the same field personnel) over the next 3 years (1985–87), a consistent and obvious deterioration in crown and stand conditions was observed. Decline of crown conditions (shifting into class 2 and 3) was immediately observed in 1985. It is interesting to note that most spruce crowns shifted from class 1 to 2 (figure 4.6) while fir rapidly shifted from 1 to 3 (figure 4.7). The mortality of spruce and fir was observed. By 1986, 9 percent of all spruce had died while 16 percent of all fir within the permanent plots had succumbed (figures 4.6 and 4.7). An alarming increase in mortality occurred over the winter of 1986–87. Forty-one percent and 49 percent of all spruce and fir, respectively, were classified as dead during the spring 1987 resurvey. This sudden mortality is likely attributable to the

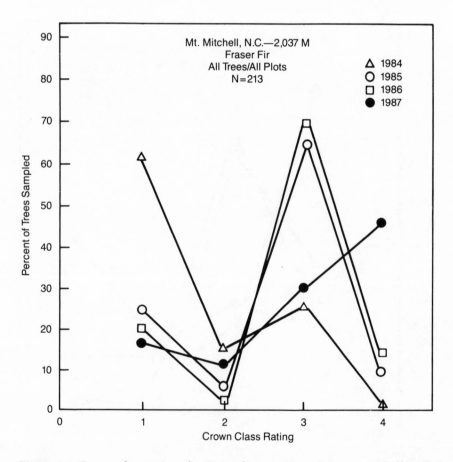

Figure 4.7. Crown class ratings for Fraser fir growing on Mount Mitchell, N.C. Surveys were conducted spring and summer of years 1984–87. Class 1 = 0–10% foliar loss, class 2 = 11–50% foliar loss, class 3 = 51–99% foliar loss, and class 4 = standing dead.

combined effects of a severe regional drought during the spring and summer of 1986 and a severe rime ice event that struck the Black Mountains on December 1, 1986. Many trees rated as class 1 in 1986 were observed as standing dead in 1987. It appears that the combined effects of coincidental drought and ice had serious consequences for this already stressed and deteriorating ecosystem.

The aspects in which stands were situated appeared to have a significant influence on the incidence and rate of tree-crown class rating change. Spruce and fir growing on the west face consistently exhibited greater crown thinning and mortality than those situated on an east aspect (figures 4.8 and 4.9). In addition, those trees growing at high altitude (above 1800 meters) consistently were rated in poorer condition than individuals growing at lower elevations

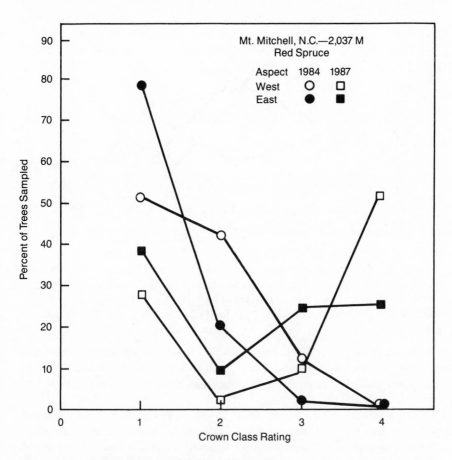

Figure 4.8. Crown class ratings for red spruce separated by east/west aspects. Surveys were conducted spring and summer of years 1984–87. Class 1 = 0– 10% foliar loss, class 2 = 11–50% foliar loss, class 3 = 51=99% foliar loss, and class 4 = standing dead.

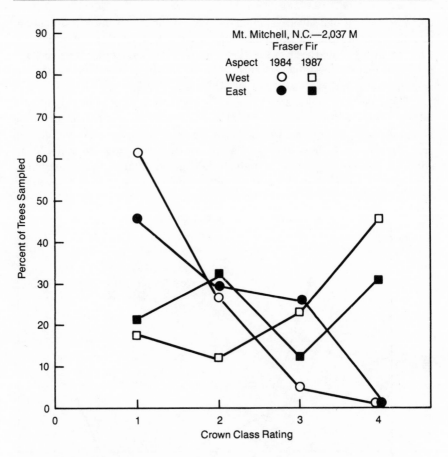

Figure 4.9. Crown class ratings for Fraser fir separated by east/west aspects. Surveys were conducted spring and summer of years 1984–87. Class 1 = 0– 10% foliar loss, class 1 = 11–50% foliar loss, class 3 = 51–99% foliar loss, and class 4 = standing dead.

(figures 4.10 and 4.11). Many predominantly fir stands growing at high altitude on the west face had deteriorated so badly by 1987 (greater than 50 percent mortality) that visits to continue overstory assessments will probably be discontinued.

Stand elevation and aspect as well as species composition had a significant effect on tree diameter and stand basal area. High-elevation stands (especially those on west slopes) composed of dense, small-diameter Fraser fir are probably a result of natural and human-activity-induced disturbances (Pyle et al. 1985). Logging, fire, and BWA infestations have created numerous mid-age stands of fir at high elevation in the Black Mountains. Low-elevation (east face) stands of

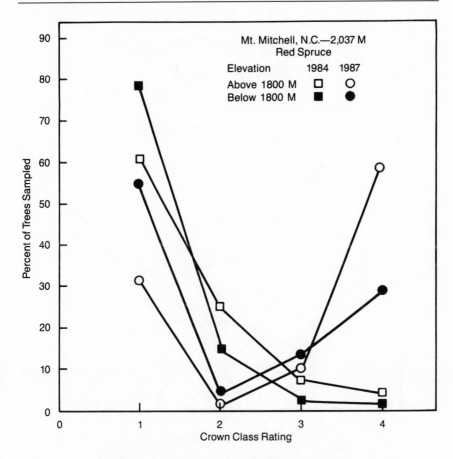

Figure 10. High altitude (above 1800 m) vs. low altitude (below 1800 m) crown class ratings for red spruce. Surveys were conducted spring and summer of years 1984–87. Class 1 = 0–10% foliar loss, class 2 = 11–50% foliar loss, class 3 = 51–99% foliar loss, and class 4 = standing dead.

relatively undisturbed older red spruce exhibit far greater basal areas on a per tree basis though the stand density is much lower.

Although smaller-diameter class red spruce (10–30 cm DBH) and Fraser fir (10–20 cm DBH) represent a significant proportion of the forest resources studied, the trend toward basal area shifting into visible decline and "dead" crown class rating categories appears to be consistent (figures 4.12 and 4.13).

Fewer red spruce trees are observed above 1800 meters elevation than below. In addition, the individuals exhibit smaller basal areas (figures 4.10, 4.12, and 4.13). However, regardless of red spruce size or elevation, the trend toward crowns deteriorating into higher classes was observed throughout 4 years.

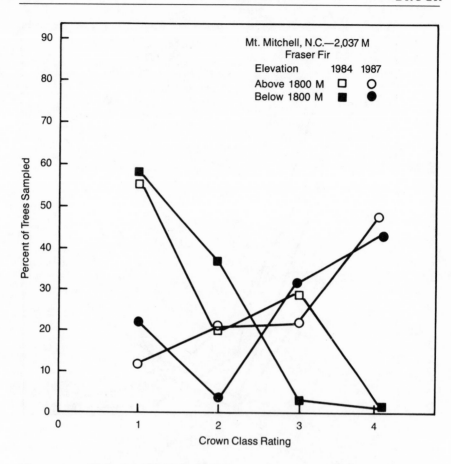

Figure 4.11. High altitude (above 1800 m) vs. low altitude (below 1800 m) crown class ratings for Fraser fir. Surveys were conducted spring and summer of years 1984–87. Class 1 = 0–10% foliar loss, class 2 = 11–50% foliar loss, class 3 = 51=99% foliar loss, and class 4 = standing dead.

In contrast, Fraser fir basal areas did not significantly differ between high- and low-altitude plots (below 1800 meters) (figures 4.14 and 4.15). This can be accounted for by the observation that Fraser fir individuals exhibit a far narrower average range in DBH size classes (for example, 7–39 cm DBH) as compared with red spruce (9–72 cm DBH). It was consistently observed, however, that high-altitude individuals (regardless of species) demonstrated the greatest shift toward greater crown thinning and mortality.

From the beginning of our initial observations (Bruck 1984) in 1983, it was noted that with the exception of a narrow band of blowdown along the Black Mountain ridge crest (usually above 1950 meters), crown thinning and mortality were greater on west-facing (windward) slopes than on east-facing aspects.

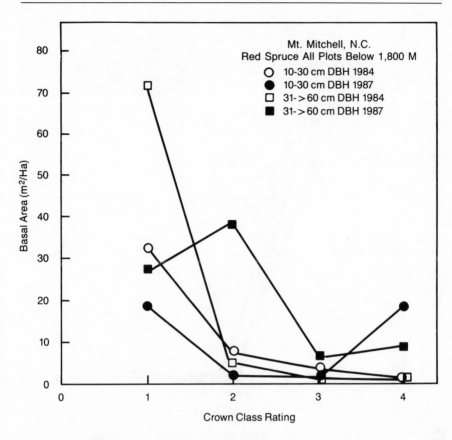

Figure 4.12. Basal area: summary of crown class ratings for 1984–87 stratified by red spruce diameter classes, at altitudes below 1800 m. Class 1 = 0.10% foliar loss, class 2 = 11–50% foliar loss, class 3 = 51–99% foliar loss, and class 4 = standing dead.

There was a consistent shift of trees from "healthy" crown classes toward "diseased" classes between 1984 and 1987. However, individuals growing on west-aspect plots were always observed to have greater crown deterioration. By 1987 almost half the standing basal area of spruce on west-facing aspects was dead (figure 4.16). A far lower percentage of spruce basal area (and fewer individuals) was as greatly affected on east-facing slopes. Fraser fir exhibited an even greater overall deterioration, again most commonly observed on west aspects (figure 4.17). By 1987, more than half the standing basal area in plots on west-facing slopes had died, while approximately 30 percent of fir growing on east slopes had succumbed (figure 4.17).

After 4 years of study we conclude that regardless of cause-and-effect mechanisms, the physical condition and structure of the boreal montane ecosystem on and around Mount Mitchell have greatly deteriorated. We recognize that this

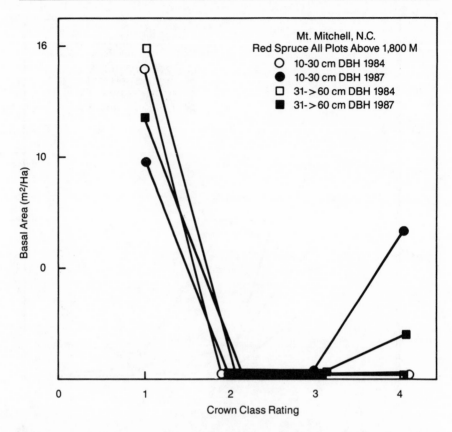

Figure 4.13. Basal area: summary of crown class ratings for 1984–87 stratified by red spruce diameter classes at altitudes above 1800 m. Class 1 = 0–10% foliar loss, class 2 = 11–50% foliar loss, class 3 = 51–99% foliar loss, and class 4 = standing dead.

conclusion is drawn from a limited data set. However, it is probably biased toward the conservative side due to the fact that all plots were initially (1984) in a good state of physical appearance. Numerous areas along the Black Mountain range were in an advanced state of deterioration prior to our survey.

The incidence of decline appears to be patterned in a mosaic. That is, at a given elevation and aspect, great diversity in forest conditions may be observed—ranging from greater than 90 percent mortality to class 1 dominant stands within 50 meters of one another. It was for this reason that in our limited survey, stands reflecting a high degree of crown integrity were chosen. We question the utility of permanent plot networks to assess the status of "whole" spruce-fir ecosystems and suggest that random transects (to assess ground conditions) combined with remote sensing techniques would better suit this purpose.

The spruce-fir ecosystems of the Black Mountain range have been and con-

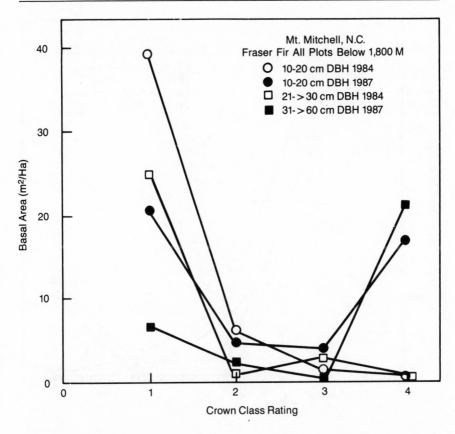

Figure 4.14. Basal area: summary of crown class ratings of 1984–87 stratified by Fraser fir diameter classes, at altitudes below 1800 m. Class 1 = 0.10–10% foliar loss, class 2 = 11–50% foliar loss, class 3 = 51–99% foliar loss, and class 4 = standing dead.

tinue to be subject to a myriad of natural and anthropogenic stresses. Color slides dating back to the mid–1950s and aerial photographs from the 1970s indicate that although certain quantifiable damages were present during recent times, the rate of acceleration in decline symptomatology and mortality has increased greatly over the past 4 years. One could speculate that the southern spruce-fir ecosystem is already "living on the edge." We speculate that numerous insults of meteorological, natural biotic, and/or human origin have had an impact on the boreal montane forests for quite some time. The high deposition of acidic, nutrient, and toxic substances principally due to cloud deposition may also be a predisposing or contributing factor to the apparent increased rate of stand deterioration (Schutt and Cowling 1985).

The future of the southern spruce-fir ecosystem is uncertain at best. As the mosaic of damage in these forests becomes manifest, the openings in crown

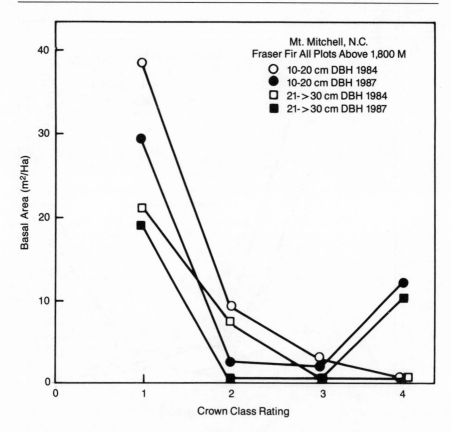

Figure 4.15. Basal area: summary of crown class ratings from 1984–87
stratified by Fraser fir diameter classes at altitudes above 1800 m. Class 1 =
0–10% foliar loss, class 2 = 11–50% foliar loss, class 3 = 51–99% foliar loss,
and class 4 = standing dead.

structure and loss of stand integrity permits wind, ice, and solar energy penetration, thus changing everything from soil microbial populations and understory
herbaceous cover to the reproductive dynamics of the trees themselves. In addition, if current trends indicating potential global warming are confirmed, the
continued presence of the unique boreal forests in the southern highlands is
threatened.

Observation and Quantification of Spruce and Fir Needle Necrosis. On
June 6, 1987, a cloud event fully enshrouded all peaks above 1500 meters in the
Black Mountain range of North Carolina. Beginning approximately 48 hours
after the event, observations were made of newly flushing foliage of mature red
spruce and Fraser fir trees. A limited number of cloud samples were collected
from this event at the Tower I site on Mount Gibbs of the EPA Mountain Cloud

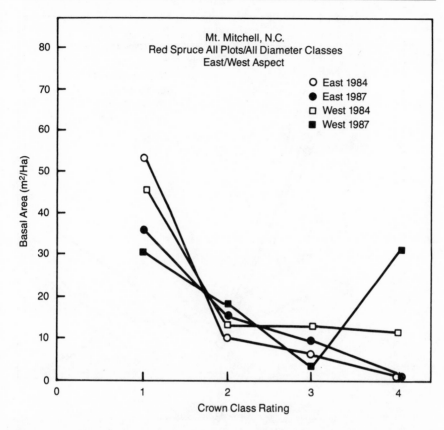

Figure 4.16. Basal area summary of crown class ratings of east vs. west aspect red spruce for 1984–87. Class 1 = 0–10% foliar loss, class 2 = 11–50% foliar loss, class 3 = 51=99% foliar loss, and class 4 = standing dead.

Chemistry Program, indicating a range in hourly cloud values from pH 2.6 to 3.1. A limited sampling of needle pH from droplets in situ was undertaken following this cloud event.

Because the long-duration cloud event occurred at the same time as the needle flush of most red spruce and Fraser fir, certain unique symptoms were observed. After the initial observations of needle necrosis on abaxial branch surfaces by removing branches from 1.0 to 3.5 meters, a survey was conducted from the summit of Mount Mitchell (2037 meters) to the Tower II Mountain Cloud Chemistry site located at 1740 meters on the southeast-facing Commissary Ridge. This stratified survey was accomplished by selecting 10 spruce and 10 fir randomly at 50-meter increments in elevation gradient down to the Tower II site, and removing branches from 1.0 to 3.5 meters from each tree.

Approximately 48 hours after the event, distinct abaxial needle surface necrosis was observed principally in areas with high air circulation and hence

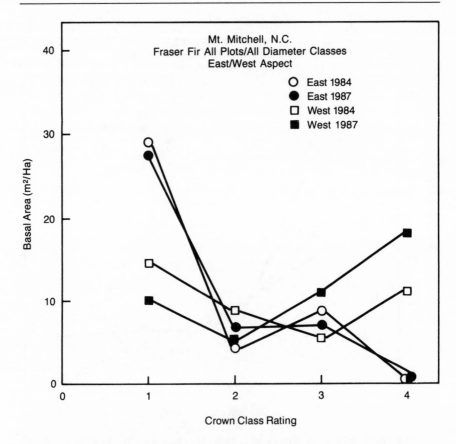

Figure 4.17. Basal area summary of crown class ratings of east vs. west aspect Fraser fir for 1984–87. Class 1 = 0–10% foliar loss, class 2 = 11–50% foliar loss, class 3 = 51–99% foliar loss, and class 4 = standing dead.

cloud deposition. Trees and specific branches that were protected by other foliage appeared to have less noticeable necrosis. In addition, there appeared to be an altitudinal gradient, with a greater number of individual trees and branches exhibiting needle necrosis at higher altitude. Although a limited number of in situ needle pH readings were taken (thirty), the lowest pH reading observed 12 hours following the cloud event was reported at pH 2.54. The mean reading of pH was 2.72. Six areas were observed down the southeast face of Commissary Ridge 5 days after the initial appearance of symptoms. A total of 10 trees near trailside were chosen at random at each of the 6 sites. The appearance of symptomatic trees and the severity of needle necrosis were distinctly stratified by altitude (table 4.1). From 1980 meters to the summit, virtually all observed trees showed needle necrosis, and at least some needles on all observable foliage had damage. Between 1940 meters and just above the Mountain Cloud Chemistry Program

Table 4.1. *Needle damage assessment of spruce and fir needles, following a 15 hr cloud event on June 6, 1987.*

Site evaluation (m)	Spruce exhibiting necrosis[a]	Fir exhibiting necrosis[a]	Spruce: necrosis laterals (%)[b]	Fir: necrosis laterals (%)[b]
2030	10	10	100	100
1980	10	10	100	100
1940	8	10	75	80
1900	7	3	60	75
1850	7	2	40	30
1760	5	0	20	0

[a]Of 10 observations.
[b]Visual estimate of those trees having symptoms.

Tower II at 1760 meters, there was a steady decrease in observed symptoms on both spruce and fir foliage (table 4.1).

The specific symptoms of needle necrosis were interesting to observe. Proximal (closest to the branch) tissues were usually bright green and undisturbed, fading into an area of ruptured cells that appeared yellow. The tissue most distal to the branch appeared to be shrunken and necrotic as if the water had been dehydrated from the needle tips. Through the services of the U.S. Environmental Protection Agency, Research Triangle Park, North Carolina, an X-ray scanning electron microscope study was initiated on necrotic needle tissue. An elemental gradation was observed between healthy and necrotic needle tissue. Damaged tissues from both the interface zone and the necrotic region showed highly elevated sulfur content. These preliminary data indicate a potential acid concentration effect following the evaporation of cloud deposition moisture onto the newly flushing spruce and fir needles.

Reciprocal Soil Transplants. On May 12, 1986, 600 two-year-old seedlings of red spruce, procured from the United States Forest Service in Durham, New Hampshire, were transplanted into reciprocal soil transplant plots on Mount Gibbs (2004 meters) and on Commissary Ridge (1740 meters) on Mount Mitchell. The plots measured 1 meter square by 8 cm deep. A total of 6 plots was excavated at each site, with 3 containing homogenized soil from the site itself and three containing homogenized and transplanted soil reciprocally from the opposite site. A total of forty-nine 3-year old red spruce seedlings per plot was transplanted into each reciprocal soil plot at a spacing of 8 × 8 cm. The original purpose and objective of this experiment was to observe the recolonization of these amycorrhizal seedlings by the natural soil flora from the respective sites on Mount Mitchell.

By November 1986 symptoms began to develop on certain seedlings in the reciprocal transplant plots. Approximately 30 percent of all seedlings transplanted into the high-elevation soil were exhibiting needle necrosis and terminal dieback. No destructive sampling was done at this time, however. After the winter of 1986, all the transplant plots located at the high-elevation site experienced severe frost heaving over the winter months. Although many of the seedlings were still alive, most of them had heaved out of the ground, exposing their root systems. The high-elevation site plots were therefore terminated. None of the low-elevation plots exhibited significant damage, and the 6 plots containing 49 seedlings in each plot of high-elevation homogenized and low-elevation homogenized soil, respectively, were continued for observations.

On October 2, 1987, all seedlings remaining in the low-elevation plot were removed along with soil samples and returned to the Forest Pathology Laboratory at North Carolina State University. Observations made at this time indicated a high incidence of mortality on seedlings transplanted into the high-elevation (Gibbs Peak) soil currently reciprocally transplanted to the low-levation (Commissary Ridge) site (table 4.2). Only 6 percent of the seedlings remaining in the homogenized low-elevation soil were dead, compared with 93 percent of the seedlings planted at the low-elevation site in the Gibbs Peak soil. The average needle necrosis observed on the Commissary Ridge seedlings was 12 percent, and that of seedlings planted in the Gibbs Peak soil, 92 percent. In addition, both root and shoot dry weights of all seedlings were significantly suppressed from seedlings grown on the high-elevation soils. Soil chemical analyses were carried out on field-moist subsamples using procedures 5.1, 5.2, and 5.3 of Robarge and Fernandez (1986).

No clear trends are evident in the data when the foliar analysis of the needles is expressed on a concentration basis (table 4.3). The mean elemental concentrations of copper (Cu), iron, zinc (Zn), and aluminum were higher in seedlings grown in the Mount Gibbs soil but the elemental concentrations of the macronutrients phosphorus (P), calcium, magnesium, and potassium were approximately the same for seedlings grown in either soil.

There was, however, a definite difference in the needle mass per tree for the seedlings grown on the two different soils. A large difference in needle mass per tree can distort comparisons of elemental concentrations in needle tissue. Total

Table 4.2. *Growth and mortality dynamics of transplanted red spruce seedlings growing on high and low elevation soils. Commissary ridge site: reciprocal soil transplant 18 months.*

Soil type	Average dry root weight (g)	Average dry shoot weight	Average necrosis (%)	Dead (%)
Commissary Ridge (1740 m)	1.42	1.85	12	6
Gibbs (2004 m)	0.63	1.03	92	93

Table 4.3. *Elemental concentration of needles from soil transplant plots. Needles from all the trees within one plot were combined to form one bulk sample per plot. Needle mass indicates the average needle mass per tree within a plot.*

Plot ID No.	N	Needle mass (g/tree)	Element (mg/kg)									
			P	Ca	Mg	K	Cu	Fe	Mn	Zn	Pb	Al
TSP1	12	0.256	2500	4080	1730	4510	11.9	570	484	91	<1	481
TSP2	16	0.645	2050	2440	1170	6300	7.0	131	806	50	<1	102
TSP3	14	0.346	2260	3160	1320	7080	15.5	307	794	76	<1	247
Mean	14	0.415	2270	3230	1410	5960	11.5	336	695	72	<1	277
HSP1	19	1.353	1900	2580	1220	6960	6	176	459	44	<1	142
HSP2	10	0.756	1910	2770	1210	6590	8.4	164	621	52	<1	150
HSP3	11	0.915	2310	2850	1280	7110	8.4	121	816	48	<1	93
Mean	13	1.008	2040	2730	1240	6890	7.6	154	632	48	<1	128

Note: TSP = transplanted Mount Gibbs soil. HSP = homogenized low-elevation soil.

elemental uptake per tree, therefore, was calculated to compare nutrient uptake between plots.

When the results for the foliar analysis were expressed as total elemental uptake per tree, definite differences between seedlings grown on the Mount Gibbs soil and the low-elevation soil became clearer. The uptake per tree of phosphorus, calcium, magnesium, potassium, copper, and manganese for seedlings grown in the low-elevation soil was two to three times that for seedlings grown in the Mount Gibbs soil (table 4.4). The differences for iron and aluminum observed in table 4.3 were no longer evident, and the trend observed for zinc was reversed.

The differences in total elemental uptake per tree in table 4.4 are most probably due to differences between the Mount Gibbs and low-elevation soil.

Table 4.4. *Total elemental uptake per tree from soil transplant plots. Values shown were calculated using the average needle mass per tree for a given plot times the respective elemental concentration derived from the bulk sample.*

Plot ID No.	Element (μ/tree)									
	P	Ca	Mg	K	Cu	Fe	Mn	Zn	Pb	Al
TSP1	640	1040	440	1150	3	146	124	23	—	123
TSP2	1320	1570	750	4060	4.5	84	520	32	—	66
TSP3	780	1090	460	2450	5.4	110	270	26	—	85
Mean	940	1340	590	2470	4.8	139	288	30	—	115
HSP1	2570	3490	1650	9420	8.1	238	621	60	—	192
HSP2	1440	2090	920	4980	6.4	124	469	39	—	113
HSP3	2110	2610	1170	6510	7.7	111	747	44	—	85
Mean	2060	2750	1250	6950	7.7	155	637	48	—	129

Note: TSP = transplanted Mount Gibbs soil. HSP = homogenized low elevation soil.

The Mount Gibbs soil has fewer nutrient cations (11 percent base saturation) than the low-elevation soil (22 percent base saturation), even though both soils have approximately the same total exchangeable acidity and exchangeable aluminum (table 4.5). The critical difference is in the amount of exchangeable calcium and magnesium present, with the low-elevation soil containing two to three times the concentration of the Mount Gibbs soil. Aluminum and calcium interactions are very important in acid soils, and the lower concentration of exchangeable calcium in the Mount Gibbs soil would increase the likelihood of aluminum toxicity to plant roots due to the decrease in calcium activity in the soil solution.

Plant response to degrees of soil acidity can be shown graphically by plotting growth or elemental uptake as a function of soil base saturation. Needle mass per tree as a function of percentage of base saturation is shown in figure 4.18. The data indicate a definite linear increase in needle mass accumulation per tree with increase in percentage of base saturation, with a change in slope at approximately 14 percent base saturation. A similar relationship is shown for elemental uptake per tree for phosphorus and magnesium in figure 4.19. The trend is even more evident when the data are plotted as a function of the soil Ca/Al ratio (cmol(+)/kg) (figures 4.20 and 4.21).

The trends shown in figures 4.18–4.21 are similar to those for agronomic crops in acid soils. Below the critical base saturation, there is a linear decrease in yield. Above the critical base saturation there is a slight increase or no increase in yield. The positive linear slopes above 14 percent base saturation in figures 4.18 and 4.19 suggest that the red spruce seedlings are still responding to a decrease in soil acidity although at a slower rate.

The results from the preliminary transplant experiment suggest that the critical level for survival of red spruce seedlings in the soils of the high-elevation boreal forests of the southern Appalachian Mountains can be set at 14 percent base saturation. The significance of this observation becomes apparent when

Table 4.5. *Soil chemical analysis for soil transplant plots. Results are expressed on an oven-fry weight loss.*

Plot ID No.	pH		Exchangeable					Total acidity	Effective CEC	B.S. %
	Water	1N KCl	Ca	Mg	K	Na	Al			
					(cmol (+)/kg)					
TSP1	3.92	3.35	0.71	0.23	0.20	—	7.5	9.5	10.6	10.7
TSP2	4.17	3.54	0.93	0.28	0.20	—	8.3	9.7	11.1	12.7
TSP3	4.05	3.53	0.93	0.33	0.17	—	9.3	10.5	11.9	12.0
Mean	4.05	3.47	0.86	0.28	0.19	—	8.4	9.9	11.2	11.8
HSP1	3.90	3.21	3.15	0.86	0.30	—	7.7	10.1	14.4	29.9
HSP2	4.15	3.41	1.05	0.37	0.20	—	8.0	10.0	11.6	13.9
HSP3	3.66	3.08	2.13	0.58	0.21	—	8.8	11.0	13.9	21.0
Mean	3.90	3.23	2.11	0.60	0.24	—	8.17	10.4	13.2	22.2

Note: TSP = transplanted Mount Gibbs soil. HSP = homogenized low elevation soil.

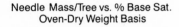

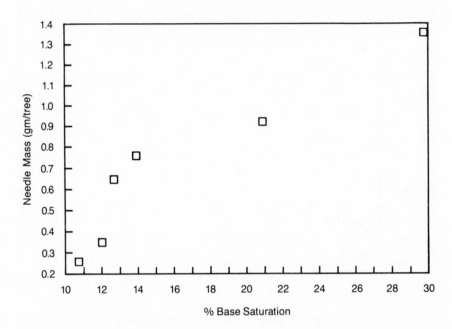

Figure 4.18. Needle mass per tree as a function of percentage of base saturation for soil transplant study.

contrasted to the range in percentage of base saturation as a function of elevation and soil depth within elevation obtained by Wells et al. (1987) for the Black Mountains of North Carolina (table 4.6). Clearly the litter layer would be favorable to seedling growth based on the above criteria, but approximately half the plots sampled would have less than the critical base saturation in the mineral soil directly under the litter layer. This is especially important because the fine root mass of red spruce and Fraser fir trees is most often located between the litter layer and the top 10 cm of the underlying mineral soil. Obviously, deeper penetration of the fine roots into the mineral soil would not alter the situation (table 4.6).

The transplanted soil from Mount Gibbs used in this experiment is from an area of known high deposition of acidic compounds of nitrogen and sulfur. The 2- to 3-fold decrease in exchangeable calcium as compared to the low-elevation soil is consistent with the hypothesis that increased deposition of acidic nitrogen and sulfur compounds leads to increases in the loss of base cations even though the total acidity of these already acid soils may not increase significantly (table 4.5). For such soils, relatively small changes in the calcium activity of the soil solution could result in a substantial increase in aluminum toxicity.

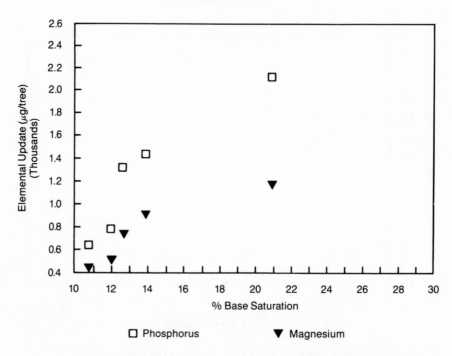

Figure 4.19. Phosphorus and magnesium total foliar elemental uptake per tree as a function of percentage of base saturation for soil transplant study.

The results from this preliminary experiment demonstrate the necessity of obtaining field data in order to form the critical link between the growing data base characterizing the high-elevation boreal ecosystem in its present condition and the effects of acidic deposition on red spruce and Fraser fir. Lack of a prior data base as extensive as that currently being generated by the U.S. Forest Service Spruce–Fir Research Cooperative precludes the direct comparison approach to determine the impact of acidic deposition on these ecosystems. Rather, indirect comparison and interpretations of the data will form the basis for assessing the impact of acidic deposition. Establishing the critical level of soil acidity for red spruce survival in situ is a vital step in this process.

Soil Lysimeter Study. In the fall of 1986, a series of soil water samplers (lysimeters) were installed on Mount Gibbs, North Carolina. The primary objective for the installation of these samplers was to attempt to study in situ the dynamics and mechanisms controlling aluminum in solution in the rooting zone of these high-elevation soils. Of particular interest are the relationship of aluminum to pH, the ratio of calcium activity to aluminum activity, the predominant

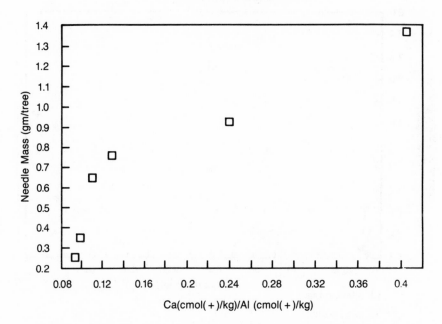

Figure 4.20. Needle mass per tree as a function of soil calcium/aluminum ratio.

form of aluminum that is present in solution, and the influence of SO_4-S on aluminum solubility in an ecosystem with high inputs of acidic sulfur compounds. With the close proximity of the collectors of the Mountain Cloud Chemistry Program, these samplers can also be used to measure NO_3-N, SO_4-S, and Ca flux out of the rooting zone and to test the hypothesis that during periods of high deposition of acidic cloudwater and/or rainwater there is a corresponding response in the chemical composition of the soil solution. Presented here are the preliminary results of analyses of the data from the samples collected in the fall of 1986.

A total of 12 lysimeters was installed just below the rooting zone for red spruce and Fraser fir (15 cm). A depth of 15 cm was typically beneath the Oa (organic) horizon within the upper portion of the A horizon.

Attempts to install the samplers without disturbing the overlying soil proved unsuccessful. The samplers were therefore installed by carefully removing the overlying litter and Oa horizon as a continuous unit and then excavating the underlying mineral soil to the desired depth. The installed sampler was then backfilled with mineral soil (excluding rocks and roots), and the overlying litter and Oa horizon were replaced in their original position. The overlying litter and Oa horizon constituted 50–75 percent of the total depth of excavation, so the

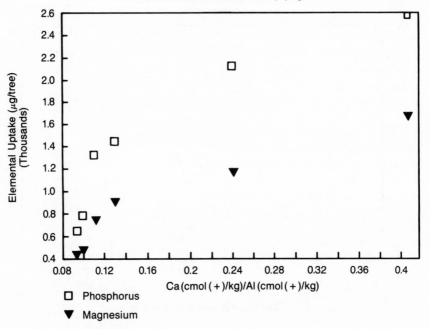

Figure 4.21. Phosphorus and magnesium total foliar elemental uptake per tree as a function of soil calcium/aluminum ratio.

Table 4.6. *Percentage of base saturation (+/− standard error) as a function of elevation and soil depth within elevation for the Eastern Spruce-fir Research Cooperative permanent plots located in the Black Mountains of North Carolina.*

Soil depth (cm)	Elevation (m)			
	1980	1830	1680	1520
LF	45 ± 18	37 ± 19	43 ± 15	32 ± 16
0–5	16 ± 6	17 ± 9	15 ± 7	13 ± 7
5–10	11 ± 4	11 ± 4	10 ± 6	10 ± 5
10–15	8 ± 3	9 ± 3	7 ± 2	8 ± 3
N	8	30	26	16

Note: LF = litter layer, variable depth. N = number of plots sampled.

actual disturbance of the surrounding mineral soil was minimal. However, any root mass in the overlying litter and soil was severed during the installation procedure.

Chemical analysis of the samples included pH, "labile-Al" (James et al. 1983), soluble calcium, magnesium, potassium, and sodium by flame atomic absorption spectroscopy, and chlorine (Cl), nitrate, and sulfate by suppressed ion chromatography. Analysis for fluorine (F) and phosphate (PO_4) yielded values below detection limits (0.1 and 0.5 mg/liter, respectively).

The first collection was made on September 19, 1986, with collections proceeding thereafter on approximately a weekly basis until November 20, 1986.

Ionic Composition. A summary of the ionic composition of the soil water collected during the period is presented in table 4.7. The predominant cation in solution was aluminum, followed by calcium, magnesium, hydrogen ion, potassium, and sodium. This trend remains even when the data are expressed on a volume-weighted basis. The predominant anion was NO_3 followed by SO_4 and Cl. The predominance of NO_3 was not expected, given the past emphasis on SO_4 in soils receiving high inputs of acidic deposition. For this set of samples collected at this time of the year it appears that it is NO_3 and not SO_4 that is the main anion accompanying movement of Al, Ca, and Mg out of the rooting zone.

Soil water samples collected during this period were devoid of color, suggesting that little or no dissolved organic matter was leaving the rooting zone. The linear relationship between the sum of negative and positive charge based on the ionic species in table 4.7 (figure 4.22) has a slope of approximately one. This suggests that whatever dissolved organic matter is present in solution is not associated to any significant degree with the ions in table 4.7. The data in figure 4.22 were plotted assuming that all "labile-Al" detected is the trivalent species.

Maximum concentrations of ions in the soil water leaving the rooting zone were observed for the October 10, 1986, collection (Julian date: 284; figures 4.23

Table 4.7. *Summary of the chemical composition of soil water collected from September 19 through November 20, 1986.*

Ionic species	Mean	s.e.	min.	max.
		(μeq/l)		
H	87	43	19	191
Al	233	155	23	713
Ca	170	111	27	501
Mg	104	61	7	325
K	52	47	5	273
Na	26	12	3	57
Cl	65	72	6	440
NO3	442	313	32	1323
SO4	148	72	29	380

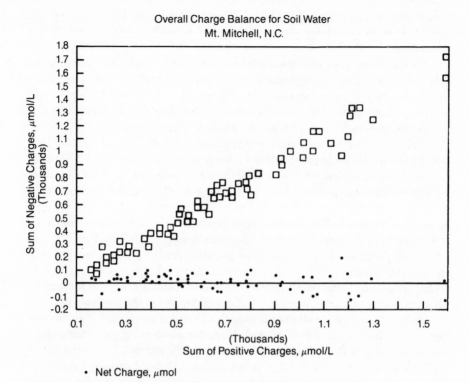

Figure 4.22. Sum of the negative charge versus the sum of positive charge, and the net charge, for each soil water sample collected during the fall of 1986.

and 4.24). The concentration of Al, Ca, Mg, H, K, and NO_3 peaked at this time, with the concentration of NO_3 exceeding that of SO_4 by approximately 8-fold. By the last collection for 1986, the concentrations of all the ions except aluminum were approximately equal to or lower than the first collection in September. The concentration of NO_3 was actually lower than that of SO_4. Throughout the sampling period the concentration of SO_4 remained essentially constant, suggesting either a constant flux of SO_4 through the rooting zone, or the presence of a solid phase controlling the activity of SO_4 in solution. The concentration of aluminum decreased after October 10, 1986, but not to the same extent as Ca, Mg, or K (figures 4.23 and 4.24).

Aluminum Equilibria. The pH of the soil water samples collected during the fall of 1986 ranged from 3.7 to 4.7. The concentration of aluminum varied as a function of pH in a manner similar to that reported by other investigators in the literature (figure 4.25). The spread in the data in figure 4.25 is probably due to expressing the results on a concentration basis and to the variability among the soil water samplers.

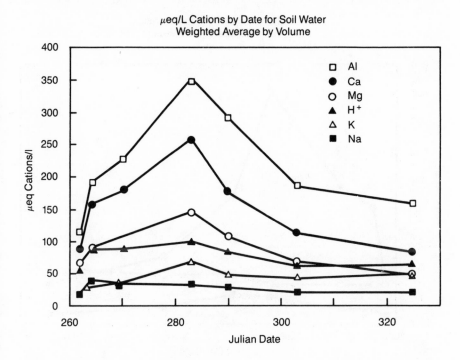

Figure 4.23. Mean concentration of cations in soil water samples as a function of date of collection.

When plotted as the negative logarithm of the concentration of aluminum (pAl) versus pH, the data appear to approach a linear relationship (figure 4.26), but not one corresponding to gibbsite as the controlling solid phase. Gibbsite is present in the mineral fraction of these soils (S. Feldman pers. comm. 1987) and is often assumed to control the activity of aluminum in the soil solution. However, gibbsite can also serve as a retention site for SO_4 with the possible formation of an aluminum hydroxy sulfate mineral as an intermediate product with sufficient levels of SO_4 input. Although such a solid phase may not be stable in these soil systems over the long term, its presence could nevertheless exert an influence on the concentration of Al, SO_4, and H in the soil solution.

The data, as presented in figures 4.26–4.28, suggest that the presence of an aluminum hydroxy sulfate phase may indeed be influencing the activity of aluminum in the rooting zone of these soils. A more definite conclusion may be reached when the data are corrected for the influence of ionic strength and replotted as activities.

The data presented here represent preliminary observations of the chemical composition of the soil water leaving the rooting zone of high-elevation spruce-fir forests. Additional collectors have since been installed at both the high-

μq/L Anions by Date for Soil Water
Weighted Average by Volume

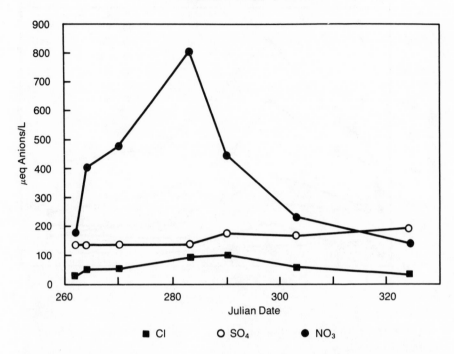

Figure 4.24. Mean concentration of anions in soil water samples as a function of date of collection.

elevation and low-elevation sites of the Mountain Cloud Chemistry Program in the Black Mountains of North Carolina. Collections have proceeded on a weekly basis since late May 1987 in an attempt to correlate observed changes in soil water composition with wet deposition at each site. Measurements of throughfall and stemflow at each site on an event basis provide information on flux of ions into the rooting zone. Since the incidence of cloud interception is significantly less at site 2 than at site 1, the deposition of acidic compounds of nitrogen and sulfur is less. Therefore, contrasting the soil water composition between the two sites offers a means to determine to what extent acidic deposition at site 1 (62 percent tree mortality versus 4 percent at site 2) has influenced the chemistry of aluminum in the rooting zone.

Throughfall and Stemflow Studies. The forest canopy plays a dual role when assessing the impact of acidic clouds on the high-elevation red spruce and Fraser fir forest ecosystem. The forest canopy can be considered as a porous medium with a large effective surface area, thereby serving as a conduit for increased deposition by cloud interception. The presence of the forest canopy

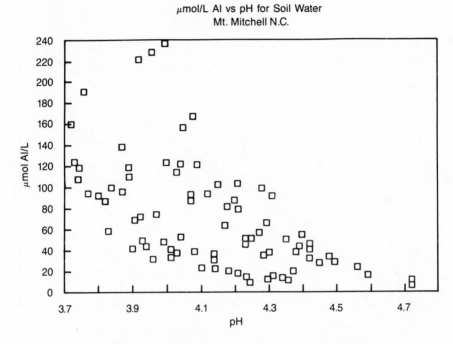

Figure 4.25. Concentration of aluminum versus pH for soil water samples collected during the fall of 1986.

can perhaps double the total deposition to the forest ecosystem above that from rainfall alone.

The large effective surface area of the forest canopy also serves as a zone of chemical interaction where the initial chemistry of the cloudwater can be modified by a number of possible reactions. These reactions range from a simple concentrating of the original composition due to evaporation of water from the needle surfaces to complete changes in the ionic composition due to interaction with the needles themselves or with material deposited from prior events. The whole reactive surface is periodically flushed by cloud or rainfall events with enough moisture to fully wet the needle surfaces, adding to the complexity of the system.

The forest canopy is a dynamic system, which makes assessing the impact of acidic cloud deposition not as straightforward as it would first appear. Interpretation of results must proceed with due scientific caution both because of a lack of our understanding of the dynamics within the forest canopy and the limitations in our techniques used to quantify processes occurring there.

Samples for 1986 were collected on an event basis. Throughfall collectors were kept covered with plastic bags between events to prevent dry deposition. Once an event started, the plastic bags were removed and throughfall collectors

pH vs. pAl for Soil Water
Mt. Mitchell, NC

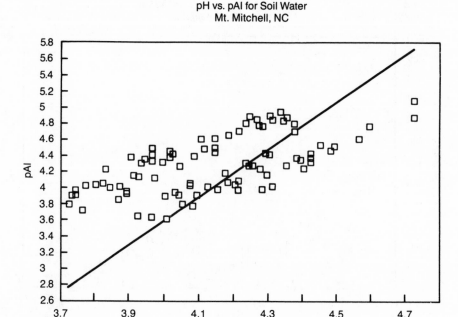

Figure 4.26. Negative logarithm of the concentration of aluminum (pAl)
versus pH for soil water samples collected during the fall of 1986. The solid
line represents the activity of aluminum controlled by gibbsite as a function of
pH.

left uncovered until the event was over. Approximately 30 minutes after an
event, the collection bottle was removed and replaced with another. The sample
was taken back to the field laboratory and weighed.

Most of the stemflow and all of the throughfall samples collected in 1986
were the result of rain events. Cloud interception, especially of low-moisture-
content orographic clouds, was insufficient to produce sufficient amounts of
sample using our collection technique. Total volume collected by throughfall
and open collectors is very similar, approximately 11 liters for 1986. Volume
collected on individual events is more variable but still similar between the two
sampling positions.

The overall results can be summarized as follows. In general, cloudwater is
10 times more acidic than rainwater. The dominant anions in cloudwater and
rainwater are NO_3 and SO_4. The dominant cations are H^+ and NH_4. Ammonia
accounts for approximately 30 percent of the positive charge between H^+ and
NH_4 combined but represents greater than 50 percent of the inorganic-N in
cloud- and rainwater.

Throughfall and stemflow are more acidic than rainwater but less acidic

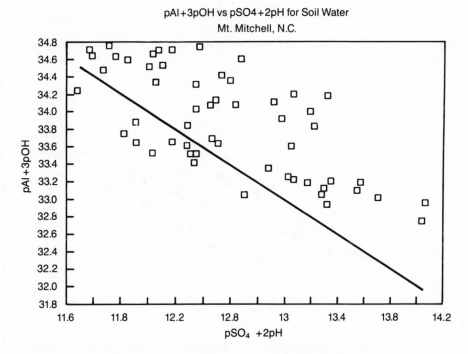

Figure 4.27. Soil solution ionic activities compared with expected relationship (solid line) if a solid of general composition $AlOHSO_4$ controls solution Al activity.

than cloudwater. Stemflow is more acidic than throughfall, probably due to the presence of dissolved organic acids. Stemflow samples were typically highly colored, whereas throughfall samples were almost always clear.

The concentration of all anions and cations measured was higher in throughfall and stemflow than in rainwater but less than or equal to the concentrations in cloudwater (tables 4.8 and 4.9). The increase in concentration is due in part to evaporation from needle surfaces during a particular rain event, but this alone can probably not account for the magnitude of the increase in concentration for these ions.

The increases in concentration for chlorine, sodium, potassium, calcium, and magnesium are probably a result of leaching from the needles. Release of potassium is normal and well documented in the literature. Release of chlorine, sodium, calcium, and magnesium is also known but to a lesser degree. It is not possible to determine from these data whether release of these elements is being enhanced by the acidity of both the cloud- and rainwater.

Release of nitrate and sulfate from the needles in amounts needed to increase the concentration of these ions to the extent demonstrated in tables 4.8 and 4.9 is highly unlikely. It is more probable that the increase in concentration

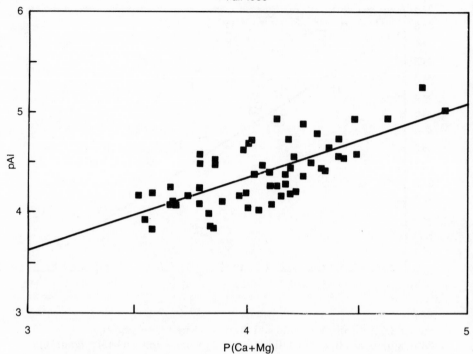

Figure 4.28. Relationship between the concentration of aluminum in solution and the concentrations of calcium and magnesium in the soil solution.

Table 4.8. *Volume weighted mean concentration per event of NH_4, Na, K, Ca, and Mg for cloudwater, throughfall, and stemflow collected in 1986.*

Collector	NH_4	Na	K	Ca	Mg
			(μmol/l)		
Red spruce					
T	23	12	63	44	15
S	41	18	88	60	18
Fraser fir					
T	37	11	63	37	14
S	53	22	75	61	18
Open	26	4	7	5	1
Cloud	241	55	8	48	27

Table 4.9. *Volume weighted mean concentration per event of H, Cl, NO$_3$, and SO$_4$ for cloudwater, throughfall, and stemflow collected in 1986.*

Collector	pH	H	Cl	NO$_3$	SO$_4$
			(μmol/l)		
Red spruce					
T	3.90	127	21	54	120
S	3.71	193	35	78	164
Fraser fir					
T	4.00	101	18	47	114
S	3.65	226	34	88	139
Open	4.21	62	5	18	35
Cloud	3.24	571	28	212	318

observed for these ions is due to prior deposition from low-moisture-content cloud events and dry deposition.

Calculation of total deposition of the various ions (tables 4.10 and 4.11) removes some of the uncertainty as to the meaning of the data in tables 4.8 and 4.9. Direct measurement of deposition from cloud interception is not possible but can be inferred from the difference in total deposition measured as throughfall and the deposition from rain events. Correction using the rainfall deposition estimates is valid because the actual throughfall samples were for the most part generated by rain events. Thus correction for the deposition due to rainfall should provide an indirect estimate of deposition due to cloud interception and dry deposition.

Total deposition over 90 days for NO$_3$ and SO$_4$ under the forest canopy ranged from approximately 4 to 6 kg/ha for NO$_3$ and from 16 to 22 kg/ha for SO$_4$

Table 4.10. *Deposition of NH$_4$, Na, K, Ca, and Mg via throughfall for June 29, 1986–September 21, 1986, at Mount Gibbs, N.C.*

Collector	NH$_4$	Na	K	Ca	Mg
			(kg/ha)		
Red spruce	0.80	0.52	4.65	3.35	0.67
Fraser fir	0.93	0.34	3.46	2.02	0.46
Open	1.04	0.20	0.54	0.46	0.07
			Excess[a]		
Red spruce	0.24	0.32	4.08	2.90	0.60
Fraser fir	0.11	0.14	2.89	1.56	0.39

[a]Red spruce or fraser fir minus open.

Table 4.11. *Total deposition of H, Cl, NO$_3$, and SO$_4$ via throughfall for June 29, 1986–September 21, 1986.*

Collector	H	Cl	NO$_3$	SO$_4$
		(kg/ha)		
Red spruce	0.24	1.42	6.26	21.6
Fraser fir	0.15	0.87	3.87	15.7
Open	0.14	0.35	2.40	7.3
		Excess[a]		
Red spruce	0.10	1.07	3.86	14.3
Fraser fir	0.01	0.52	1.47	8.4

[a]Red spruce or fraser fir minus open.

(table 4.12). Correction for the deposition from rainfall indicates that between 40 and 70 percent of the deposition of NO$_3$ and SO$_4$ is due to cloud interception or dry deposition.

Total deposition of ammonium (NH$_4$) measured for this period is actually less than that measured for just rainfall, resulting in negative net deposition estimates. This indicates rapid uptake of NH$_4$ by the forest canopy. The significance of this observation lies in the fact that NH$_4$ constitutes more than half of the total inorganic-N in both rain- and cloud water (tables 4.8 and 4.9).

A similar observation can be made for the total deposition of H$^+$. The net deposition calculated for Fraser fir is approximately zero, suggesting that all of the H$^+$ deposition from dry deposition and cloudwater interception is neutralized by the forest canopy. The H$^+$ deposition reaching the forest floor appears to be due mainly to rainfall. The positive deposition calculated for red spruce may well be a result of the uncertainties in the estimates for total and rainfall deposition.

The net deposition calculated for Cl, Na, K, Ca, and Mg reflect both the natural cycling of these elements in the forest ecosystem and enhanced leaching by acidic cloud and dry deposition. Neutralization of both dominant cations (H$^+$

Table 4.12. *Total deposition compared to soil flux from rooting zone for 1986.*

Total deposition[a]	H	Ca	Mg	NO$_3$	SO$_4$
			(kg/ha)		
Red spruce	0.24	3.36	0.67	6.26	21.6
Fraser fir	0.15	2.02	0.46	3.87	15.7
Soil flux#	0.20	7.3	2.6	51.3	23.4

[a]Sampling period: June 29–September 21, 1986.
[b]Sampling period: September 19–November 20, 1986.

and NH_4) in the cloud- and rainwater requires release of other cations to balance the NO_3 and SO_4 remaining. Calculation of the total negative and positive charge available as net deposition equals 100 percent balanced when averaged between red spruce and Fraser fir. Calcium and magnesium combined account for approximately 50 percent of the total positive charge calculated for net deposition, suggesting that neutralization of acidic deposition by the forest canopy is enhancing the release of these nutrient elements (table 4.8).

Total deposition listed in tables 4.10 and 4.11 spans a period of approximately 90 days. Assuming the same percentage of cloud and rainfall deposition throughout the year yields estimates of 20 and 75 kg/(ha-yr) for NO_3 and SO_4, respectively. Direct estimates for H^+ and NH_4 are not possible, owing to absorption interactions within the canopy.

EPA's Mountain Cloud Chemistry Program. This section describes the exposure of a high-elevation fir-spruce forest in the southern Appalachians. An analysis of air pollutant measurements, made at two high-elevation sites during the summer of 1986, is presented along with information on the location, weather, and related pollutant measurements made nearby. This study was accomplished through the cooperation of the NCSU Departments of Plant Pathology, Marine, Earth and Atmospheric Sciences, and Soil Science.

Mount Mitchell is situated in western North Carolina, a relatively isolated region of the eastern United States, at approximately 35° N, 82° W. This peak, at an elevation of 2037 meters (6684 feet) above mean sea level, is the highest elevation east of the Mississippi River, but it is only one of a substantial number of peaks of elevation close to 2 kilometers in the southern Appalachians. Mount Mitchell is part of a ridgeline, called the Black Mountains, that forms the eastern cordillera of the Appalachians at this latitude. Because other high ridges, such as the Great Smoky Mountains and the Plott Balsams, lie to the west and south, one would expect little, if any, orographic enhancement of rainfall at this location.

The site is relatively isolated from major pollutant sources. There are a number of large electric power generation stations on the Tennessee River about 125 km north and west of the site. The only significant industrial factilities are paper mills in Canton, North Carolina, about 50 km wsw of the measurement location. Major population centers that may influence air pollutant concentrations at this location are shown in table 4.13, along with separation distances.

Table 4.13. *Sources of air pollutants.*

City	Population	Distance
Knoxville, Tn.	250,000	125 km
Atlanta, Ga.	500,000	165 km
Greenville, S.C.	280,000	100 km
Spartenburg, S.C.	200,000	100 km
Charlotte, N.C.	350,000	125 km

There are sources of air pollutants in every direction within 150 kilometers (93 miles) but north.

Although there were a few cloudwater collections in May, the major field season for pollutant measurements began with ozone measurements on June 13, 1986. Measurements of gases, cloudwater, and rain were made from that time through mid-September. In addition, a few particle samples were collected in early September using a dichotomous sampler, a device that sorts particles into large (1.5–2 μm) and small (less than 2 μm) size ranges.

In general, the 1986 field season was dry and hot. For example, figure 4.29 presents rainfall at the nearby town of Asheville, North Carolina, compared with the 30-year average (1951–80) for that site. Clearly, June through September registered less than half the rainfall normal for that period. Such conditions are conducive to photochemical processes and ozone formation.

In spite of the general drought conditions, high-altitude cloud deposition events were very numerous in 1986. Most of the events appeared to be associated with orographic clouds during June and July. Most of the rain experienced during the season occurred in August. Cloud events during that period were

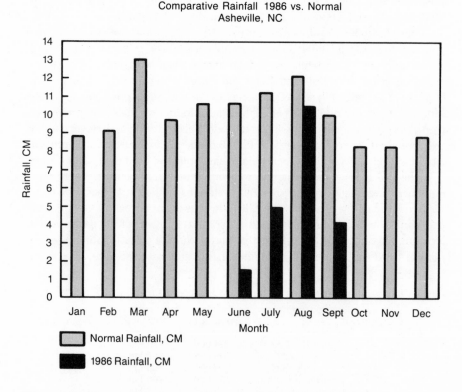

Figure 4.29. Comparison between 1986 rainfall and normal rainfall.

accompanied by rain and associated with cold front passages on August 2 and 12. In September, two stratiform cloud events were experienced, apparently associated with Atlantic low pressure systems. These cloud episodes appeared to dominate wet deposition in this particular season.

Of the major gaseous air pollutants, only ozone was measurable for the whole season. Sulfur dioxide measurements were attempted for the entire period. However, detectable concentrations were present on only three occasions. All three SO_2 observances were in the early morning hours, just as the mixing layer began to develop. For very brief, approximately half-hour, periods, SO_2 was present at concentrations of 11, 17, and 40 ppb, respectively. For the balance of the time, approximately 2700 hours, no SO_2 was detected at the 2 ppb limit. Measurements of nitrogen dioxide (NO_2) were attempted in the latter part of the season, the last 500 hours or so. No SO_2 was detected, however, again at a detection limit of about 2 ppb.

Figure 4.30 presents measured hourly ozone values for the period June 14–October 1, 1986. During the earlier portion of this period, from June 14 to 28, the area was under the influence of a high pressure system centered in West Virginia that produced warm, dry weather over the entire southeastern portion of the

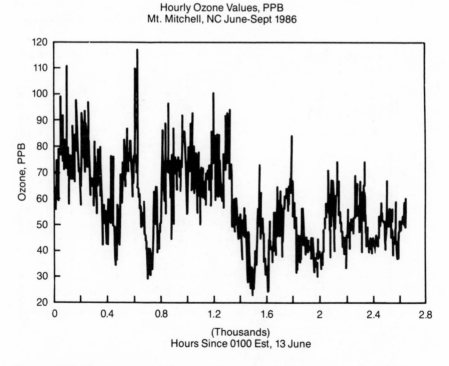

Figure 4.30. Hourly ozone values on Mount Mitchell between June and September 1986.

country. A very weak cool front passed through on the twenty-first. However, the only influence was formation of a minor cloud episode with insufficient deposition for collection.

On June 28, the area came under the influence of Hurricane Bonnie, which had moved ashore in Louisiana 2 days previously. For the next several days, the entire area was enveloped in clouds, the residuum of the tropical storm. During that period ozone values were quite low, both for Mount Mitchell and for other nearby locations, until July 4. During the next week, ozone concentrations built to substantial levels (90 ppb at peak), then dropped temporarily with the passage of another cold weather system on July 11. From that time until about August 4, ozone remained elevated (over 80 ppb). Rainfall and cloudiness from a series of cold fronts in August depressed ozone concentrations for essentially the balance of the season, even though rather little rain fell.

Ozone concentrations over 70 ppb were experienced for a period of about 7 weeks, from mid-June until early August. Thereafter, the higher values of ozone measured were in the 60 ppb range.

Ozone measurements at Mount Mitchell appeared to be modulated primarily by synoptic events and, therefore, should have some general correlation with those in nearby low-elevation locations. Available measurements for our

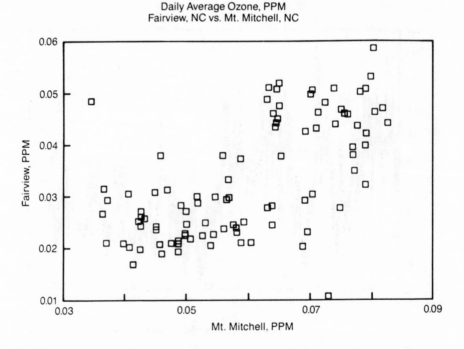

Figure 4.31. Comparison between ozone levels in Fairview and on Mount Mitchell.

locations were compared with those from other mountain sites. The ozone monitor operated at Fairview, North Carolina, by the North Carolina state government
is only 15 km from the Mount Mitchell site. Other monitors at Lenoir, North
Carolina, and Marion, Virginia, also produce relevant data. All these monitors
are relatively isolated, at considerable distance from major pollutant sources.
One monitoring location, Knox County, Tennessee, is immediately downwind
of Knoxville, a city of about 250,000.

   Figure 4.31 presents a comparison of daily average ozone values for Fairview and Mount Mitchell. Generally, these values track reasonably well with a
few exceptions. On this basis, the mountaintop experiences generally higher
ozone concentrations than does the valley location, even though higher days
tend to correlate reasonably well. Figure 4.32 is a cross-plot of the same data set
showing the degree of correlation between the two sites. Regression analysis
reveals a significant association between ozone values at the two sites; thus,
about 40 percent of the variability in Mount Mitchell daily average ozone can be
associated with Fairview ozone variability.

   Comparisons with Lenoir are quite similar. Figure 4.33 compares Mount

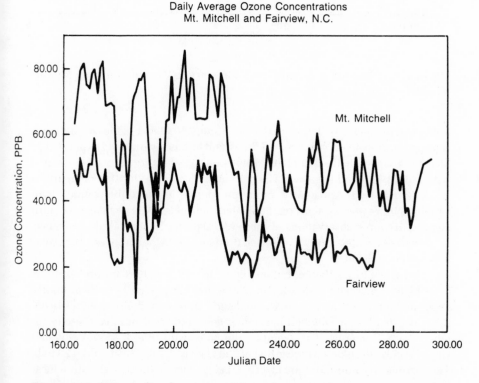

Figure 4.32. Twenty-four hr. means of "growing season" ozone concentrations
on Mount Mitchell.

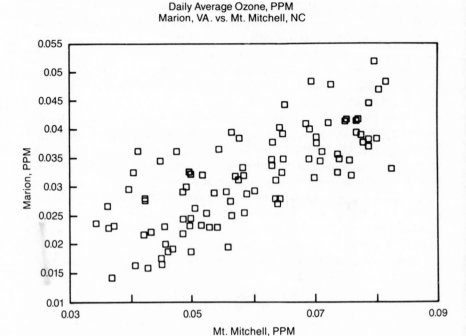

Figure 4.33. Comparison of ozone levels in Marion, Va., with those on Mount Mitchell.

Mitchell daily ozone with that of Marion, Virginia. In this case, the correlation is even better, with about 70 percent of the Mount Mitchell variability explained by Marion variance. On the other hand, Knox County values are not at all correlated with Mount Mitchell averages. The correlaton coefficient for that comparison was only $r = 0.03$. Apparently, pollution sources in Knoxville coupled with variances in weather systems on the western side of the Appalachians are sufficient to decouple observations at these two stations. In general, daily average ozone values for sites relatively isolated from sources in the southeast are reasonably well correlated.

This is not to say that mountain and valley ozone values are similar, however. Figures 4.34 and 4.35 are cumulative frequency distributions of hourly ozone measurements for Mount Mitchell and Fairview for the 1986 field season. It is clear that Mount Mitchell has higher concentrations and that these occur much more frequently. For example, the 50th percentile of Mount Mitchell values is close to 60 ppb whereas that of Fairview is slightly greater than 20 ppb. Thus, ozone concentrations are clearly much greater at the high-elevation location.

Figure 4.36 compares diurnal patterns in ozone concentration for Mount

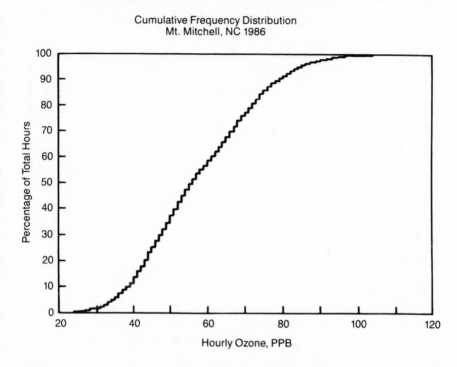

Figure 4.34. Cumulative frequency distribution of ozone levels on Mount Mitchell, N.C.

Mitchell and Fairview for the relatively high ozone period for both sites. It is clear that the major difference between these two curves is the maintenance and even increase of nighttime ozone relative to daytime values at the summit station. These patterns are almost universal; essentially on every day, the nighttime values increase at the summit and decrease at low-elevation locations. Figure 4.37 compares the diurnal patterns for the relatively low ozone period associated with the remains of Hurricane Bonnie. Again, the same diurnal patterns apply. It should be noted, however, that not all these high values occur at night. At low-elevation locations, morning ozone values tend to be at a minimum, building toward the diurnal maximum near midday. At the mountain summit, ozone values are quite consistently high during this photosynthetically active period.

The observation of high nighttime ozone values is by no means unique to this study. But the first observation of this phenomenon of which we are aware was reported for Mount Mitchell. Berry (1964) operated a Mast automatic iodometric analyzer for ozone at Mount Mitchell and at a nearby low-elevation site. This author also noted both the small diurnal variation at the summit and the nighttime maximum. Since then, there have been numerous reports of this phenomenon, both in the United States (Samson 1978) and in Europe (Broder

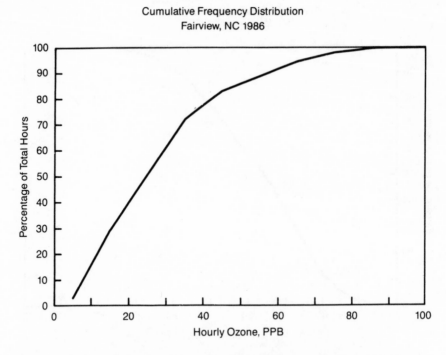

Figure 4.35. Cumulative frequency distribution of ozone levels in Fairview, N.C.

and Gygax 1986, Reiter et al. 1986). The reasons for the phenomena are not completely understood, but there does seem to be an interplay between flow separation of the nighttime jet and dry deposition of ozone at a lower elevation (Broder and Gygax 1986).

Thus, ozone concentrations on mountaintops are higher than in nearby valleys, but the significance of this difference is likely to be expressed in the early morning hours. It is also possible that there may be some issues associated with recovery. For example, ozone concentrations tend to be elevated on mountaintops for days or weeks at a time. At lower elevations, there are periods of low concentration that may permit some biochemical recovery from the insult period. It should also be noted that ozone concentrations may serve as an indicator for other pollutant products of photochemical conversions. Consequently, high ozone periods may also be high periods for nitric acid vapor, hydrogen peroxide, and sulfate aerosol.

As reported previously, the 1986 field season was a period of very low rainfall. Figure 4.38 presents weekly collection amounts of rain from the National Atmospheric Deposition Program collector at the mountain site. Figures 4.39 and 4.40 offer comparisons of cations and anions, respectively. There appears to be relatively little novelty associated with rainfall at Mount Mitchell. The rain is

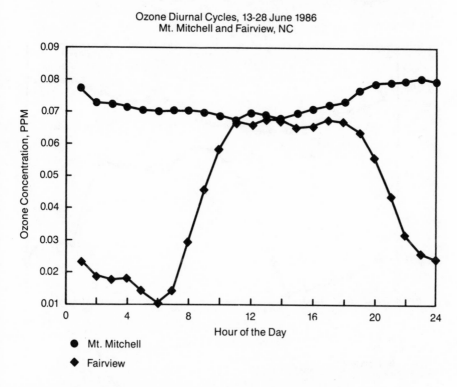

Figure 4.36. Comparison of daily ozone levels in Fairview and on Mount Mitchell, June 13–28, 1986.

not very acid even though, paradoxically, there is very little neutralization. For example, on the average, measured anions and cations in these samples balance very well, usually within 5 percent or so. The principal cation in rain is hydrogen ion; ammonium ion accounts for less than one-third of the cation balance and the metals much less. Therefore, the rain is very little neutralized by alkali even though it is quite dilute. Sulfate is the principal anion in all the samples measured; nitrate accounts for about one-fourth the anion balance, with detectable but insignificant amounts of chloride also present.

Cloudwater deposition events were considerably more common than rain in 1986. Figure 4.41 shows the range of hourly collection volumes obtained with a string-type cloudwater collector against Julian date. In all, 24 event periods were experienced at the summit site during the field season. It should be noted that a second site on the east face (low-elevation site) of the mountain experienced no events during this period. Figure 4.42 presents a plot of pH (axis on right) in cloudwater compared with hourly collection rates (axis on left) from the string sampler. It is interesting that there seems to be an association between these two parameters; the more water deposited, the higher the pH. Thus, acidity in these

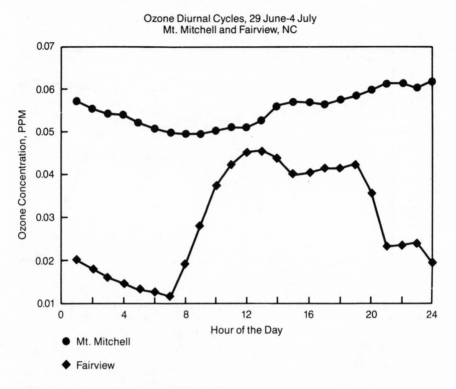

Figure 4.37. Comparison of daily ozone levels in Fairview and on Mount Mitchell, June 29–July 4, 1986.

samples seems to be controlled primarily by ion strength. The more concentrated the sample, the more acidic. On the other hand, the degree of neutralization seems to be greater with the more concentrated samples. For example, figure 4.43 presents a comparison of the ratio of hydrogen ion to the sum of sulfate and nitrate (a measure of the lack of neutralization) to the collection volume. Higher values of this ratio seem to be associated with high collection volumes, hence low ion strength.

Thus, like rainfall, the rather dilute cloudwater samples, probably mostly those associated with mixed cloud-rain events or with frontal system clouds, tend to be rather little neutralized but relatively low in acidity. By contrast, the low liquid-water content orographic clouds tend to be high in acidity but even higher in ion strength, and therefore relatively more neutralized.

The hydrogen ion concentration in clouds was more than an order of magnitude greater than that of rain. At this site, in 1986, cloud deposition dominated the wet deposition of acidity.

The dry, hot 1986 field conditions at Mount Mitchell were favorable for the development of ozone and other photochemical air pollutants such as nitric and

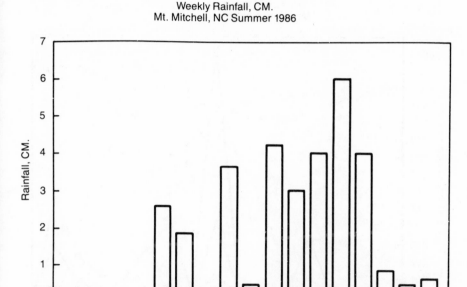

Figure 4.38. Weekly rainfall on Mount Mitchell, summer 1986.

sulfuric acids. At the same time, orographic cloud events occurred quite frequently. Since there is no climatological record of orographic events, it is not
possible to say whether these observations are unusual or not.

Ozone measured at the mountaintop and at several rural low-elevation stations correlated reasonably well on a daily average basis and tracked prevailing
synoptic weather conditions. On the other hand, the dramatic difference in
diurnal pattern between high- and low-elevation sites produced a substantial
difference in frequency distributions, biased mainly by high night values for the
mountain summit. However, morning values of ozone are also very much greater
for the mountain.

During 1986, rainfall was scarce, not very acid, but also not very much
neutralized. By contrast, orographic clouds were very frequent, very acidic, and
somewhat more neutralized than rain. These data suggest that the significantly
greater spruce-fir mortality seen at the higher elevations may be linked to
orographic cloud deposition (50 times more frequent at high altitudes) and not
rainfall.

In Situ Cloud Simulator Studies. The high-elevation spruce-fir forests of
the Black Mountains are bathed in clouds for approximately two-thirds of the

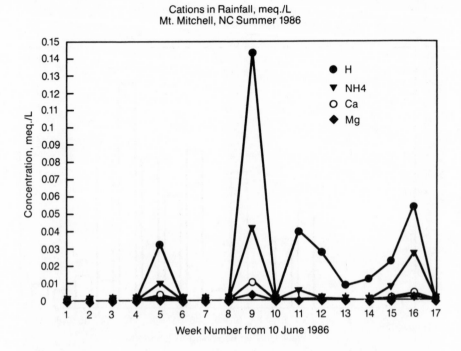

Figure 4.39. Concentrations of various cations in rainfall on Mount Mitchell, summer 1986.

summer growing season. The majority of these clouds are of an orographic nature. Cloud samples (both frontal system and orographic) taken during the summer of 1986 have ranged in pH from 5.7 to 2.3 with a mean pH of 3.3. Orographic clouds commonly form during the early morning hours (0100–0300) and affect the spruce-fir forests throughout the night before dissipating by midmorning (0700–1000). This night-time cloud cycle often leaves the needles of red spruce and Fraser fir coated with a film of moisture. This film of moisture is very quickly evaporated once the cloud has dissipated in the morning sun. It is thought that this frequent cycle of deposition and evaporation of relatively acidic cloudwater may have a damaging effect on the cuticle and epicuticular wax ultrastructure of these needles.

An experiment was designed to simulate the nighttime orographic cloud cycle under field conditions and to quantify the amount of damage to the surface of needles of red spruce. The data presented here are a summary of the preliminary results from the quantification of epicuticular wax damage in the stomatal antichamber of red spruce.

Stomatal wax plugs in plants are believed to be adaptive structures that minimize water loss while still facilitating efficient gas exchanges. It has been

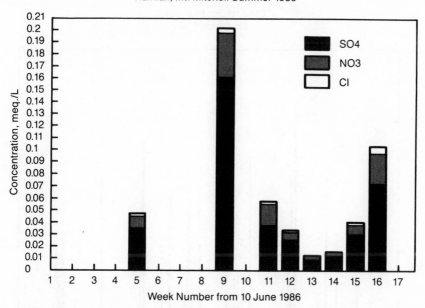

Figure 4.40. Concentrations of various anions in rainfall on Mount Mitchell, summer 1986.

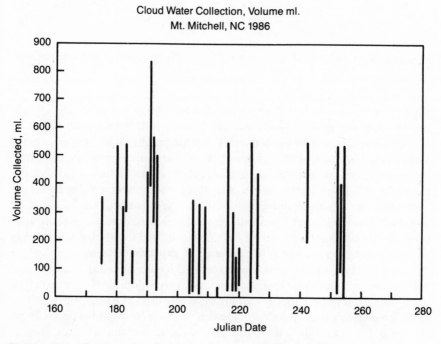

Figure 4.41. Cloudwater collected on Mount Mitchell, 1986.

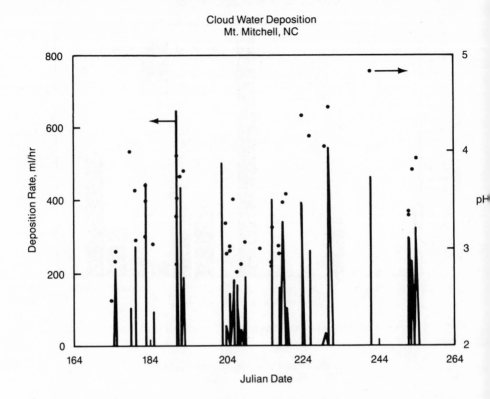

Figure 4.42. Hourly deposition rate and pH of cloudwater on Mount Mitchell.

shown that changes in the ultrastructure of wax plugs can result in a reduction in gas exchange and growth (Gunthardt and Wanner 1982, Jeffree et al. 1971).

On the east face of Mount Mitchell (1770 meters), in a relatively healthy red spruce forest four greenhouse-shaped plastic chambers were constructed with each chamber evenly partitioned into five subchambers. Within each subchamber five evenly spaced holes were dug in the ground, and 6-inch clay pots, each containing one 5-year-old red spruce seedling, were sunk until the tip of the pot was flush with the ground. The seedlings used in the experiment were randomly selected 1 month before the start of the experiment and transferred into the clay pots containing organic soil collected near the experimental site.

Four fluid reservoirs were constructed, each corresponding to one of four pH treatments. Simulated cloudwater solutions were prepared by adjusting the pH of deionized water with a 3:1 mixture of sulfuric (H_2SO_4) and nitric (HNO_3) acids. To each solution the following background ions were added (in mg/liter): Ca^{2+}, 0.8; K^+, 0.14; Mg^{2+}, 0.07; and Cl^-, to simulate cloudwater collected in 1986. The simulated cloudwater acidity levels were pH 2.5, 3.5, 4.5, and 5.5.

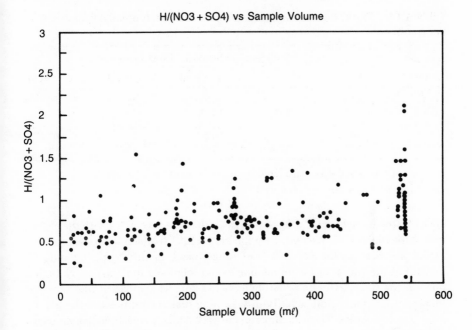

Figure 4.43. Comparison of the ratio of hydrogen ion to the sum of sulfate and nitrate to the collection volume.

Reinforced polyvinyl chloride (PVC) tubing and a series of PVC pipe manifolds could be used to distribute the solutions to each of the subchambers. Stainless steel fog nozzles were attached to the ends of the PVC tubing.

At 0230 hours the chamber tops were placed on the chambers and the seedlings were misted for 5 minutes. At 0400 the seedlings were again misted for 5 minutes. This misting series was repeated at 0500, 0600, and 0700. After the final misting period the chamber tops were immediately removed to enhance air circulation and therefore facilitate the rapid evaporation of the droplets. After the needles were completely dry the seedlings were removed from the ground and rotated counterclockwise to the next adjacent hole. This procedure helped to ensure an even mist distribution over the seedlings during the course of the experiment. This series of hourly 5-minute mists thoroughly covered the needles with a film of moisture but not to the point where the droplets were dripping off.

At the end of the experiment approximately 20 two-year-old needles were randomly chosen from each of the seedlings for examination. As shown in table 4.14, 91 percent of the wax plugs on the control needles were classified as relatively undamaged. These plugs consisted of a network of long, thin, tubular wax crystals. The margins of the wax plugs that come in direct contact with the needle surface exhibited very little if any cracking.

The wax plugs on the pH 5.5 mist-treated needles showed a moderate in-

Table 4.14. *Percentage of wax plugs occurring in each damage class per treatment.*

| Treatment | Damage class (amount of wax fusion) | | | | |
	1(0–25%)	2(26–50%)	3(51–75%)	4(76–90%)	5(91–100%)
C	91%	3%	3%	3%	0%
5.5	61%	13%	17%	9%	0%
4.5	25%	31%	28%	16%	0%
3.5	14%	21%	42%	11%	12%
2.5	35%	22%	14%	13%	16%

Note: Total of 390 wax plugs rated.

crease in the percentage of plugs fused in all the damage classes with the exception of the highest (5). Wax plugs on the pH 4.5 mist-treated needles exhibited a further increase in the percentage of plugs fused. In this treatment only 25 percent of the wax plugs were rated as being relatively undamaged. Note that none of the plugs exhibited complete wax fusion. The apparent trend toward increased incidence and severity of wax plug fusion continues in the pH 3.5 mist-treated needles. The wax plugs on the pH 2.5 mist-treated needles showed a slight increase in the percentage of wax plugs in the two highest damage classes, but an unexpected increase in the percentage of needles rated as relatively undamaged. The surfaces of pH 2.5 mist-treated needles show a large amount of wax fusion, with cracks developing in and around the plugs. Some needles showed completely fused wax plugs.

Although the results are preliminary, these data suggest that there is a trend toward an increase in the incidence and severity of wax fusion as the acidity of the mist treatments increases. An exception to this trend occurred in the pH 2.5 mist treatment. The increase in the percentage of undamaged plugs may be due to biological variation among the needles sampled, or to the relatively small sample size taken.

Recent research on Norway spruce exposed to "fog type" sprayings with a pH 3.5 solution has shown a fusion of the wax plug similar to the observations made in this experiment (Rinalla et al. 1986). Other work with Norway spruce exposed to pH 3.0 mists has suggested that the cracks found in and around the wax plug are in fact due to acidic mist, but wax fusion is a result of ozone exposure (Magel and Ziegler 1986). If wax plug fusion on red spruce is in fact a result of exposure to acidic mists, as this experiment suggests, then damage to the wax plug of native red spruce trees may be possible under the present ambient field conditions. This condition not only may result in a reduction in gas exchange, but large gaps between the wax plug and the needle surface could potentially result in excessive water loss during the winter months. Physiological stresses such as these resulting from acidic orographic clouds could be contributing to the forest decline on Mount Mitchell.

Evidence for the Causes of the Growth Reduction in Southern Pine

Analyses of the Piedmont growth reduction suggest that diameter growth rates of pine in natural stands decreased between the fourth and fifth survey periods. Unfortunately, the data collected in these surveys do not permit an identification of specific causes. Forest Inventory and Analysis personnel believe that most of the decline can be attributed to one or more of 14 causes. Five of these proposed causes—stand aging, increased density, hardwood competition, loss of old-field sites, and climatic changes—are products of natural stress and are receiving special attention. Each has sound scientific merit in its support and warrants special investigation. Air pollution is included as a sixth possibility. A discussion follows of these 6 possible factors.

Aging of Stands. Diameter growth of individual trees in a stand generally declines as the stand matures. Little change in average stand age occurs in fully regulated forests, but natural pine stands in the Piedmont are not regulated. Sheffield et al. (1985) report that the percentages of trees by age class are greatly disproportionate. In a regulated forest these percentages would be approximately equal. The abandonment and reversion to pine of millions of acres of farmland between 1945 and 1965 are largely responsible for this age distortion. The current age structure suggests that a high proportion of remeasured pines are in older stands now than when previously measured.

Increased Stand Density. Diameter growth rates of individual trees in a stand tend to decrease as stand density increases. In simple terms, increased stand density translates into more competition for sunlight, water, and nutrients for individuals. Average stand density in the Southeast has been increasing since the initial forest surveys in the 1930s, but changes have been most rapid in the upland areas during the past two decades. In natural pine stands of the Georgia Piedmont and Mountain Survey Units, basal area in trees of 1-inch diameter at breast height and larger increased from 61 ft^2/acre in 1961 to 96 ft^2/acre in 1982.

Increased Hardwood Competition. Hardwood trees and shrubs can adversely affect the growth of pines, even when they are in the understory of relatively old pine stands. Improved fire protection in the Southeast over the past several decades has favored the survival and development of increasing numbers of hardwood trees and shrubs in pine stands. In pine forests on old-field sites, hardwood numbers have increased in what is seen as natural succession. In the Georgia Piedmont and mountains, hardwood accounted for one-fourth of the basal area (1.0 inches DBH and larger) in natural pine stands in 1982, compared with 17 percent in 1961. Numbers of hardwood stems per acre increased from 240 to 390 over the same period.

Loss of Old-field Sites. The proportion of the Piedmont pine resource occurring on land that was farmed as recently as 1945 is higher than the proportion in Coastal Plain forests in the Southeast. Abandonment of farmland between 1945 and 1965 was concentrated in the Piedmont. Pine tree growth on these acres benefited from the relative absence of non-pine competing vegetation. Also, soil nutrient levels may often have been higher than those typical of forest sites. Over the past 20 years, relatively little farmland has been abandoned, and the majority of new natural stands have been established on cutover forest land. Trees in these stands do not have the initial growth advantage of trees in old fields.

Climatic Changes. Of the climatic factors that affect tree growth, rainfall is probably the most obvious. Available soil moisture is the variable that directly influences tree growth. Unfortunately, specific climatic data have not yet been assembled to relate its influence on the southern pine resources.

Atmospheric Deposition. Recent awareness of the forest declines in Europe attributed to air pollution has resulted in atmospheric deposition being included as a possible contributing factor. A significant difference between the Piedmont growth decrease and European forest decline is absence of any of the visible foliage and crown symptoms of damage associated with the European decline. The only symptom of the Piedmont decline in common with the European problem is reduction in radial growth, which is also a universal symptom of natural stress.

There is a possibility that ozone and other pollutants may have contributed to the decline. The southern region has had a degradation in ambient air quality that is a logical issue of concern.

Summary. A hypothesis was formulated after serious and deliberate consideration of the various natural stress factors. The single most prevalent hypothesis is that the severe drought in this region of the country over the past few years together with the loss of old-field site conditions and a significant increase in hardwood competition has caused a measurable decrease in average pine growth rates on drought-prone Piedmont sites in the Southeast.

It seems clear that nonatmospheric factors are closely related to the decreases in growth observed in the Piedmont and have not acted independently. Stand age, stand density, hardwood competition, and effects of old-field conditions are likely to have had some type of an impact on growth. There is a belief that if air pollution has had some influence, it is most likely through some type of amplification of stresses caused by competition, drought, and other natural stresses. Present data are insufficient to estimate the probability that atmospheric deposition has been or is a causal factor in the growth decline of pines in the southern region.

Summary

The recent and widespread phenomena of forest declines have prompted a great deal of scientific inquiry into their nature and causes. Perhaps the most intriguing characteristics of the declines observed in North America and in central Europe are the number of species affected and the geographic extent over which symptoms have been observed. In the known scientific record, this type of syndrome affecting entire ecosystems and many species appears to be unique.

Because these declines are occurring in areas of high pollutant deposition, much research has been aimed at determining whether air pollutants are being benignly assimilated and cycled through terrestrial ecosystems or whether these toxic, nutrient, and acidic compounds cause significant perturbations to above- and below-ground plant tissues leading to mortality. This chapter has reviewed the current understanding of the decline in the boreal montane ecosystems in the southern Appalachian Mountains and of recently observed growth losses in southern commercial pine forests.

The review is divided into five sections. The first provided evidence of the forest declines in the eastern United States. The second section provided a framework for understanding stresses on forests and the resulting declines, and the third discussed the role of natural factors in causing forest declines. The fourth section reviewed the evidence that human activities can contribute to forest declines. The fifth section reviewed recent data specifically related to the spruce, fir, and pine problems in southeastern United States.

The investigations reviewed in the fifth section indicate that the decline of spruce and fir trees in the Black Mountains of North Carolina has been increasing over the past 4 years. In 1984, 272 healthy red spruce and 213 Fraser fir in 16 stratified plots were selected for observation. By 1986, 9 percent of all spruce and 16 percent of the fir had died. By the spring of 1987, 41 percent of the spruce and 49 percent of the fir had died. Drought and ice appear to have severely affected these already stressed trees.

Following a highly acid cloud event of 16 hours' duration that occurred during the spring flush of spruce and fir foliage, specific needle necrosis was observed. Analysis of the damaged needles indicated high sulfur content in the tissues. Experiments designed to study the effects of simulated acid fog indicated significant stomate wax plugging at low pH values. Cuticular erosion was also observed.

Soil was transplanted from a high-altitude site to a low-altitude site and subsequently planted with 3-year-old red spruce seedlings. A significant differential effect in seedling necrosis and mortality was observed. Of the seedlings planted in the transplanted high-elevation soil, 93 percent died after one-and-a-half years, compared with 6 percent of the trees planted in the homogenized low-elevation soil. The high-elevation soil had fewer available nutrients (11 percent base saturation) than the low-elevation soil (22 percent base saturation) even though both soils had the same total exchangeable acidity and aluminum. The critical difference was in the amount of exchangeable calcium and magnesium present, with the low-elevation soil containing two to three times as much as high-elevation soil.

The predominant cation in soil solution samples collected in the fall of 1986 with soil lysimeters installed below the root zone of red spruce and Fraser fir was aluminum, followed by calcium, magnesium, hydrogen, potassium, and sodium. The predominant anion was nitrate, followed by sulfate and chloride. The concentration of nitrate showed a strong temporal trend, with peak concentrations occurring during mid-October. Sulfate concentrations remained relatively constant during the sampling period. A preliminary review of the data suggested that an aluminum hydroxy sulfate species may be controlling the activity of aluminum in soil solution. This could contribute to the availability of soluble aluminum to the root systems of fir and spruce, possibly leading to toxicity.

The EPA Mountain Cloud Chemistry Program air quality data indicated elevated ozone concentrations on Mount Mitchell compared with low-elevation monitoring sites. Average daily ozone concentrations were observed to be consistently double those of Marion, Virginia, just 8 kilometers (5 miles) away but 1525 meters (5000 feet) lower. During 1986 and 1987, ozone levels at or near the summit of Mount Mitchell were in excess of 75 parts per billion (ppb) approximately 33 percent of all hours during the growing season. During the 1986 and 1987 growing seasons, high-ozone events in excess of 80 ppb frequently occurred within a matter of 6–12 hours of orographic cloud events with pH acidities of 2.7 or lower. Acidity of cloud moisture during 1986 and 1987 ranged from pH 2.5 to 3.4.

Recently analyzed data for Mount Mitchell indicate the following average annual wet deposition rates: sulfate, 122 kg/ha; nitrate, 65 kg/ha; ammonium, 15 kg/ha; and base cations, 5 kg/ha. At the present time there is no definitive proof that acid deposition is limiting the growth of forests in either Europe or the United States. Detecting responses in mature forest trees is made extraordinarily difficult by several factors, including the complexity of competition, climate variations, and the potential interactions among anthropogenic pollutants such as acid deposition, gaseous pollutants, and trace metals. Moreover, there are no unimpacted (control sites) to compare with sites affected by acid deposition. For these reasons, assessing the possible impacts (cause and effect) of atmospheric deposition on forest productivity is difficult and will require techniques other than those involving comparisons between affected and unaffected sites.

The standard tenets of Koch's postulates—applied by forest pathologists for more than a hundred years to determine the cause of forest injury—may have little or no application in assessing the forest damages now occurring in eastern North America. I contend that the scientific and policy making communities will have to find new approaches to deal with the problem of "imperfect knowledge" in assessing whether further controls should be adopted to reduce anthropogenic pollutant deposition on spruce-fir ecosystems.

References

Abrahamsen, G. 1980b. Acid precipitation, plant nutrients, and forest growth. In *Ecological Impact of Acid Precipitation*, ed. D. Drablos and A. Tollan, pp. 58–63. Proceedings of the International Conference, Norwegian Interdisciplinary Res. Prog. Acid Precipitation, Effects Forest Fish. Sandfjord, Norway. March 11–14, 1980.

F. Adams. 1984. *Agronomy #12, Second Edition*. Madison, Wisc.: Soil Science Society of America, pp. 240–272.

Anderson, F., T. Fagerstrom, and S. I. Nilsson. 1980. Forest ecosystem responses to acid deposition-hydrogen ion budget and nitrogen/tree growth model approaches. In *Effect of Acid Precipitation on Terrestrial Ecosystems*, ed. C. Hutchinson and M. Havas, pp. 319–334. New York: Plenum Press.

Babich, H., and G. Stotsky. 1979. Environmental factors that influence the toxicity of heavy metal and gaseous pollutants to microorganisms. *Crit. Rev. Microbiol.* 8:99–115.

Bauch, J. 1983. Biological alterations in the stem and root of fir and spruce due to pollution influences. In *Workshop in the Effects of Accumulation of Air Pollutants in Forest Ecosystems*. Hingham, Mass.: D. Reidel.

Berry, C. R. 1964. Differences in concentrations of surface oxidant between valley and mountaintop conditions in the southern Appalachians. *Journal of the Air Pollution Control Association* 14:238–239.

Broder, B., and H. A. Gygax. 1986. Terrain-induced effects on the ozone, temperature and water vapor daily variation in the upper part of the PBL over hilly terrain. *Tellus* 37B:259–271.

Bruck, R. I. 1984. Decline of montane boreal ecosystems in central Europe and the southern Appalachian Mountains. *TAPPI Porc.* 1984:159–163.

Bruck, R. I., and W. P. Robarge. 1986. Decline of boreal montane systems in the Southern Appalachian Mountains. Final report to the U.S. Environmental Protection Agency.

Bruck, R. I. and W. P. Robarge. 1988. Change in forest structure in the boreal montane ecosystem of Mt. Mitchell, N.C. USA. *European Journal of Forest Pathology* 18:357–366.

Cole, D. W., and D. W. Johnson. 1977. Atmospheric sulfate additions and cation leaching in a Douglas fir ecosystem. *Water Resources Research* 13:313–317.

Cole, D. W., and M. Rapp. 1981. Elemental cycling in forest ecosystems. In *Dynamic Properties of Forest Ecosystems*, ed. D. E. Reichle, pp. 187–238. Cambridge, Mass.: University Press.

Cowling, E. B., and L. S. Dochinger. 1980. Effects of acidic precipitation on health and productivity of forests. In *Effects of Air Pollutants on Mediterranean and Temperate Forest Ecosystems*, ed. P. R. Miller, pp. 165–73. U.S. Forest Serv. Gen. Tech. Rep. PSW-43. Upper Darby, Pa.

Cronan, C. S. 1980a. Controls on leaching from coniferous forest floor microcosms. *Plant Soil* 56:301–22.

Feldman, S. Personal communication.

Foy, C. D. 1973. Manganese and plants. In *Manganese*, pp. 51–76. Natl. Academy Sci., Natl. Res. Counc., Washington, D.C.: National Academy of Sciences, National Research Council.

Foy, C. D., 1974b. Effects of aluminum on plant growth. In *The Plant Root and its Environment*, ed. E. W. Carson, pp. 601–42. Charlottesville, Va.: University Press of Virginia.

Foy, C. D. 1981. Effect of nutrient deficiencies and toxicities in plants: acid soil toxicity. *Manuscript for Handbook of Nutrition and Food*. Boca Raton, Fla.: CRC Press.

Foy, C. D., R. L. Channey, and M. C. White. 1978. The phytiology of metal toxicity in plants. *Annual Review of Plant Physiology* 29:511–66.

Friedland, A. J. Personal communication.

Friedland, A. J., A. H. Johnson, T. G. Siccama, and D. L. Mader. 1983b. *Soil Sci. Soc. Am. J.* 48:422–25.

Friedland, A. J., A. H. Johnson, and T. G. Siccama. 1984a. Trace metal content of the forest floor in the Green Mountains of Vermont: Spatial and Temporal patterns. *Water Air Soil Pollution* 21:161–70.

Friedland, A. J., R. A. Gregory, L. Karenlampi, and A. H. Johnson. 1984b. Winter damage to foliage as a factor in red spruce decline. *Canadian Journal of Forest Research* 14:963–965.

Gunthardt, M. S., and Wanner, H. 1982. Veranderungen bei Spaltoffnungen und der Wachsstruktur mit zunchmendem Nadelatter fei Pinus cembra L. und Picae abies (L.) Karten an der Waldgrenze. *Botanica Helvetica* 92:47–60.

Hadfield, J. S. 1968. Evaluation of diseases of red spruce on the Chamberlin Hill scale, Rochester Ranger District, Green Mountain National Forest. Unpublished manuscript No. A-68-8-5230. U.S. Department of Agricultural Forest Service, Amherst Field Office, Amherst, Mass.

Hepting, G. H. 1971. Diseases of Forest and Shade Trees of the United States. U.S. Department of Agriculture, Forest Service Handbook No. 386, 658 pp.

Hepting, G. H. 1961. Climate and Forest disease. *Ann. Rev. Phytopathology* 1:31–47.

Humphreys, F. R., M. J. Lambert, and J. Kelly. 1975. The occurrence of sulfur deficiency in forests. In *Sulfur in Australian Agriculture*, ed. K. D. McLachlan, pp. 154–62. Sidney, Australia: Sidney University Press.

James, B. R., C. J. Clark, and S. J. Riha. 1983. An 8-hydroxquinoline method for labile and total aluminum in soil extracts. *Soil Science of America Journal* 47:893–897.

Jeffree, C. E., Johnson, R. P., and Jarvis, P. G. 1971. Epicuticular wax in the stomatal antechamber of Sitka spruce and its effects on the diffusion of water vapor and carbon dioxide. *Planta (Berl).* 98:1–10.

Johnson, A. H., T. G. Siccama, R. S. Turner, and D. G. Lord. 1983. Assessing the possibility of a link between acid precipitation and decreased growth rates of trees in the northeastern U.S. In *The Effects of Acidic Deposition on Vegetation*. American Chemical Society Meeting, Las Vegas, Nev. Ann Arbor, Mich.: Ann Arbor Science (in press).

Johnson, A. H., T. G. Siccama, R. S. Turner, and D. G. Lord. 1984. Assessing the possibility of a link between acid precipitation and decreased growth rates of trees in the northeastern U.S. In *Direct and indirect effects of acidic deposition on vegetation*, ed. R. A. Linthurst. pp. 31–53. *Acid Precipitation Series*, vol. 5. Boston, Mass.: Butterworth.

Johnson, A. H., T. G. Siccama, D. Wang, R. S. Turner, and D. G. Lord. 1981. Recent changes in patterns of tree growth rate in the New Jersey pinelands: A possible effect of acid rain. *Journal of Environmental Quality* 10(4):427–30.

Johnson, A. H., T. G. Siccama, and A. J. Friedland. 1982. Spatial and temporal patterns of lead accumulation in the forest floor in the northeastern U.S. *Journal of Environmental Quality* 11:577–80.

Johnson, D. W., and D. W. Cole. 1977. Sulfate mobility in an outwash soil in western Washington. *Water Air Soil Pollution* 7:489–95.

Kokorina, A. L. 1977. Effect of soil acidity on yield and chemical composition of herbage under different soil moisture regimes (Rue). *Rationyi Zhurnal* 55:363:88–91.

Lord, D. G. 1982. Root and foliar composition of declining and healthy red spruce in Vermont and New Hampshire. M.S. thesis, University of Pennsylvania.

Lovett, G. M., W. A. Reiners, and R. K. Olson. 1982. Cloud droplet deposition in subalpine balsam fir forests: Hydrological and chemical inputs. *Science* 218:1303–05.

Magel, E., and Ziegler, H. 1986. Einflub von ozon und saurem nebel auf die struktur der stomataren wachspfropfen in den nadelin von Picae abies (L.) *Karst. Forst. Cbl.* 105:234–238.

Manion, P. D. 1981. *Tree Disease Concepts*. Englewood Cliffs, N.J.: Prentice-Hall.

Matzner, E., and B., Ulrich. 1981. Effect of acid precipitation on soil. In *Beyond the Energy Crisis—Opportunity and Challenge*, ed. R. A. Fazzolare and C. B. Smith, pp. 555–564. New York: Pergamon Press.

Mayer, R., and B. Ulrich. 1977. Acidity of precipitation as influenced by the filtering of atmospheric sulfur and nitrogen compounds—its role in the elemental balance and effect on soil. *Water Air Soil Pollution* 7:409–416.

Mohnen, V. A. 1988. Personal communication.

Mollitor, A. V., and D. J. Raynal. 1982. Acid precipitation and ionic movements in Adirondack forest soils. *Soil Sci. Soc. Am. J.* 36:137–41.

Prinz, B., G. H. M. Krause, and H. Stratmann. 1982. Forest Damage in the Federal Republic of Germany, Landesanstalt für Immissionsschultz des Landes Nordrheinwestfalen, Essen. LIS Rep. No. 28, 144 pp.

Pyle, C., M. P. Schafale, and T. R. Wentworth. 1985. History of Disturbance in Spruce-Fir Forests of the SARRMC Intensive Study Sites—Mt. Rogers National Recreation Area, Black Mountains, and Great Smoky Mountains. SARRMC—Southern Appalachian Spruce-Fir Ecosystem Assessment Project. 67 pp.

Rehfuess, K. E. 1981. Uber die Wirkungen der sauren Niederschlage in Waldokosystemen. *Forstwissenschaftlliches Centralblatt* 100:363–81.

Rehfuess, K. E., C. Bosch, and E. Pfannkuch. 1982. Nutrient imbalances kin coniferous stands in southern Germany. International Workshop on Growth Disturbances of Forest Trees. IUFRO/FFRJ-Jyvaskyla, Finland, October 10–13.

Reiter, R., R. Sladkovic, and H. J. Kanter. 1986. *Meteorol. Atmos. Phys.* 37:27–47.

Reuss, J. O. 1977. Chemical and biological relationships relevant to the effect of acid rainfall on the soil-plant system. *Water Air Soil Pollution* 7:461–78.

Rinalla, C., P. Raddi, R. Gellini, and V. DiLonardo. 1986. Effects of simulated acid deposition on the surface structure of Norway spruce and silver fir needles. *Eur. J. For. Path.* 16:440–446.

Robarge, W. P., and I. Fernandez. 1986. Quality Assurance Methods Manual for Laboratory and Analytical Techniques. Test Group F. Forest Response Program. U.S. Forest Service and U.S. Environmental Protection Agency. NCSU Acid Deposition Program, Raleigh, N.C. 201 pp.

Roberts, T. M., T. A. Clarke, P. Ineson, and T. R. Gray. 1980. Effect of sulfur deposition on litter decomposition and nutrient leaching in coniferous forest soils. In *Effects of Acid Precipitation on Terrestrial Ecosystems*, ed. T. C. Hutchinson and M. Havas, pp. 381–93. New York: Plenum Press.

Roman, J. R., and D. J. Raynal. 1980. Effects of acid precipitation on vegetation: Actual and Potential Effects of Acid Precipitation in the Adirondack Mountains. Rep. ERDA 80-28. New York State Energy Res. Dev. Authority, 4-1-4-3.

Rosenqvist, I. Th. 1977. suf Jord/Surt Vann., Ingenioforlaget. Oslo: Norway:

Sampson, P. J. 1978. Nocturnal ozone maxima. *Atm. Environ.* 12:951–955.

Schütt, P., and E. B. Cowling. 1985. Waldsterben, a general decline of forests in central Europe: Symptoms, development, and possible causes. *Plant Disease* 69:548–60.

Shafer, S. R. 1984. Effects of acid rain on soilborne plant pathogens. Ph.D. Thesis, North Carolina State University, Raleigh.

Sheffield, R. M., N. D. Cost, W. A. Bechtold, and J. P. McClure. 1985. Pine growth reductions in the Southeast. Resour. Bull. SE-83. U.S. Dept. Agric. Forest Service, Southeast. Forest Exp. Sta.

Siccama, T. G., M. Bliss, and H. W. Vogelmann. 1982. Decline of red spruce in the Green Mountains of Vermont. *Bull. Torrey Bot. Club* 109:162–8.

Sollins, P., et al. 1980. The internal element cycles of an old growth Douglas-fir ecosystem in western Oregon. *Ecol. Mon.* 50:261–85.

Tanaka, A., and Y. Hayakawa. 1974. Comparative studies on plant nutrition. Crop tolerance to soil acidity: (1) Tolerance to low pH, (2) Tolerance to high levels of Al and Mn, and (3) Tolerance to soil acidity. *J. Soil Sci. Pl. Nutr.* 21:305–8.

Tomlinson, G. H. 1983. Air pollution and forest decline. *Environ. Sci. Technol.* 17:246a–56a.

Ulrich, B. Personal communication.

Ulrich, B. 1981a. Bodenchemische und Umwelt-Aspekte der stabilitat von Waldokosystemen. Library Environment Canada Translation No. OOENV TR-2038.

Ulrich, B. 1981b. Eine okosystemare Hypothese uber die Ursachen des Tannensterbens (Abies alba Mill.) *Forstw. Cbl.* 100:228–36.

Ulrich, B. 1982. Dangers for the forest ecosystem due to acid precipitation: Necessary countermeasures—Soil liming and exhaust gas purification. U.S. Environ. Prot. Agency translation TR-82-0111. North Carolina State University, Raleigh.

Ulrich, B., R. Mayer, and P. K. Khanna. 1980. Chemical changes due to acid precipitation in a loss-derived soil in central Europe. *Soil Science* 130:193–99.

U.S. Environmental Protection Agency. Personal communication.

Vitousek, P. M., J. R. Gosz, C. C. Grier, J. M. Melillo, W. A. Reiners, and R. L. Todd. 1979. Nitrate losses from disturbed ecosystem. *Science* 204:469–74.

Wells, C. G., W. P. Robarge, L. P. Zelazny, and S. B. Feldman. 1987. Soil and Plant Tissue Properties Associated with Stand Characteristics of Spruce fir in the Southern Appalachians. Terrestrial Effects Task Group V. Team review. Summaries. National Acid Precipitation Assessment Program. March 8–13, 1987, Atlanta, Ga. Pp. 114–121.

Will, G. M., and T. C. Youngberg. 1978. Sulfur status of some central Oregon pumice soils. *Soil Sci. Soc. Amer. J.* 42:132–34.

Youngberg, C. T., and C. T. Dyrness. 1965. Biological assay of pumice soil fertility. *Soil Sci. Soc. Am. Proc.* 29:182–7.

Zoettl, H. Personal communication.

5

Decline of Red Spruce in the High-Elevation Forests of the Northeastern United States

ARTHUR H. JOHNSON AND
THOMAS G. SICCAMA

Background and Evidence of Decline

The recent widespread and obvious mortality and growth reduction in high-elevation red spruce in the Northeast have led to a focused scientific research effort designed to determine the causes (U.S. EPA/U.S. Forest Service 1986). Since the earliest reports (Siccama et al. 1982), air pollution has been suspected by many as a possible contributor to the unusual mortality, largely because spruce decline has occurred where the trees are subject to relatively high exposures to polluted cloudwater and ozone; because no obvious biological causes have been immediately apparent; and because a concomitant deterioration in the health of conifers in the Federal Republic of Germany appeared to have many similar characteristics, which many German forest scientists suspected could be linked to air pollution.

At the same time, theories about air pollution involvement in the red spruce decline have met with skepticism because obvious signs of air pollution injury have been lacking; because the types and levels of air pollution that act as subtle stresses to spruce have been unknown; because major, synchronized episodes of mortality in red spruce and several other species have occurred at times and in places where air pollution clearly could not have been a factor; and because data

The research reported here was supported by three grants from the A. W. Mellon Foundation and the following grants and contracts: EPA-381347NAEX, GM-11-3, USDA 23-085, USDA 23-114, USDA 23-157, and USDA 23-172.

Significant contributions to our research beyond those acknowledged in the text have been made by A. Friedland, J. Scott, E. Miller, J. Battles, S. Andersen, R. Gregory, L. Karenlampi, D. Vann, E. Fox, M. Saunders, T. Schwartzman, R. Miller, J. Dushoff, M. Arthur, R. Zanes, D. Lord, G. Boyagian, B. Bovee, W. Silver, and many others who gave their time as field assistants over the past seven years.

on the many known, natural causes of tree mortality and growth reduction have not been collected and analyzed.

Against this backdrop of competing ideas, many research projects have been initiated. Their goals and rationales have been summarized recently by Hertel et al. (1987). The many hypotheses on how air pollution might be involved in forest decline have been reviewed in several publications (for example, Kline and Perkins 1987, McLaughlin 1985, Schutt and Cowling 1985).

Given the lack of consensus about the most important factors involved in the current episode of spruce decline, this chapter summarizes the data available that we think will eventually contribute to discovering the causes of the recent spruce decline. The issue of air pollution involvement is not settled. The chapter presents the most logical questions to ask regarding possible air pollution involvement in light of empirical studies and experimental studies where effects of air pollutants applied at reasonable levels have been observed.

Background of Declines

Episodes of widespread and synchronized mortality in forest species in the northeastern United States have been recognized for over a century and have been extensively studied by forest pathologists and entomologists. Numerous cases of widespread, unusually rapid mortality have occurred in areas where or at times when air pollution cannot logically be considered a cause. For example, such episodes have been reported in Hawaii and elsewhere around the Pacific rim (see, for example, Mueller-Dombois 1983, 1987). Many episodes of hardwood decline were noted in North America in the early twentieth century, and a major episode of red spruce mortality occurred in New York and New England in the 1870s and 1880s (Johnson et al. 1986, Weiss et al. 1985). A number of scientists view cases of synchronized canopy mortality as periodic natural phenomena that occur infrequently enough to be difficult to recognize as normal processes of stable ecosystems (see, for example, Pastor et al. 1987). One study (Kozin 1982) provides evidence that over the past 300 years, long periods of disintegration of parts of virgin boreal forests of the Soviet Union have been associated with cycles of solar activity.

In this century, many insect and disease problems have had an impact on forests of the Northeast. Mortality from introduced diseases and insects (such as beech bark disease, Dutch elm disease, chestnut blight, and gypsy moth) have and will continue to have a major effect on shaping forest composition (Weiss and Rizzo 1987). Aside from situations where a single cause has been identified, in many cases widespread tree mortality in New York, New England, and eastern Canada has occurred from a combination of stress factors. These multiple stress diseases are called "declines" (see, for example, Houston 1981, Manion 1981). In a review of declines occurring in major eastern U.S. forest types, Weiss and Rizzo (1987) note that oaks (*Quercus* sp.), white ash (*Fraxinus americana* L.), white and yellow birch (*Betula papyrifera* Marsh., *Betula alleghaniensis* Britt.), American beech (*Fagus grandifolia* Ehrh.), and sugar maple (*Acer saccharum* Marsh.)

have undergone major declines during the twentieth century. Several other species (mostly hardwoods) have been subject to periods of unusually severe mortality related to natural stress factors. Various combinations of drought, frost, waterlogging, climatic warming, insect defoliation, old age, poor site conditions, and secondary organisms (insects and fungal diseases) have been determined to be key stress factors.

Decline of sugar maple is currently being studied in Ontario, Quebec, and Vermont and has been thought by some to be aided by air pollution. McIlveen et al. (1986) have reviewed the historically documented occurrences of maple decline, with special emphasis on Ontario. They suggest that the current situation there fits the pattern of previous maple declines, where combinations of insect defoliation, drought, and poor site conditions, followed by secondary organisms, served as causal agents. They also indicate that, at present, effects of regional-scale air pollution have not been demonstrated.

Air Pollution Effects on Trees

From the extensive record of controlled and empirical studies of the interactions of air pollutants and plants, many ways in which air pollution could serve as a stress contributing to declines have been suggested (see, for example, Kline and Perkins 1987, McLaughlin 1985, Woodman and Cowling 1987). Support for the idea that air pollution can injure trees comes from many studies conducted near smelters and industrial districts and short-term exposure studies involving acidified rain, mist, and toxic gases. Many alterations of plant structure and function have been noted in the presence of high doses of pollutants. Most of the current questions, however, involve chronic effects of lower pollutant doses because levels of most pollutants measured in the high-elevation forests of the northern Appalachians are usually below the levels that have caused adverse effects in short-term experiments (up to months in duration). The notable exception to that generalization is ozone.

A growing body of information clearly indicates that ambient levels of ozone in the northeastern United States constitute a stress to vegetation. The studies of Reich and Amundson (1985), Skelly et al. (1983), and Wang et al. (1986) (see also Bormann 1985), using charcoal-filtered air and controlled ozone additions, suggest that ambient levels of ozone in the Northeast may be causing reduced net photosynthesis and growth of at least several species of forest trees, and that these effects can occur without visible symptoms.

Because of the current political environment in which the research is being conducted, rigorous proof of a connection between air pollution and reduced forest health is being sought (see, for example, Woodman and Cowling 1987). "Confirmation" of air pollution as a factor causing a measured forest response requires that there be a consistent association between symptoms and the suspected pollutant, that the same symptoms observed in the forest be produced when healthy trees are exposed to the pollutant at controlled levels similar to the exposures known to occur in the forest, and that the same pattern of genetically controlled variation in susceptibility observed in the forest be duplicated when clones of the same trees are exposed to the suspected causal agent under controlled conditions. The data needs for such a proof are enormous even when considering a single species. Interim steps require that the pollutant(s) most likely to cause adverse effects be identified and that their role in the decline complex be specified for rigorous testing under the rules just noted. The research effort designed to determine if air pollution has contributed to the recent spruce decline is currently identifying which pollutants are present at what levels and testing hypotheses about how ambient levels of pollutants might act to promote declines.

As pointed out by Wallace (1978), the assignment of cause and effect in multiple stress diseases can be very difficult. The data demands in the case of red spruce decline are extensive because of the uncertainty involved in extrapolating pollutant effects from controlled experiments with seedlings to mature trees in a complex forest, the observation that air pollution stress and natural stresses can cause similar symptoms, and the possibility of multiple interactions among normal stresses in the forest and airborne chemicals. Over the next several years, therefore, judgments about the probability that air pollution is an important factor will have to rely on simpler arguments and on scientific consensus.

The rules of proof noted by Woodman and Cowling (1987) are designed to show conclusively the existence of an air pollution effect. Because current data are not nearly sufficient to judge cause and effect using their criteria, a less rigorous examination is carried out here of the existing possibilities to find the most promising directions for continuing research on the causes of spruce decline.

Red Spruce Mortality in High-Elevation Forests of the Northeast

Hertel et al. (1987) reported a consensus among scientists who have studied or are currently studying spruce decline that a high degree of mortality and reduction in growth has occurred in the northern Appalachians and Adirondacks above about 900 meters that is not readily explained by current models of usual stand dynamics and forest growth. An elevation of 900 meters is a somewhat arbitrary but useful division for discussion purposes.

At high elevations, the most prominent symptom is the loss of foliage due to the dieback of the terminal and lateral apices. Recent personal observations are that the upper surfaces of older needles are often slightly chlorotic, and that relatively few year classes (three to five) of needles are retained on declining red spruce. No unusual, systematic decrease in balsam fir or white birch importance or growth has been detected. At low elevations, mortality, growth reductions, and reproduction are as expected in second-growth forests except on some islands along the Maine coast (Jagels 1986), where chlorosis, loss of old needles, stork's nest crowns, and extensive crown thinning have been observed in some stands.

The difference between high- and low-elevation stands is fairly consistent across New York, Vermont, and New Hampshire. Hornbeck and Smith (1985) and Federer and Hornbeck (1987) studied 3001 red spruce tree cores from an extensive network of largely previously logged, low-elevation sites across New England and the Adirondacks. They concluded that the decrease in diameter growth rates they observed after 1960 were predictable based on accepted models of very even-aged, pure stands of second-growth spruce (Meyer 1929). In the Bartlett Forest of New Hampshire, Leak (1987) observed that on suitable types of substrate at and below 650 meters, red spruce basal area increased, often substantially, between 1931–32 and 1984. Thus, recent quantitative studies suggest mostly expected changes in red spruce growth and population dynamics in the lower elevation forests.

Gradients in the species and environmental conditions on the mountains of the Northeast have been studied by several investigators (Adams et al. 1920, Foster and Reiners 1983, Harries 1966, Holway et al. 1969, McIntosh and Hurley 1964, Scott and Holway 1969, Siccama 1974). Figure 5.1 shows the elevational gradients on Whiteface Mountain, New York, in 1986–87. Sugar maple is the most important canopy species below 700 meters, and balsam fir (*Abies balsamea* [L.] Mill.) is most important above 1000 meters. Red spruce is a minor

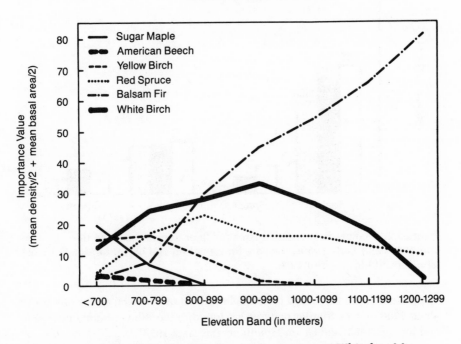

Figure 5.1. Elevational Gradients in Species Importance at Whiteface Mt., N.Y. Source: Battles et al. 1988.

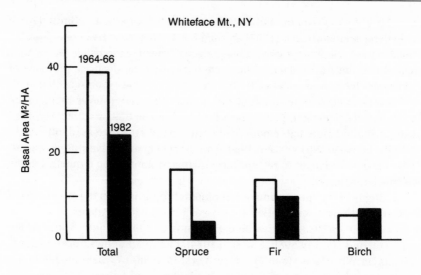

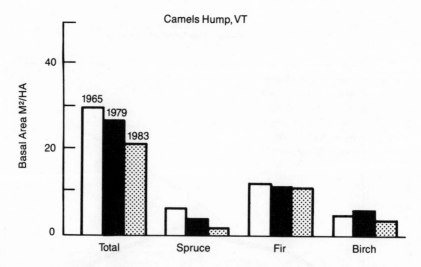

Figure 5.2. Changes in Live Basal Area of Major Species between the Mid-1960s and Early 1980s. Data are for transition and boreal forests. Source: Johnson and McLaughlin 1986.

component of the hardwood forest (below about 800 meters) and is rare above about 1200 meters. Similar patterns are found in the White Mountains (Foster and Reiners 1983) and Green Mountains (Siccama 1974).

The patterns of vegetation on the mountains of the Northeast are complex mosaics that result from natural gradients of climate and soil characteristics and the effects of natural and human disturbances including fire, windthrow, land-slides, and logging. Smaller scale disturbances (from less than 1 to about 20 hectares or so) within each vegetation type have significant effects on species composition and the age structure of the forest (see, for example, Foster and Reiners 1983).

A 1982 survey of the condition of red spruce in the canopy at sites at or above 650 meters on Whiteface Mountain (New York), Mount Mansfield (Vermont), and Mount Washington (New Hampshire) showed a relationship between eleva-tion and severely declining and dead spruce (Johnson and McLaughlin 1986, Johnson and Siccama 1983). In that survey, the percentage of dead stems deter-mined in prism surveys was rather evenly distributed among diameter classes up to 40 cm, with a higher percentage dead in size classes above 40 cm (Johnson and McLaughlin 1986).

The condition of spruce in high-elevation forests changed substantially between the mid–1960s and the early 1980s (figure 5.2, table 5.1) (see also Scott et al. 1984, Siccama et al. 1982), as evidenced by data collected at Whiteface Mountain and four of the Green Mountains (Vermont). Reductions of live spruce density and basal area of 50 percent or more were common above 900 meters in

Table 5.1. *Changes in red spruce density and basal area above 760 m between 1965 and 1979 at four mountains in Vermont.*

Size (cm) Class	Density (stems/ha)				Basal area (m²/ha)			
	1965[a]	1979	Change (%)	Dead[b]	1965	1979	Change (%)	Dead
Boreal forest (>880 m)								
	Camels Hump							
<2	4211	2866	−32	—	—	—	—	—
2–9	206	151	−27	39	0.52	0.30	−42	0.09
≥10	125	60	−52	73	6.32	3.51	−44	7.00
	Other mountains (pooled data)[c]							
2–9	256	73	−71	73	0.48	0.15	−69	0.17
≥10	145	76	−48	66	3.17	3.22	+2	1.98
Transition forest (760–880 m)								
	Camels Hump							
<2	8748	4333	−50	—	—	—	—	—
2–9	495	145	−70	102	0.89	0.14	−84	0.18
≥10	91	54	−41	65	5.73	3.42	−40	8.55
	Other mountain (pooled data)							
2–9	352	45	−87	54	0.79	0.08	−90	0.11
≥10	124	9	−93	9	2.03	0.26	−87	0.10

Source: Siccama et al. 1982.

[a]Camels Hump studied in 1965, other mountains in 1964.

[b]Dead stems measured in 1979 only.

[c]Jay Peak, Bolton Mt., and Mt. Abraham; stems < 2 cm dbh were not tallied.

spruce 10 or more cm diameter-at-breast-height (dbh) and in spruce less than 10 cm dbh (table 5.1).

A resurvey of the 56 transects at Mount Washington, Mount Mansfield, and Whiteface Mountain (see Johnson and McLaughlin 1986) shows the progress of the spruce decline between 1982 and 1987 (figure 5.3) (Silver et al. 1989). Above 900 meters, the majority of trees that in 1982 were in crown class 3 (greater than 50 percent foliage lost from the upper portion of the live crown) died, while the percentage of spruce in crown classes 1 and 2 was unchanged.

Below 900 meters the pattern was different. Severely declining trees (crown class 3) were rare at lower elevations in both surveys. A decrease was noted in crown class 1 trees, with a corresponding increase in dead spruce (class 4 and 5). This pattern suggests relatively rapid death of individual trees that appeared healthy in 1982. Spruce beetle (*Dendroctonus rufipennis* Kirby) damage has been observed in large spruce along some of the transects (Weidensaul, pers. comm.). As this beetle causes rapid death, presumably at least some of the low-elevation spruce mortality is due to the insect.

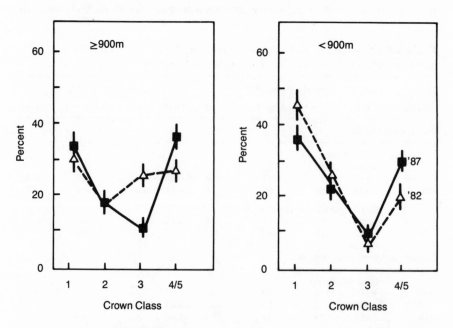

Figure 5.3. Changes in Crown Condition between 1982 and 1987 at Five Sites in Northern Appalachia. The data are from 56, 100 m transects on Mt. Mansfield, Vt., Mt. Washington, N.H., and Whiteface Mt., N.Y. Crown class 1 is 0–10% loss of foliage from the upper portion of the live crown; crown class 2 is 11–50%; class 3 is 50–99%; class 4 is dead/intact; and 5 is dead, broken above dbh. Source: Silver et al. 1988. Methods are given by Johnson and McLaughlin 1986.

Figure 5.4 shows the changes in the percentage of dead spruce greater than or equal to 10 cm dbh in the canopy at the study sites between 1982 and 1987 as a function of elevation. Above 900 meters, approximately 20 percent of the spruce alive in 1982 died by 1987. This rate (4 percent per year) is substantially faster than the general rate reported for low-elevation spruce (0.5 percent per year [Hertel et al. 1987]). The most rapid mortality observed at the northern study

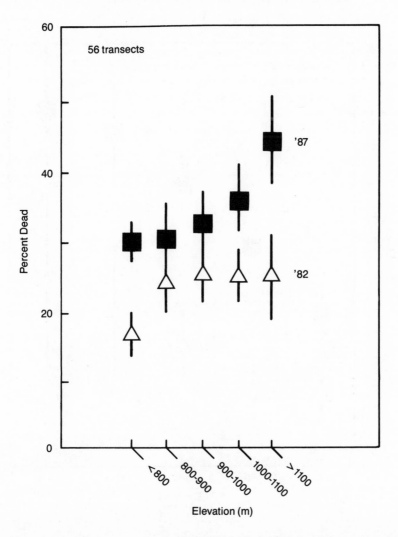

Figure 5.4. Change in the Percentage of Dead Spruce between 1982 and 1987 as a function of elevation. Sites are the same as in figure 5.3. Source: Silver et al. 1989.

sites was at Whiteface Mountain (figure 5.5), with elevational patterns and the extent of the mortality varying from site to site (figure 5.6).

Based on the data in figure 5.3, it appears that vigorous spruce (class 1 and 2) are not beginning to decline and that the trees currently in crown class 3 will die. Thus, at least a temporary stabilization in the high-elevation spruce population may be near. Figures 5.2–5.6 and table 5.1 reveal that spruce as a component of the canopy above 900 meters will have decreased in abundance by 50–80 percent over two decades at least at the range of sites studied.

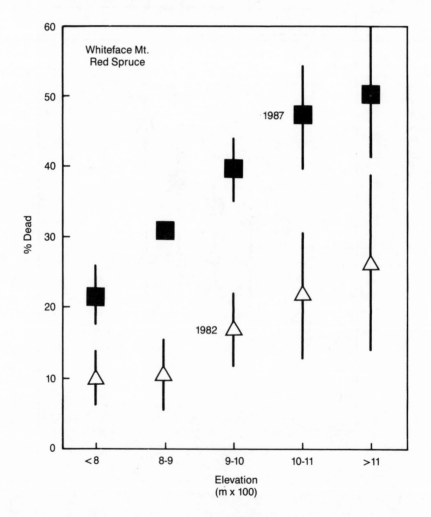

Figure 5.5. Percentage of Dead Spruce Stems at Whiteface Mt. in 1982 and 1987 as a Function of Elevation. Data are from 20,100 m transects on SW and SE slopes. Source: Silver et al. 1989.

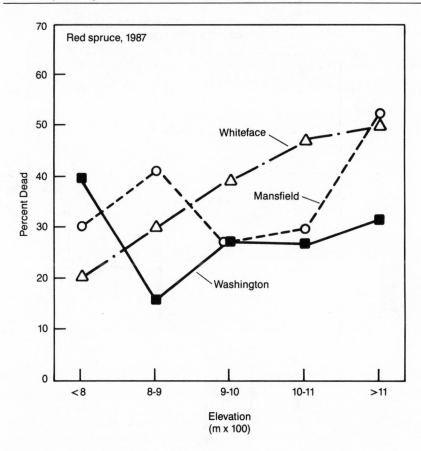

Figure 5.6. Percentage of Dead Spruce Stems in 1987 at Mt. Mansfield, Vt., Mt. Washington, N.H., and Whiteface Mt., N.Y., as a function of elevation. Source: Silver et al. 1989.

Natural Factors

Relationships between Mortality and Elevation, Age, and Aspect

Interacting Influences. In 1986 and 1987, a network of 331 systematically located, 10-meter-diameter permanent plots was established on Whiteface Mountain to better define spatial patterns in the spruce decline. Details of the sampling design and methods are given by Battles et al. (1988). From that survey, which included more than 7000 trees, some of the relationships among mortality and site and stand variables emerge more clearly.

Figure 5.7 shows the percentage of dead spruce as a function of elevation, with the upper three elevational bands showing a significantly greater proportion of dead stems than the lower three bands. The increase above 1000 meters

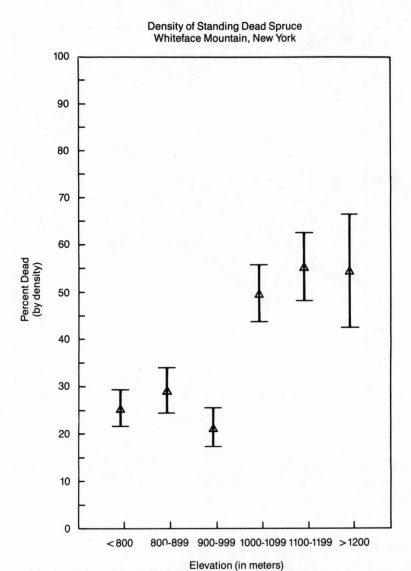

Figure 5.7. Percentage of Dead Spruce in 1986–87 in 331 Permanent Plots at Whiteface Mt., N.Y. Source: Battles et al. 1989. Calculations are based on the number of live and dead stems greater than 5 cm diameter at breast height per unit area.

arises primarily from two factors—the age and exposure of the trees. Figure 5.8 indicates a greater degree of mortality in the larger size classes at both high and low elevations, as well as greater mortality in all size classes at the higher elevations. The latter probably indicates a real association between elevation and spruce decline. Figure 5.9 indicates that there was a greater percentage of spruce in larger size classes above 1000 meters. If we assume that the increase in dead stems in larger size classes is real (that it does not reflect only the tendency for large dead trees to remain standing longer than small dead trees), the distribution of size classes helps account for the increased number of dead stems above 1000 meters.

At Whiteface Mountain, mortality has been greatest over 1000 meters on the northwest exposure (figure 5.10). Figure 5.11 shows that mortality has been greater in almost all size classes on the upper northwest-facing slopes compared with other aspects, suggesting a real relationship between aspect and mortality. The northwest aspect has the most pronounced development of fir waves (for example, Sprugel 1976), owing most likely to the severer impact of the prevailing (northwesterly) winter winds, clouds, and storms that cause mechanical defoliation of ranks of exposed, mature balsam fir. This natural form of disturbance

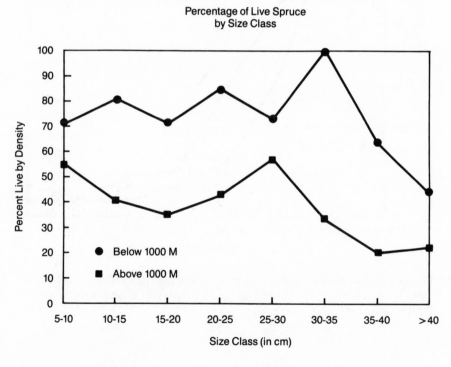

Figure 5.8. Distribution of Live Spruce by Diameter Classes above and below 1000 m at Whiteface Mt., N.Y. Source: Battles et al. 1988.

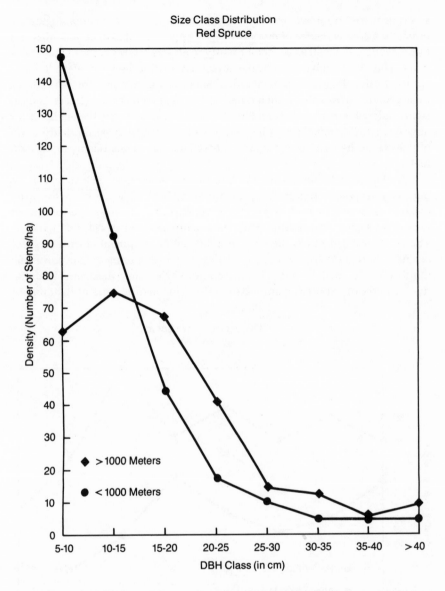

Figure 5.9. Distribution of Spruce by Size Class for Stands above and below 1000 m on Whiteface Mountain. Data include live plus dead stems. Source: Battles et al. 1988.

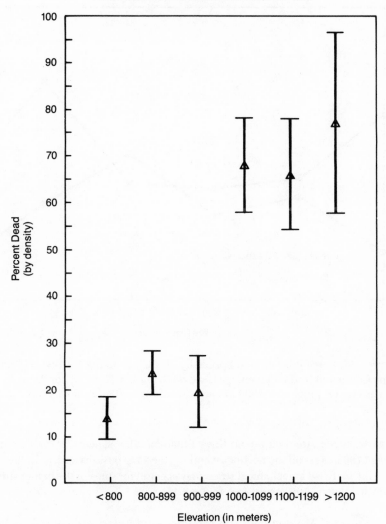

Density of Standing Dead Spruce
Northwest Face, Whiteface Mountain, New York

Figure 5.10. Percentage Dead Spruce Greater than 5 cm Diameter at Breast
Height as a Function of Elevation on Whiteface Mountain, Northwest Face.
Source: Battles et al. 1988.

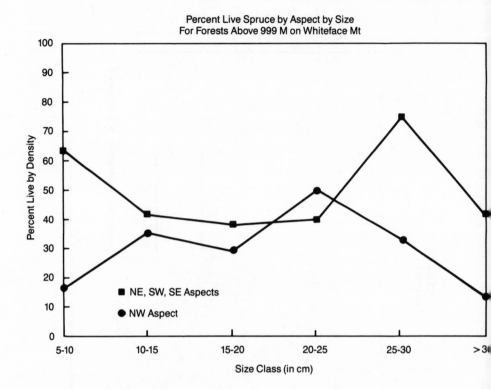

Figure 5.11. Percentage of Live Spruce by Size Class on the Northwest-Facing Slope Compared to the Percentage Dead on the Other Three Major Aspects. Data are for all plots over 1000 m elevation. Source: Battles et al. 1988.

suggests greater stress on the northwest than on other aspects. Fir waves occur above 1100 meters on the northwest and southeast exposures and, as in the case of fir, the increased mechanical stress may account for some of the higher spruce mortality above 1100 meters (see figure 5.12).

Summary. Age and poor site conditions are factors generally thought to control, in part, the severity of other declines. At Whiteface and elsewhere the upper slopes have thinner, more acidic soils, considerably shorter growing seasons, and greater wind stress in winter; they are in general poorer sites than the lower elevations. Thus, the association between elevation and mortality suggests that site conditions in this case are related to the severity of decline. Likewise, tree age on Whiteface Mountain appears to be a factor related to the spatial patterns of decline severity.

The implication of age and site conditions as determinants of decline severity cannot be subjected to experiment to further support their role as causes.

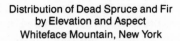

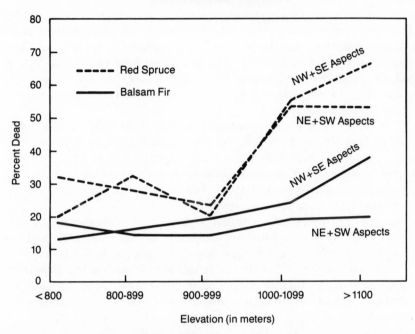

Figure 5.12. Distribution of Dead Spruce and Fir on Aspects Having Fir
Waves (NW and SE) Compared to Dead Spruce and Fir on Aspects with No
Fir Waves (NE and SW). Source: Battles et al. 1988.

Hence acceptance of those as causal factors will depend on how consistently
they are associated with the severity of mortality at the range of spruce decline
sites now being studied.

Relationship between Climate and Red Spruce Decline

Climate stress in the form of drought or frost (freezing) injury has been
considered a component of other declines in the Northeast (Manion 1981, Weiss
and Rizzo 1987), and climate stress has long been regarded as a likely contributor
to forest diseases. Hepting (1963) discussed some of the indirect ways in which
climate change (such as long-term warming) can alter important life-supporting
factors in a forest environment and lead to disease. But, except in the case of
severe climatic events, rigorous proof of climatic influence on mature trees is
very difficult given current experimental capabilities. Johnson et al. (1986, 1988)
and Cook et al. (1987) used tree-ring and other empirical studies to infer climate
involvement in spruce decline, as described in this section.

Important Characteristics of Tree-Ring Series. The current episode of high-elevation spruce mortality has been accompanied by a pronounced reduction in ring width of vigorous dominant and codominant trees that lasted at least 20 years and was unreversed as of the early or mid–1980s (figure 5.13). The reduction in tree-ring width occurred in stands of widely different age structure and disturbance history, and the high frequency variance is strongly correlated across virtually all high-elevation sites in our sample of the Catskill, Adirondack, Green, and White mountains. This large degree of correlation suggests that regional-scale factors (such as climate) have an important influence on the year-to-year variation in growth.

The disturbance histories and age structure of the high-elevation forests reported here are distinctly different from those studied by Hornbeck et al. (1986) and Meyer (1929). Federer and Hornbeck (1987) suggest that the decrease in growth they observed in mostly lower elevation spruce fit the expected patterns for pure, highly even-aged stands of second-growth red spruce, as reported by Meyer (1929), and further suggest that spruce in the lower elevation forests are behaving as if in even-aged stands that reached breast height about 1915. Their application of Meyer's growth curves may be reasonable for stands that had been cut at the turn of the century, a major period of logging in the Northeast.

However, red spruce reported on in this section reached breast height on average in about 1850 (with a wide range of variation [50 to 200 years]) in multiaged stands known to be uncut and unburned, as well as in stands known to

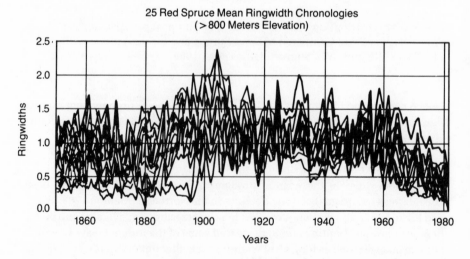

Figure 5.13. Tree-ring Width Series for 25 High-Elevation Sites in the Catskill, Adirondack, Green, and White Mountains. Data are average annual ring-width values normalized to the 131-year mean. Each site is represented by two cores from 12–20 canopy trees selected on the basis of a systematic sampling scheme (Johnson and McLaughlin 1986).

have been logged at the end of the last century (see figure 5.14). In short, the stands are very different from one another and far more heterogeneous and older than the pure, highly even-aged, old-field stands described by Meyer (1929). The data of Conkey (1984) suggest that after major disturbances, ring width in dominant and codominant red spruce in uneven-aged, high-elevation stands increases dramatically within a few years, then decreases gradually, and fluctuates around a constant mean value for well over a century or until the next major disturbance. Our personal observations imply that the prolonged period of constant ring width might result from the fact that spruce tend to be considerably taller than the co-occurring fir or birch and often form a supercanopy.

Overall, there is no theoretical or empirical basis for applying Meyer's data to high-elevation forests and, contrary to the impression given by Zedaker et al. (1986) and Van Deusen (1987), a simultaneous decrease in ring width across the sites studied is unexpected. Related arguments are presented by McLaughlin et al. (1987). Whether or not there has been a ring-width decline and what it might be due to have been controversial. But interestingly, it makes little difference to the arguments presented here regarding the cause of the ring-width decline. Other characteristics of the tree-ring series show clear evidence of unusual behavior after 1960 and serve as an indicator of the beginning of the spruce decline.

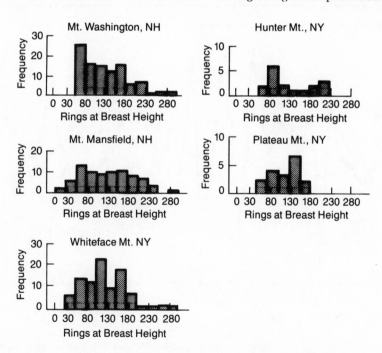

Figure 5.14. Distribution of Ages of Red Spruce Cored for the Tree-Ring/Climate Studies of Johnson et al. (1988) and Johnson and McLaughlin (1986).

Our analyses showed that in high-elevation red spruce, the year-to-year variation in tree-ring width was associated with temperatures in the prior December (or January) and in the previous August (or July). Smaller-than-expected rings occurred after warm, late summers and/or after cold, early winters from 1856 to about 1960, but not from 1961 to 1981 (Cook et al. 1987, Johnson et al. 1988).

Figures 5.15 and 5.16 show that there was an abrupt shift in the early 1960s, when neither August nor December was strongly correlated with ring width, but November (year prior to ring formation) and current July temperatures were. Several factors appear to contribute to these patterns. Most important is the breakdown in the association between ring width and previous August (or July) temperature. This accounts for most of the lack of fit between modeled and actual standardized ring widths from about 1962 to 1974. It is difficult to pinpoint an exact year when the climate/ring-width relationship changes, and it appears that this may vary slightly across sites. Most sites seem to change between 1960 and 1965, whereas the high-elevation Adirondack sites appear to model well until 1966 or 1967 (Cook 1987). Interestingly, the years 1977–1981

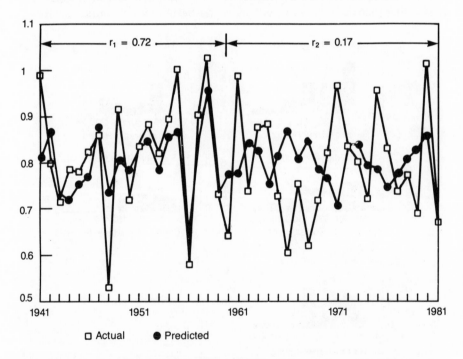

Figure 5.15. Actual Standardized Ring Widths (□) and Those Predicted by the August-December Temperature Model (●). Values are regional averages, and correlation coefficients (r_1, r_2) are for climate-predicted vs. observed values for 1941–60 and 1961–81.

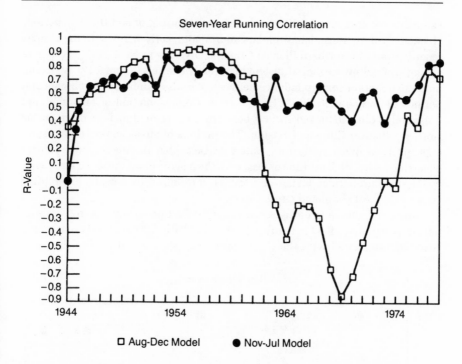

Figure 5.16. Seven-year Running Correlation Coefficients for Actual vs. Climate: Predicted Standardized Ring Widths. Squares show the forecasting ability of the August-December temperature model and filled circles show the forecasting ability of the November-July temperature model. Data are regional averages as in figure 5.15.

are predicted reasonably well by the August-December model, suggesting perhaps that the trees have assumed a "normal" response to climate after a period of shock.

Temporal Relationships. The changes in tree-ring/climate relationships suggest that during the 1960s and 1970s, spruce may have been subject to stress severe enough to alter the physiology of mature trees. As will be shown, this was also a time of widespread red spruce death. Historical reports of spruce mortality dating back to 1959 indicate that many high-elevation spruce were dying during this period. Given the unfavorable effect of warm Augusts and cold Decembers on ring width, we analyzed long-term climate records to see if there were an association between anomalies or trends in those climatic factors during recent and past episodes of spruce mortality or injury.

Although it is not known how warm, late summers and cold, early winters contribute to small rings, or how anomalies in those parameters might lead to mortality, the historical evidence suggests a connection. Episodes of severe and extensive red spruce mortality tended to occur during or shortly after periods characterized by warm, late summers and/or cold, early winters. Using climate records from the Northern Plateau Climatic Division in New York, reviews of episodes of red spruce mortality, and historical materials (figure 5.17), we examined the temporal correspondence between periods of unfavorable climate and periods of spruce mortality. Figure 5.18 is a representation of favorable and unfavorable climate for red spruce, based on long-term climate patterns in the Northern Plateau Climatic Division. The periods of stress simulated in figure 5.18 (positive index values) are based on December or August temperatures occurring 5 percent of the time or less, with the magnitude of the stress simulated by the standard normal deviate or Z score (for example, the rarest temperatures have the highest simulated stress value).

Since red spruce tree-ring series show distinct persistence or autocorrelation, the carry-over effect in figure 5.18 of climatically favorable and unfavorable years was simulated with the statistical relationship that describes the auto-

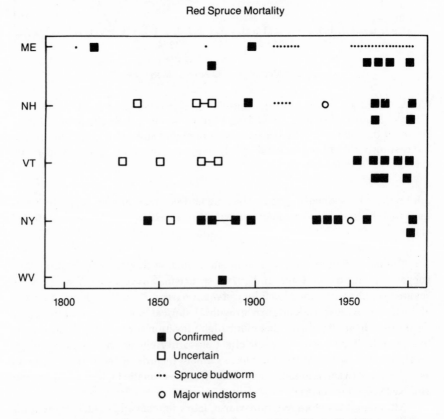

Figure 5.17. Years When Widespread or Severe Mortality in Red Spruce was Reported.

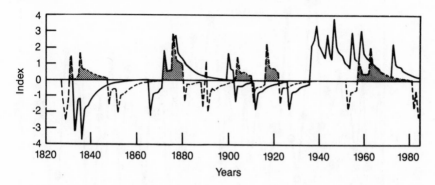

Figure 5.18. Periods of Climate Stress (Positive Values) and Periods of Favorable Climate (Negative Values) for Red Spruce Based on August (solid) and December (dashed) Temperatures. Methods are explained by Johnson et al. 1988. The anomalously warm August of 1871 is obscured.

regressive persistence in the standardized tree-ring series of high-elevation stands (see Johnson et al. 1988).

In this representation of climate stress, the most pronounced periods of unfavorable temperatures occurred from 1870 to 1880, from 1935 to 1950, and from 1956 to 1980. The 1870s are particularly notable as they contain the only two years that had both unusually warm Augusts and unusually cold Decembers. Field reports indicate that the three major periods of red spruce mortality began in or about 1871 (Hopkins 1901), 1938 (Bermingham/McIntyre 1938), and probably around 1962. The earliest periods of mortality involved spruce beetles, and some speculated that climate was involved (for example, Hopkins 1891, New York State 1891). The episode of the late 1800s, as noted previously, was responsible for very heavy spruce mortality in the Adirondacks (an estimated 50 percent of the mature spruce died). Reports from New Hampshire and Vermont suggest very heavy mortality there as well (Hopkins 1901). The spruce mortality of the 1930s was reported only in New York as far as we know. Further discussion of the historical aspects of spruce mortality is reviewed by Johnson et al. (1986) and Weiss et al. (1985).

In summary, periods of severe and region-wide red spruce mortality are temporally consistent with periods of especially warm Augusts and/or cold Decembers. Figure 5.19 shows that the most recent unfavorable climatic period was characterized by warm Augusts in the 1950s with cold Decembers in the late 1950s and early 1960s (see also Namias 1970).

Winter Damage as an Initiating and Synchronizing Stress. Many recorded observations suggest that severe winter damage (from dessication, freezing, or

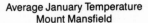

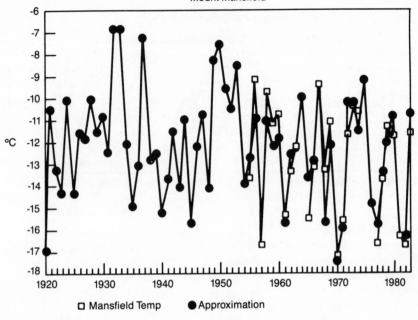

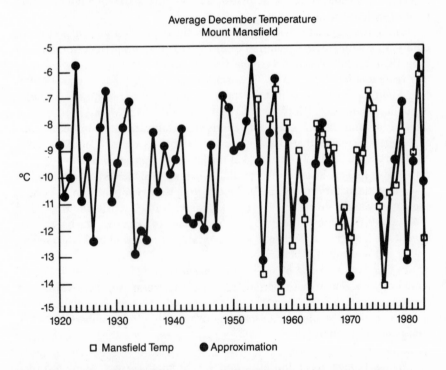

Figure 5.19. Winter Temperatures at Mt. Mansfield, Vt. (approximately 1350 m) during the 1960s. Values prior to 1954 were determined from the regression of Mount Mansfield temperatures on Burlington (Vt.) temperatures.

both) was the key factor that triggered the recent spruce decline. Figure 5.20 shows that reports of severe winter damage to foliage (usually confined only to the newest foliage), sometimes accompanied by extensive bud mortality and tree death, have been widespread across New England since about 1959 (Friedland et al. 1984a, Johnson et al. 1986, Weiss et al. 1985).

Reports citing severe browning episodes were particularly frequent and widespread in the early 1960s. Patches of high-elevation spruce in New Hampshire were judged to have been killed by severe winter conditions by 1962 or 1963 (Kelso 1965, Stark unpublished field notes, Tegethoff 1964, Wheeler 1965). By 1963, studies or repeated visits to affected sites were initiated in some places to follow the progress of the injured trees (see, for example, Stark unpublished, Tegethoff 1964). Curry and Church (1952) showed that similar browning events in the late 1940s had triggered an episode of dieback and red spruce death in the Adirondacks.

Based on the concurrence of the synchronous downturn in ring width across the study sites, the many reports of spruce mortality in 1962–65, and the change in the tree-ring/climate relationship in the early 1960s, we suspect that the frequent winter damage served as a synchronizing and initiating factor that started a sequence of decline. This assertion conforms with the fact that the recent (post–1960) foresters' reports from which figure 5.17 is drawn often attribute mortality in spruce to severe winters.

The effect of several closely spaced damaging winters might also explain the shift in tree-ring width shown in figure 5.13. Perhaps severe winter damage to

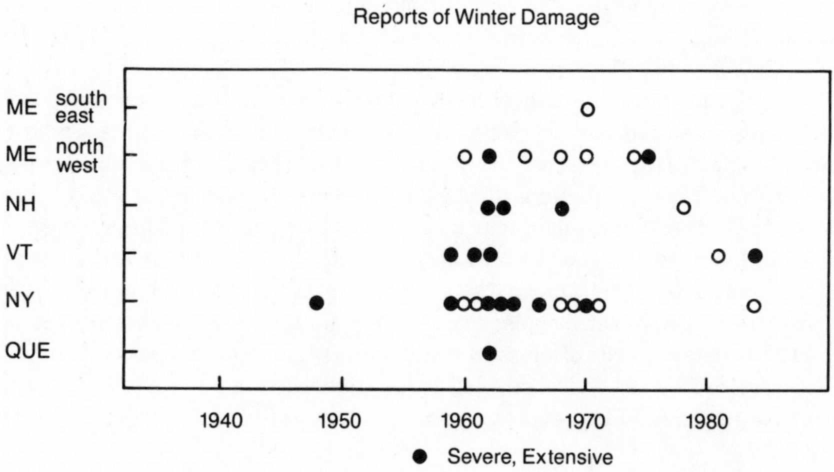

Figure 5.20. Reports of Winter Damage to Red Spruce. Reports are based on findings of red needles in the spring, bud and twig death, and tree death attributed by the observers to winterkill.

foliage and buds shifted carbon allocation in favor of producing new shoots and foliage and away from wood production.

Spatial Consistency. If climatic stress is an important determinant of spruce decline, the regional pattern of mortality, symptoms, and tree-ring characteristics should be spatially consistent with climatic anomalies. In the eastern United States, the 1960s were characterized by unusually cold winters (Namias 1970). Climate records for the Northeast show a strong gradient in the departure from normal temperatures, with the most unusual temperatures in western New England and New York (Namias, 1970) and near-normal temperatures in central and eastern Maine.

At present there are insufficient published data for a statistically rigorous analysis of the spatial consistency between winter temperature conditions and spruce decline. However, we expect such data to be available very soon.

Drought. During the 1960s widespread drought in the Northeast was prolonged and especially severe (see, for example, Cook and Jacoby 1977), and the potential for drought as an initiating stress in other declines has been noted (Burgess et al. 1984, Houston 1981, Manion 1981). Information relating to the importance of the mid–1960s drought in the high-elevation forests is conflicting. Spatial patterns do not strongly suggest drought as a major factor, since precipitation during the drought increased with elevation (Johnson and McLaughlin 1986, Siccama 1974), and mortality has been greatest at the highest elevations. However, high-elevation soils are thinner and hold less water. The available evidence argues clearly that there were major episodes of high-elevation spruce mortality in progress by 1962 or 1963, two years prior to the severe drought (see, for example, Tegethoff 1964) that had the same symptoms and characteristics reported more recently.

On the other hand, although the Palmer drought index is not correlated with the standardized ring widths or with residuals from the temperature-based regression models, the tree-ring analyses often show strong negative residuals from the temperature-based predictions during the peak drought years of 1965–67 (figure 5.15), indicating that drought may be a stress adding to the decline syndrome. Additionally, the timing of the breakdown in the climate/tree-ring relationships in the Adirondacks (1967 or so) (Cook 1987) corresponds more with the drought period than with the earlier episodes of winter damage. It is also noteworthy that the 1870s were a decade of stressful temperature (figure 5.18) and severe drought (Cook and Jacoby 1977). The climatological similarities of the two severest periods of spruce mortality are intriguing and worthy of further study.

Summary. Climatic influences are thought to play important roles in the decline of other tree species, and the available evidence indicates that climatic events, most likely damage to foliage and buds related to winter conditions, are factors in the current spruce decline. Major periods of red spruce mortality, including the recent one, are temporally and spatially consistent with unfavorable winter temperatures. A climate-driven mechanism that has been shown previously to initiate a period of dieback and decline in red spruce—severe winter damage to foliage and buds—is documented in field reports, but the exact mechanisms that contribute to the foliar and bud damage are unknown. Drought may be another climatic factor helping to synchronize or initiate the mortality, and its effect may be less regional and more site specific. The evidence implicating drought is weaker than that implicating winter damage.

Influence of Insects and Diseases

Very commonly, insects and diseases contribute to declines as factors responsible for the death of trees subject to other types of stress. In some cases, defoliating insects can initiate a period of decline (see, for example, Weiss and Rizzo 1987). Currently, permanent plot studies are being carried out to systematically categorize symptoms on declining red spruce and the factors responsible. Many symptoms are recognized as related to known insects and diseases, but many symptoms are of unknown origin (Hertel et al. 1987).

Hopkins (1901) and Weiss et al. (1985) reviewed the occurrence of insects and diseases in episodes of spruce mortality during the past 100 years or so. The spruce beetle, *Dendroctonus rufipennis* Kirby, has been implicated in many cases of spruce mortality and is present in some locations in the current episode of decline (McCreery et al. 1987). It is rare in the declining stands on Whiteface Mountain (Weidensaul pers. comm.). *Armillaria mellea* (Vahl:Fr.), the shoestring root rot fungus, is present in declining spruce at lower elevation but rare at higher elevations (in about 10 percent of the severely declining trees in a study done by Carey et al. 1984). Other fungi (such as *Fomes pini* and Cytospora canker) that are considered to be secondary in their action are present and have been identified in declining spruce in some places over the past two decades (see, for example, Hadfield 1968). At lower elevations, dwarf mistletoe (*Arceuthobium pusillum* Peck) has been identified as a factor associated with growth loss and mortality (McCreery et al. 1987).

Early observers of spruce mortality at high elevation in New Hampshire and western Maine (Kelso 1965, Stark unpublished, Tegethoff 1964, Wheeler 1965) looked for but found no evidence of insects or fungal disease as a cause of the spruce mortality they studied through repeated visits in the early 1960s. They attributed the mortality and the visible decline in spruce to the effect of severe winter conditions. Overall, the current evidence suggests a variety of insects and diseases as contributors to decline, with different combinations present in different areas.

Summary of Natural Factors

To date, many of the naturally occurring factors implicated in declines of other species are shown to be temporally and/or spatially associated with the current red spruce decline. Tree age, elevation and aspect (indicators of more stressful sites), winter damage, drought, insects, and fungal diseases have been identified thus far. With the use of Manion's scheme (1981), age and site factors might be logically assigned as predisposing factors, winter damage and drought might be considered as inciting or triggering factors, and the fungal diseases might be considered as contributing factors. Spruce beetles, where present, might be contributing factors or in some cases primary factors in tree death. If the forthcoming findings of research from the several intensively studied sites are consistent with the findings presented here, those natural factors will likely be designated as causes of the decline.

Possibility of Air Pollution Involvement

Although the available information shows that the spruce decline fits the pattern of other declines where air pollution has not been suspected as a contributor, such evidence does not rule out air pollution involvement. Many possible roles for air pollution have been identified and summarized (see, for example, Friedland et al. 1984a, Johnson and Siccama 1983, Klein and Perkins 1987, McLaughlin 1985, Schutt and Cowling 1985). At present, there are no descriptions of symptoms attributable solely to air pollution, so it is logical to ask questions about the interaction of airborne chemicals and site conditions, winter damage, drought, insects, and disease. Could air pollution at current and past levels have changed site conditions or plant function in a way that would make spruce more susceptible to the natural stresses associated with decline? Could the recovery period after acute climatic stress require enough energy for repair that a background of air pollution stress, for which the plants could ordinarily compensate, becomes significant? Might the resources used for repair of low-level air pollution damage be slowing down repair and replacement of needed roots or foliage? A few empirical studies and a few experimental studies in which realistic levels of pollutants were used provide a sound basis for determining the direction of future investigations.

Airborne Chemicals and Site Quality

Acids and heavy metals are constituents of cloudwater and precipitation regarded as chemicals that might have adverse effects on the quality of forest soils. They are deposited at relatively high rates in the subalpine forests largely because of the interception of cloudwater, which has much higher levels of dissolved chemicals than does rain (see, for example, Weathers et al. 1986). Table 5.2 summarizes some data sets on deposition of chemicals in cloudwater and precipitation in forests of the Northeast.

Trace Metals. Friedland et al. (1984b, c), Johnson (1982), and Reiners et al. (1975) have documented a rather rapid accumulation of lead, copper, zinc, and organic matter in the forest floor in montane forests of New England. The forest floor in the subalpine coniferous forests is usually deep (more than 10 cm), and

Table 5.2. *Deposition of chemicals in precipitation and clouds to high-elevation forests in the Northeast compared to deposition in a lower elevation forest at Hubbard Brook, N.H.*

Ion	Hubbard Brook (N.H.)[a] (1963–1974)	Mt. Moosilauke (N.H.)[b] (1220 m) (kg ha⁻¹ yr⁻¹)	Whiteface Mt. (N.Y.)[c] (1060 m)
N^+	1.0	3.9	1.03
NH_4^+	2.9	20.5	9.1
SO_4^{2-}	38.4	202.7	60.2
NO_3^-	19.7	124.9	27.4

Ion concentrations in stratiform cloudwater (μeq/l) at Whiteface Mt. (N.Y.)[d]

H^+	280	(pH 3.6)
NH_4^+	89	
SO_4^{2-}	140	
NO_3^-	110	

[a]1963–74 data from Likens et al. 1977.
[b]1980–81 data from Lovett et al. 1982.
[c]E. K. Miller, A. J. Friedland, and A. H. Johnson unpublished 1985–86.
[d]Mean of 28 samples collected at Whiteface Mt. summit, Castillo et al. 1983.

since red spruce are particularly shallow rooted, changes in the chemistry or processes regulating nutrient availability and turnover in that key ecosysem compartment could be important.

Trace metal accumulations in the forest floor in the fir-spruce forest at Camels Hump (Vermont) are not now in the range where alterations of biological activity have been noted in controlled experiments; they lie in the range where experimental results yield no significant changes in the measured indicators of biological activity. Camels Hump metal levels are among the highest measured values (see, for example, Friedland et al. 1984b, c) in the Northeast, and although those levels will likely increase, perhaps to problematic levels, there is no indication in these or other data available now that trace metal deposition is a likely factor in the current spruce decline. However, effects on special-purpose organisms that are crucial in nutrient cycles (ammonifiers, nitrifiers, and mycorrhizae) cannot be judged by the data and this leaves open an area for further study.

Soil Acidification. At sufficient input levels, the leaching of base cations by mobile anions resulting from atmospheric inputs can probably increase soil acidity in some types of soils (see, for example, Ulrich et al. 1980, Van Breeman et al. 1983). Soil acidification would be expected to result in lower soil fertility and increased availability of potentially toxic elements such as aluminum, and as such could add to site-related stress. (See Binkley and Richter [1987] for a summary of the processes that govern soil acidification and the factors that make soils susceptible or resistant to acidification by atmospheric sources.)

Owing to the high rates of sulfuric and nitric acid deposition on high-elevation spruce/fir forests, an understanding of long-term changes in soil pH and base status is helpful in understanding the impact of acid deposition to date. In 1930–32, Heimburger (1934) measured soil profiles throughout the Adirondack Mountains and determined pH of the organic horizons and dilute acid-extractable calcium in organic and mineral horizons. Those measurements were repeated in 1984 by Andersen (unpublished) and the results are shown in figures 5.21–5.24. While E horizons (leached layers at the surface of the mineral soil) tended to acidify, there was no systematic change in the pH of highly acid organic horizons (pH < 4.0) over five decades, and no systematic change in the extractable calcium in mineral B horizons of acid soils. Higher pH organic horizons tended to show a decrease in pH and extractable calcium, but for the most part, high-elevation sites do not appear to be highly susceptible to acidification

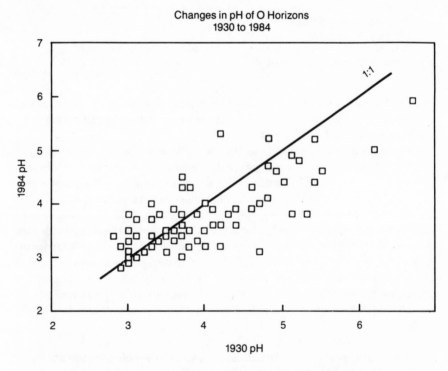

Figure 5.21. Changes in pH of Organic Horizons of Adirondack Forest Soils between 1930–32 and 1984. Values are shown for 32 sites that were relocated with very high confidence (Andersen unpublished). The original study was carried out by Heimburger (1934), and the 1984 measurements were made using the same collection and analytical procedures.

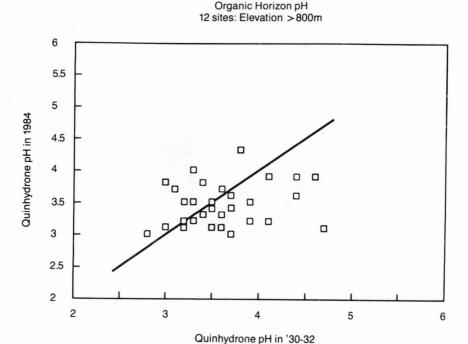

Figure 5.22. Changes in pH in Organic Horizons in High-Elevation Adirondack Forest Stands with Appreciable Numbers of Red Spruce and Balsam Fir (Andersen unpublished).

at current and past rates of acid deposition. Estimates of the quantity of calcium lost from the organic and E horizons in the past 50 years suggest that calcium depletion can be accounted for by tree uptake alone (Andersen unpublished). In light of those data, the likelihood that acid deposition has significantly increased stress related to acidity in the high-elevation forests is small.

Foliar Leaching and Foliar Nutrition

Related to the issue of site quality and soil acidification is the ability of plants to obtain adequate base cations for normal functions. Experiments with several different species have shown that acidic precipitation increases the leaching of cations from foliage. Although no detrimental effects of foliar leaching have been observed in controlled experiments, the association of magnesium and potassium deficiency with declining Norway spruce in West Germany (see, for example, Huettl chapter 2) and the high potential for foliar leaching above 900 meters make the levels of cations in red spruce foliage of interest.

Friedland et al. (1987) sampled healthy mature red spruce at high and low

Mineral & Organic Ca Comparison
32 Sites

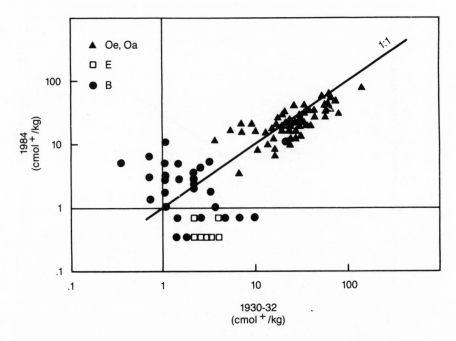

Figure 5.23. Changes in Dilute-Acid-Extractable Calcium (0.2N HCl) in Organic and Mineral Horizons between 1930–32 and 1984 at Sites Relocated with the Most Confidence (Andersen unpublished).

elevations in the Green and Adirondack mountains and reported that foliar magnesium levels were lower in their high-elevation sites than all other reported values for red spruce, and in the range of moderate deficiency according to Swan (1971).

Studies of foliar calcium, magnesium, and potassium carried out using standard sampling procedures at Whiteface Mountain (Schwartzman and Miller unpublished) show some interesting patterns. Of the three elements measured, only potassium was related to crown class (figure 5.25). Interestingly, levels appear to be sufficient, even in severely declining trees (table 5.3), and the lower foliar levels are not related to low levels of available potassium in the soil. This contrasts with the case of Norway spruce, where potassium and magnesium deficiencies are associated with soils having very low available levels. The lower levels of foliar potassium in declining trees may be related to increased foliar leaching (due to poorer foliar integrity?) or to decreased uptake capability (due to impaired root systems?).

Although magnesium is only weakly associated with crown class, figure

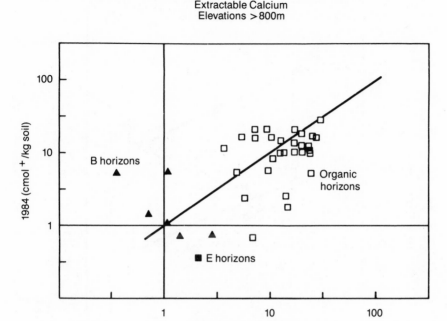

Figure 5.24. Changes in Dilute-Acid Extractable Ca at High-Elevation Adirondack Sites (Andersen unpublished).

5.26 shows that foliage is not, in general, well supplied with magnesium. According to the values of Swan (1971), red spruce at high elevations shows a slight magnesium deficiency. Foliar and soil levels decrease with increasing elevation, so the respective roles of foliar leaching and soil supply are obscured. At present, no conclusions can be drawn regarding the role of foliar nutrient status in the spruce decline, or the effect of air pollution on foliar nutrient status.

Airborne Chemicals and Climate or Winter Stresses

Some recent research has been designed to test the effect of pollutants on the resistance of seedlings to climatic stresses. In a growth chamber experiment, Norby et al. (1987) used potted red spruce seedlings from Maine planted in soil from Camels Hump (Vermont) and Acadia National Forest (Maine) coupled with ozone and acid mist plus acid rain treatments. In that experiment, mist and rain, acidified to realistic values (pH 3.6 mist and pH 4.1 rain) and applied at realistic rates, were associated with a large and statistically significant increase in drought stress. The authors attributed the effect to an increase in shoot-to-root ratio brought about by the application of mist and rain, and speculated that this might be the result of the increased nitrogen applied.

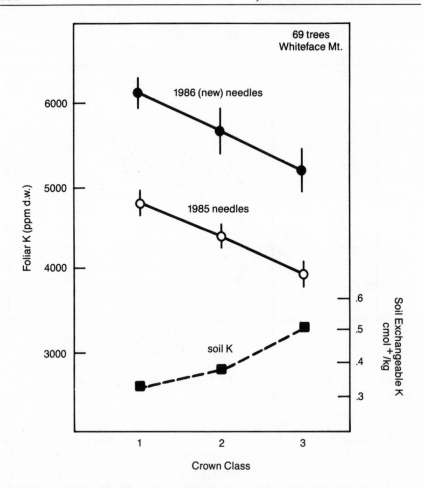

Figure 5.25. Foliar and Soil Potassium as a Function of Crown Class at Whiteface Mountain, N.Y. Foliage was sampled from the south side of the 4th–7th whorl in October 1986. Soil potassium (K) is 2N NH_4 Cl-extractable K. At each tree, soils were sampled at 0–15 cm (n = 4 per tree) and 15–30 cm (n = 4 per tree) depths and bulked into one sample for each depth. Values reported are the arithmetic mean of the two depths.

As explained previously, the tree-ring analyses for intervals prior to 1960 and from 1961 to 1980 and the timing of reports of mortality in 1962 or 1963 did not strongly suggest drought as the primary factor initiating the current spruce decline. But the tree-ring analyses cannot account for threshold responses to acute stress, so a particularly severe drought (and the mid–1960s drought was extreme) could be an important stress. As with all experimental work to date, it is not known if tests done with seedlings are applicable to the mature trees in the

Table 5.3. *Suggested provisional standards for the evaluation of the results of foliar analyses species: red spruce foliar concentration expressed as percent dry matter.*

Element	Range of acute deficiency	Range of moderate deficiency	Transition zone from deficiency to sufficiency	Range of sufficiency for good to very good growth	Range of luxury to excess consumption
K	Below 0.19	0.19–0.30	0.30–0.40	0.40–1.10	1.10 and up
Mg	Below 0.04	0.04–0.06	0.06–0.08	0.08–0.17	0.17 and up
Ca	Below 0.05	0.05–0.08	0.08–0.12	0.12–0.30	0.30 and up

Notes: Modified from Swan 1971. These suggested standards are essentially judgments; they are based both on the results of greenhouse studies and on experience gained from the use of foliar analysis in field studies.

field. Nonetheless, because the treatment levels were realistic, additive or interactive effects between drought and pollution stress appear to be possible.

The many reports of red-brown needles and bud death in spring during the early 1960s, the change in tree-ring/climate relationships that began about that time, and the beginning of region-wide reporting of dead and dying spruce in the early 1960s suggest that winter damage is a significant cause of the spruce decline. Thus, the effects of air pollutants on resistance to winter stresses comprise an important area for investigation. Many factors, some reasonably well understood and others much less so, contribute to a plant's ability to resist damage in winter (see, for example, Weiser 1970).

Davison et al. (1987) reviewed the types of damage that plants are subject to in winter, and the ways in which air pollution might alter resistance to those types of damage. In winter, conifers must contend with photooxidation of chlorophyll (resulting in bleaching of the needles), desiccation (caused by water loss through cuticles on bright, sunny days when water in the conducting tissues is frozen), and freezing injury (usually called "frost damage"). The latter two stresses are associated with the red-brown color observed in spring, and it is unclear which type of injury has been most frequently observed in red spruce during the past two decades.

As for evidence that air pollution might affect resistance to winter damage, electron microscopy in polluted and unpolluted areas in Finland suggested that conifers exposed to pollution were more susceptible to winter damage (Davison et al. 1987), but the exact reasons remain unknown.

Since several pollutants attack constituents of cell membranes and since changes in cell membranes occur during hardening against freezing injury, Davison et al. (1987) suggest there is a sound theoretical basis for suspecting that air pollution might interfere with the development of freezing resistance. Their experiments with ozone fumigations (120 ppb ozone for six hours per day, for 70

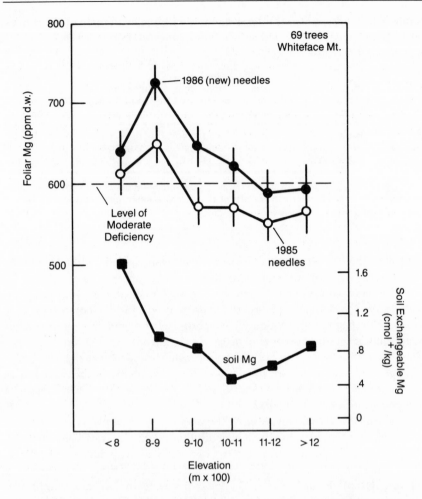

Figure 5.26. Foliar and Soil Magnesium (Mg) Levels as a Function of Elevation at Whiteface Mountain, N.Y. Methods are the same as for figure 5.25.

days) indicated that ozone at those dose rates was associated with increased
cuticular water loss from previous year's needles in one out of eight Norway
spruce clones tested. Four clones showed an increase in freezing-related damage
when exposed to the ozone treatments.

Weinstein et al. (1987) used red spruce seedlings and exposed them to
filtered air, the ambient levels of ozone in Ithaca (New York), and twice ambient
levels. Ozone at ambient and twice ambient doses was associated with increased
respiration, decreased pigment concentrations, and continued high photosynthesis rates during the hardening period (October), whereas plants exposed
to charcoal-filtered air showed the expected decreased photosynthesis rates in

the fall. The elevated rates of photosynthesis in fall and examination of ultra-structural characteristics suggested to the authors that hardening had been delayed by ozone treatments. The data of Burgess et al. (1984) indicate that the summertime ozone doses received by the trees at Whiteface Mountain have probably been at least within the ambient range used in Weinstein et al.'s experiments, suggesting a possible way in which ozone might be linked to winter damage and the spruce decline. Again, the application of experimental results to mature trees on the mountains is suspect, but ozone stress clearly becomes a plausible candidate for more detailed research as a potentially important factor leading to spruce decline.

Weinstein et al. (1987) also tested the effect of nitric acid additions in artificial mist and rain on red spruce seedlings. Using reasonable misting periods and reasonable water application rates, they observed no increase in winter damage at even unrealistically high rates of nitrogen input (pH 2.5 nitric acid), thus implying that increased nitrogen inputs via cloudwater deposition do not increase the likelihood of winter damage, as suggested by Friedland et al. (1984a).

Summary

Experimental work now in progress and future research may show that realistic levels of acids, ozone, or other airborne chemicals affect spruce in ways that may have contributed to declining vigor. The record of experimental results from which to choose the most appropriate targets for research suggests that ozone should be given a high priority. As noted in the beginning of this chapter, a clear record has been established with seedlings and saplings of many species that suggests ozone at ambient levels can reduce net photosynthesis and growth, and recent results with red spruce suggest a possible link to winter damage.

Most of our information on treatments and effects comes from experiments done with small trees, a few years old. It will therefore be difficult to bridge the gap between experimental conditions and mature trees growing in complex ecosystems. New experimental technologies may be needed to satisfactorily refine our understanding of air pollution effects on forests. For example, chambers fitted on branches designed to control the environment around small portions of large trees may be valuable for studying the effects of cloud-like mists and gaseous pollutants for comparison with the results of carefully controlled open-top and growth chamber experiments.

Impact of Red Spruce Decline on the Forest

As data on spruce and fir seedlings and saplings have not yet been analyzed, it is difficult to predict the future of the stands. Our observations at Whiteface Mountain suggest that there are patches of young spruce less than 1 meter tall scattered throughout the high-elevation forests, and occasional 2- or 3-year-old seedlings, especially near streams. Balsam fir and white birch reproduction have been quite vigorous in the gaps left by the dying spruce, and over the lifetime of the birch and fir (100–150 years) the species composition of the canopy and the processes that regulate energy and nutrient flow will shift accordingly. Figures 5.27 and 5.28 show the distribution of live-tree diameters for spruce and the other associated species above and below 1000 meters at Whiteface Mountain. These patterns suggest a declining spruce population above 1000 meters and a stable one below (see, for example, Leak 1976).

Since red spruce are much longer lived than fir or birch (300+ years) and exteremly shade tolerant, the survivors of the current decline may be sufficient to allow the population of understory spruce to increase and ultimately replace some of the fir and birch in the canopy as the latter die off a century or so from now. The shade tolerance and longevity of spruce should allow that species to undergo serious declines periodically but to return in future years as an important canopy species. The age distribution of spruce in the virgin, high-elevation stand in New Hampshire reported by Leak (1976) suggests this may have been the case after the decline of the late 1800s. Thus if the spruce decline has only natural causes, it may be only part of a normal cycle in an ecosystem that is stable when viewed over periods of several centuries.

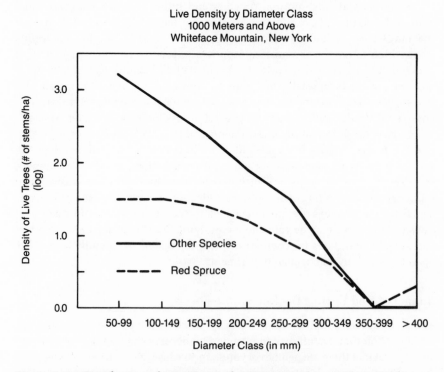

Figure 5.27. Distribution of Live Spruce by Diameter Class at Elevations Greater than 1000 m at Whiteface Mountain, N.Y., Compared to Disttribution of Co-occurring Species by Diameter Class (Battles et al. 1988).

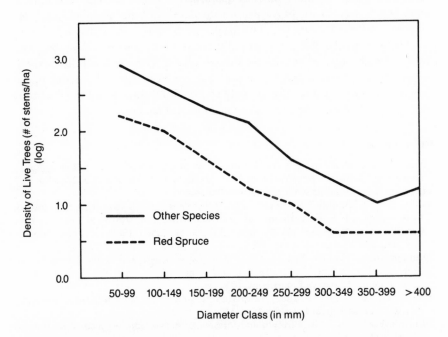

Figure 5.28. Distribution of Live Spruce by Diameter Class at Elevations Less than 1000 m at Whiteface Mountain, N.Y., Compared to the Distribution of Co-occurring Species by Diameter Class (Battles et al. 1988).

Recent studies (Eckert et al. 1987) suggest, as did Wright earlier (1955), that the montane red spruce populations are not very genetically diverse relative to other species. Wright predicted that environmental change (such as a climatic warming) might cause a high degree of mortality in the montane spruce as a result of the lack of diversity. This extensive mortality appears to have been realized. As an alternative to the species returning as an important part of the canopy, therefore, perhaps spruce are being phased out as new episodes of stress periodically remove them from the forest.

The complexity of natural forests of the Northeast has made it difficult to identify that even a large change has occurred. About 20 years passed before spruce mortality had progressed enough to be recognized as a region-wide phenomenon. That complexity, combined with our current lack of understanding of the dose rates of airborne chemicals and the physiological and biochemical effects of those chemicals, hinders a clear assessment of whether air pollution has had a major role in the decline of red spruce, and perhaps in other declines.

Those who wish to judge causes of the recent spruce decline must realize there are many natural stresses that will, in all likelihood, be agreed upon by forest scientists as causal factors. The degree to which those stresses alone could have caused the mortality cannot be quantified. There is evidence that current levels of airborne chemicals can exacerbate natural climatic stress, but applying experimental results obtained with seedlings to large trees under field conditions is tenuous. The effects of air pollution stress on red spruce can be quantified through ongoing and future studies, and if partially controlled experiments using large trees growing in the field can be successfully carried out, it may be possible to bridge the gap between seedling studies and the declining mature trees that have been the focus of empirical studies.

References

Adams, C. C. G. P. Burns, T. L. Hankinson, B. Moore, and N. Taylor. 1920. Plants and animals of Mt. Marcy, New York. Parts I–III. *Ecology* 1:71–94, 204–233, 274–288.

Andersen, S. B. Unpublished. University of Pennsylvania.

Battles, J., A. H. Johnson, T. G. Siccama, and W. L. Silver. 1988. Recent changes in spruce-fir forests of New York, Vermont and New Hampshire. Proceedings of a U.S.-German Conference on Forest Decline. Burlington, Vt. October 19–24, 1987. U.S. Forest Service, Broomall, Pa.

Bermingham, M. Unpublished reports on red spruce disease and mortality from W. G. Howard, 1938, H. L. McIntyre, 1938, and W. Smith, 1948. New York Department of Environmental Conservation, Albany, N.Y.

Binkley, D., and D. Richter. 1987. Nutrient cycles and H^+ budgets of forest ecosystems. *Advances in Ecological Research* 16:1–51.

Bormann, F. H. 1985. Air pollution and forests: An ecosystem perspective. *Bioscience* 35:434–441.

Burgess, R. L., M. B. David, P. D. Manion, M. J. Mitchell, V. A. Mohnen, D. J. Raynal, M. Schaedle, and E. H. White. 1984. Effects of Acidic Deposition on Forest Ecosystems in the Northeastern United States: An evaluation of Current Evidence. New York State College of Environmental Science and Forestry, Syracuse, N.Y.

Carey, A. C., E. A. Miller, G. T. Geballe, P. M. Wargo, W. H. Smith, and T. G. Siccama. 1984. *Armillaria mellea* and decline of red spruce. *Plant Disease* 68:794–795.

Castillo, R. A., J. E. Jiusto, and E. McLaren. 1983. The pH and ionic composition of stratiform cloud water. *Atmospheric Environment* 17:1497–1505.

Conkey, L. E. 1984. Dendrochronology and forest productivity: Red spruce wood density and ring width in Maine. In *Research in Forest Productivity, Use and Pest Control*, ed. M. M. Harris and A. M. Spearing, pp. 69–75. USDA Forest Service General Tech Report NE-90.

Cook, E. R. 1987. The use and limitations of dendrochronology in studying effects of air pollution on forests. In *Effects of Atmospheric Pollutants on Forests, Wetlands and Agricultural Ecosystems*, ed. T. C. Hutchinson and K. M. Meema, pp. 227–290. Springer Verlag, Berlin.

Cook, E. R., and G. C. Jacoby. 1977. Tree-ring-drought relationships in the Hudson Valley, New York. *Science* 198:399–401.

Cook, E. R., A. H. Johnson, and T. J. Blasing. 1987. Modelling the climate effect in tree rings for forest decline studies. *Tree Phys.* 3:27–40.

Curry, J. R., and T. W. Church. 1952. Observations on winter drying of conifers in the Adirondacks. *J. Forestry* 50:114–116.

Davison, A. W., J. D. Barnes, and C. J. Renner. 1987. Interactions between air pollution and cold stress. Proceedings of Second International Symposium on Air Pollution and Plant Metabolism, April 6–9, 1987. Neuherburg, Federal Republic of Germany.

Eckert, R. T., M. E. Dermeritt, and D. M. O'Malley. 1987. Genetic variation in red spruce pollution response. Spruce-Fir Research Cooperative Progress Report 5/1/87. U.S. Forest Service, Broomall, Pa.

Federer, C. A., and J. W. Hornbeck. 1987. Expected decrease in diameter growth of even aged red spruce. *Can. J. Forest Research* 17:266–269.

Foster, J. R., and W. A. Reiners. 1983. Vegetation patterns in a virgin subalpine forest at Crawford Notch, White Mountains, New Hampshire. *Bull. Torrey Bot. Club* 110:141–153.

Friedland, A. J., R. A. Gregory, L. Karenlampi, and A. H. Johnson. 1984a. Winter damage to foliage as a factor in red spruce decline. *Can. J. Forest Research* 14:963–965.

Friedland, A. J., A. H. Johnson, and T. G. Siccama. 1984b. Trace metal content of the forest floor in the Green Mountains of Vermont: Spatial and temporal patterns. *Water Air Soil Pollution* 21:161–170.

Friedland, A. J., A. H. Johnson, T. G. Siccama, and D. L. Mader. 1984c. Trace metal profiles in the forest floor of New England. *Soil Science Society of America Journal* 49:422–425.

Friedland, A. J., G. J. Hawley, and R. A. Gregory. 1987. Red spruce foliar chemistry in northern Vermont and New York, USA. *Plant and Soil* 105:189–193.

Hadfield, J. S. 1968. Evaluation of diseases of red spruce on the Chamberlain Hill sale, Rochester Ranger District, Green Mt. Nat. Forest. File Report A-68-8 5230. Amherst, Mass.: USDA-Forest Service Northeastern Area, State and Private Forestry, Amherst FPC Field Office.

Harries, H. 1966. Soils and vegetation in the alpine and subalpine belt of the Presidential Range. Ph.D. dissertation, Rutgers University, New Brunswick, N.J.

Heimburger, C. C. 1934. Forest Type Studies in the Adirondacks. Ph.D. Thesis, Cornell, University, Ithaca, N.Y.

Hepting, G. H. 1963. Climate and forest declines. *Ann. Review Phytopathology* 1:31–50.

Hertel, G., S. J. Zarnoch, T. Arre, C. Eager, V. Mohnen, and S. Medlarz. 1987. Status of the Spruce-Fir Research Cooperative research program. Proceedings of the Eightieth Annual Meeting of the Air Pollution Control Association, New York, N.Y., June 21–26, 1987.

Holway, J. G., J. T. Scott, and S. Nicholson. 1969. Vegetation of the Whiteface Mt. region of the Adirondacks. In *Vegetation–Environment Relations at Whiteface Mt. in the Adirondacks,* ed. J. G. Holway and J. T. Scott, pp. 1–44. Rept. No. 92, Atmospheric Sciences Research Center, State University of New York at Albany.

Hopkins, A. D. 1891. Forest and shade tree insects. II. West Virginia Experiment Station Third Annual Report. pp. 171–180.

Hopkins, A. D. 1901. Insect enemies of the spruce in the Northeast. U.S. Department of Agriculture, Division of Entomology Bulletin No. 28, new series, pp. 15–29.

Hornbeck, J. W., and R. B. Smith. 1985. Documentation of red spruce growth decline. *Canadian Journal of Forest Research* 15:1199–1201.

Hornbeck, J. W., R. B. Smith, and C. A. Federer. 1986. Growth declines in red spruce and balsam fir relative to natural processes. *Water Air Soil Pollution* 31:425–430.

Houston, D. B. 1981. Stress triggered tree diseases, the diebacks and declines. Washington, D.C. USDA Forest Service NE-INF-41-81.

Jagels, R. 1986. Acid fog, ozone and low elevation spruce decline. IAWA Bull. (in press).

Johnson, A. H. 1982. Spatial and temporal patterns of lead accumulation in the forest floor of the northeastern U.S. *Journal of Environmental Quality* 11:577–580.

Johnson, A. H., and S. B. McLaughlin. 1986. The nature and timing of the deterioration of red spruce in the northern Appalachians. Report of the Committee on Monitoring and Trends in Acidic Deposition. National Research Council. National Academy Press, Washington D.C., pp. 200–230.

232   JOHNSON AND SICCAMA

Johnson, A. H., and T. G. Siccama. 1983. Acid Deposition and Forest Decline. *Environ. Sci. Technology.* 17:294a–305a.

Johnson, A. H., A. J. Friedland, and J. Dushoff. 1986. Recent and historic red spruce mortality: Evidence of climatic influence. *Water Air Soil Pollut.* 30:319–330.

Johnson, A. H., E. R. Cook, and T. G. Siccama. 1988. Relationships between climate and red spruce growth and decline. *Proceedings of National Academy of Science* 85:5369–5373.

Kelso, E. G. 1965. Memorandum 5220,2480, July 23, 1965. Amherst, Mass.: U.S. Forest Service Northern FPC Zone.

Klein, R. M., and T. D. Perkins. 1987. Cascades of causes and effects of forest decline. *Ambio* 16:86–93.

Kozin, E. K. 1982. On the cyclic development of virgin forests in the Sikhote-Alin'. (English abstract of an article in Russian).

Leak, W. B. 1976. Age distribution in virgin red spruce and northern hardwood. *Ecology* 56:1451–54.

Leak, W. B. 1987. Fifty years of compositional change in deciduous and coniferous forest types in New Hampshire. *Can. J. For. Res.* 17:388–93.

Likens, G. E., F. H. Bormann, R. S. Pierce, J. S. Eaton, and N. M. Johnson. 1977. *Biogeochemistry of a Forested Ecosystem.* New York: Springer-Verlag.

Lovett, G. M., W. R. Reiners, and R. K. Olson. 1982. Cloud droplet deposition in a subalpine balsam fir forest: Hydrological and chemical inputs, *Science* 218:1303–1304.

Manion, P. D. 1981. *Three Disease Concepts.* Englewood Cliffs, N.J.: Prentice Hall.

McCreery, L. R., M. M. Weeks, M. J. Weiss, and I. Millers. 1987. Cooperative survey of red spruce and balsam fir decline and mortality in New York, Vermont and New Hampshire: A progress report. In Proceedings Integrated Pest Management Symposium for northern forests March 24–27, 1986. Cooperative Extension Service, University of Wisconsin, Madison.

McIlveen, W. D., S. T. Rutherford, and S. N. Linzon. 1986. A historical perspective of sugar maple decline within Ontario and outside Ontario. Rep. ARB-141-86-Pyto, Canadian Ministry of the Environment, Ottawa.

McIntosh, R. P., and R. T. Hurley. 1964. The spruce fir forests of the Catskill Mountains. *Ecology* 45:314–326.

McLaughlin, S. B. 1985. Effects of air pollution on forests: A critical review. *Journal of Air Pollution Control Association* 35:512–534.

McLaughlin, S. B., D. J. Downing, T. J. Blasing, E. R. Cook, and H. S. Adams. 1987. An analysis of climate and competition as contributors to the decline of red spruce in high elevation Appalachian forests. *Oecologia* 72:487–501.

Meyer, W. H. 1929. Yields of second-growth spruce and fir in the Northeast. USDA Tech. Bull. No. 142. 52 pp.

Mueller-Dombois, D. 1983. Tree-group deaths in North American and Hawaiian forests: A pathological problem or a new problem for vegetation ecology? *Phytocoenologia* 11:117–137.

Mueller-Dombois, D. 1987. Natural dieback in forests. *Bioscience* 37:575–585.

Namias, J. 1970. Climatic anomaly over the United States during the 1960's. *Science* 170:741–743.

New York State. 1891. Seventh Report of the Forest Commission. Albany, N.Y.

Norby, R., G. E. Taylor, S. B. McLaughlin, and C. A. Gunderson. 1987. Drought Sensitivity of Red Spruce Seedlings Affected by Precipitation Chemistry. In Proceedings of the Ninth N. American Biology Workshop, ed. C. G. Tauer and T. C. Hennessey, June 1986, Department of Forestry, Oklahoma State University, Stillwater (in press).

Pastor, J., R. H. Gardner, V. H. Dale, and W. M. Post. 1987. Successional changes in nitrogen availability as a potential factor contributing to spruce declines in boreal North America. *Canadian Journal of Forest Research* (in press).

Reich, P. B., and R. G. Amundson. 1985. Ambient levels of ozone reduce net photosynthesis in tree and crop species. *Science* 230:566–570.

Reiners, W. A., R. H. Marks, and P. M. Vitousek. 1975. Heavy metals in subalpine and alpine soils of New Hampshire. *Oikos* 26:264–75.

Schutt, P., and E. B. Cowling. 1985. Waldsterben, a general decline of forests in central Europe: Symptoms, development and possible causes. *Plant Disease* 69:548–58.

Schwartzman, T. N., and R. G. Miller. Unpublished data. University of Pennsylvania.

Scott, J. T., and J. G. Holway. 1969. Comparison of topographic and vegetation gradients in forests of Whiteface Mt., N.Y. pp. 44–88 of Vegetation-Environment Relations at Whiteface Mt. in the Adirondacks, ed. J. G. Holway and J. T. Scott. Rep. No. 92, Atmospheric Sciences Research Center, State University of New York, Albany, N.Y.

Scott, J. T., T. G. Siccama, A. H. Johnson, and A. R. Breisch. 1984. Decline of red spruce in the Adirondacks, New York. *Bull. Torrey Bot. Club* 111:438–444.

Siccama, T. G. 1974. Vegetation, soil and climate on the Green Mountains of Vermont. *Ecol. Monogr.* 44:325–349.

Siccama, T. G., M. Bliss, and H. W. Vogelmann. 1982. Decline of red spruce on the Green Mountains of Vermont. *Bull. Torrey Bot. Club* 109:163–168.

Silver, W., T. G. Siccama, and A. H. Johnson. 1988. Changes in red spruce populations in montane forests of the Appalachians. Unpublished.

Skelly, J. M., Y. Yang, B. I. Chevone, S. J. Long, J. E. Nellessen, and W. E. Winner. 1983. Ozone concentrations and their influence on forest species in the Blue Ridge Mountains of Virginia. In *Air Pollution and Productivity of the Forest*, pp. 143–59. Washington, D.C.: Izaak Walton League.

Sprugel, D. 1976. Dynamic structure of wave-regenerated *Abies balsamea* forests in the northeastern United States. *J. Ecol.* 64:889–911.

Stark, D. Unpublished notes on red spruce disease, mortality and winter injury 1957–1977. State of Maine Department of Conservation, Entomology Laboratory, Augusta.

Swan, H. S. D. 1971. Relationships between nutrient supply, growth and nutrient concentrations in the foliage of white and red spruce. Woodlands Rep. WR/34, February 1971. Pulp and Paper Research Institute of Canada. 27 pp.

Tegethoff, A. C. 1964. High Elevation Spruce Mortality. Memorandum 5220, September 25, 1964. Amherst Mass.: U.S. Forest Service Northern FPC Zone.

Ulrich, B., R. Mayer, and P. K. Khanna. 1980. Chemical changes due to acid precipitation in a loess-derived soil in central Europe. *Soil Science* 130:193–199.

U.S. Environmental Protection Agency and U.S. Forest Service. 1986. Responses of Forests to Atmospheric Deposition. National Research Plan for the Forest Response Program. Washington, D.C.

Van Breemen, N., J. Mulder, and C. T. Driscoll. 1983. Acidification and alkalization of soils. *Plant and Soil* 75:283–308.

Van Deusen, P. 1987. Testing for stand dynamics effects on red spruce growth trends. *Canadian Journal of Forest Research* (in press).

Wallace, H. R. 1978. The diagnosis of plant diseases of complex etiology. *Ann. Rev. Phytopathology* 16:379–402.

Wang, D., D. F. Karnosky, and F. H. Bormann. 1986. Effects of ambient ozone on productivity of *Populus tremuloides* Michx. grown under field conditions. *Canadian Journal of Forest Research* 16:47–55.

Weathers, K. C., G. E. Likens, F. H. Bormann, J. S. Eaton, W. B. Bowden, J. L. Andersen, D. A. Cass, J. N. Galloway, W. C. Keene, K. D. Kimball, P. Huth, and D. Smiley. 1986. A regional acidic cloud/fog water event in the eastern United States. *Nature* 319:657–658.

Weinstein, L., R. J. Kohut, J. S. Jacobson. 1987. Research at Boyce Thompson Institute on

the effects of ozone and acidic precipitation on red spruce. Proceedings of Eightieth
Annual Meeting, Air Pollution Control Association June 21–26, 1987, New York, N.Y.

Weiser, C. J. 1970. Cold resistance and injury in woody plants. *Science* 169:1269–1278.

Weiss, M. J., and D. M. Rizzo. 1987. Forest declines in major forest types of the eastern
United States. Proceedings of Workshop on Forest Decline and Reproduction: Regional
and Global Consequences. Krakow, Poland. March 23–28, 1987.

Weiss, M. J., L. R. McCreery, I. R. Millers, J. T. O'Brien, and M. Miller-Weeks. 1985.
Cooperative survey of red spruce and balsam fir decline in New York, New Hampshire
and Vermont—1984. Interim Report. USDA Forest Service, Forest Pest Management.
Durham, N.H.

Wheeler, G. S. 1965. Memorandum 2400, 5100 July 1, 1965. Laconia N.H.: U.S. Forest
Service Northern FPC Zone.

Woodman, J. N., and E. B. Cowling. 1987. Airborne chemicals and forest health. *Environ.
Sci. Tech.* 21:120–26.

Wright, J. W. 1955. Species crossability in spruce in relation to distribution and taxonomy.
Forest Science 1:319–349.

Zedaker, S. M., D. M. Hyink, and D. W. Smith. 1986. Growth declines in red spruce. *J.
Forestry* (Jan. 1987):34–36.

6

Assessment of Crop Losses from Air Pollutants in the United States

W A L T E R   W .   H E C K

Introduction

Overview of Air Pollution Effects on Crops

Air pollution effects on crops from point sources have long been recognized. Early research was directed primarily at recognizing injury symptoms and assessing losses in productivity in accordance with the severity of symptom development. Through the years many chemicals have been identified as atmospheric contaminants that can injure or damage crops at some range of exposure concentrations and times (see table 6.1).

Historical Overview. Historically, vegetation injury has been one of the first recognized manifestations of an air pollution problem. Sulfur dioxide (SO_2) and fluoride [generally as hydrogen fluoride (HF)] gases were first found to cause injury to vegetation in Europe about the mid-nineteenth century. Industry in the United States largely ignored SO_2 problems related to smelter operations in the early part of the century. This often resulted in denuding land of vegetation for several miles around the source; Ducktown (Copper Hill), Tennessee, still shows marked effects from early smelter operations. As our knowledge of the severe effects of SO_2 on vegetation grew, there was a concerted effort to control the release of SO_2 or to disperse it over a wider area. Today, with the installation of emission control systems and the use of high stacks, extensive injury to vegetation near sources of SO_2 rarely occurs. However, the growth of high-capacity power stations and the use of tall stacks have increased the distribution of lower SO_2 concentrations over a large area, so that SO_2 is now considered a regional problem and not a source-associated local problem. In addition, the long-distance transport of SO_2 permits chemical reactions to take place and SO_4^{2-} is formed, which is a major contributor to acidic precipitation.

Table 6.1. *Phytotoxic air pollutants in order of importance to crop systems.*

Pollutant	Primary or secondary	Form	Major source(s)
O_3	Secondary	Gas	Atmospheric transformation (associated with automotive emissions, NO_2, hydrocarbons)
SO_2	Primary	Gas	Power generation, smelter operations
NO_2	Primary and secondary	Gas	From direct release and atmospheric transformation (high-temperature combustion, from NO); fertilizer production
HF	Primary	Gas-particulate	Superphosphate, aluminium smelters
$C{=}C^a$	Primary	Gas	Combustion, natural
PAN-Oxid.[b]	Secondary	Gas	Atmospheric transformation (automotive emissions, NO_2, hydrocarbons)
NO	Primary	Gas	Combustion, natural
Cl_2	Primary	Gas	Spills, manufacture
HCl	Primary	Gas	Burning of plastics
Toxic elements	Primary	Particulate	Smelters, combustion processes
NH_3	Primary	Gas	Feedlots, natural
SO_4^{2-}	Secondary	Aerosol	Atmospheric transformation (SO_2)
NO_3^-	Secondary	Aerosol	Atmospheric transformation (NO_2)
H_2S	Primary	Gas	Paper production, natural, geothermal
CO_2	Primary	Gas	Combustion, natural
UV-β	Primary	Radiation	Natural, stratospheric O_3 depletion

Source: W. W. Heck, 1982. Future directions in air pollution research. In *Effects of Gaseous Air Pollution in Agriculture and Horticulture,* ed. M. H. Unsworth and D. P. Ormrod, pp. 411–435. London: Butterworth Scientific.

Note: This list is not meant to be complete but represents the most important air pollutants with respect to terrestrial plant systems. Several of those low on the list have been poorly studied and may be more important than currently thought.

[a]Ethylene.

[b]Peroxyacetyl nitrate-oxidant.

Although fluoride (discussed here as HF) injury to vegetation was described by the turn of the century, HF was not a major problem until the expansion around 1940 of industries such as aluminum smelting and superphospate production. At that time, fluoride symptoms were characterized and a spectrum of sensitivities both between and within species was understood. Fluoride accumulates in plant tissues, and foliar analysis is an acceptable diagnostic tool. The wide distribution of fluoride in nature complicates the diagnosis of the air pollution effect of HF. The problem is also of concern because pasture grasses and other forage species, appearing perfectly normal, can have a sufficiently elevated fluoride content to harm grazing animals seriously.

The earliest recognized biological effect of photochemical air pollution was injury to vegetation in the Los Angeles area. This was first identified in 1944 and was evident over large segments of southern California, plus the San Francisco Bay area, by 1950. It is now recognized that components of photochemical oxidants (primarily ozone [O_3]) injure and damage crops in most of the United States, both rural and urban areas in the East and in large areas around metropolitan centers in the West if not generally over the West. Ozone, peroxyacetyl nitrate (PAN), and nitrogen dioxide (NO_2) are phytotoxic components of the photochemical complex, and other components such as hydrogen peroxide are suspected of being important phytotoxicants.

Ethylene was identified as a phytotoxic component of illuminating gas near the turn of the century. It is a major petrochemical, a by-product of combustion (especially in automobile exhaust) and of plant metabolism. It is the most phytotoxic of hydrocarbon gases and contributes to the formation of photochemical oxidants. Ethylene from anthropogenic sources probably contributes to crop losses, but definitive research results are not available.

Airborne pesticides, chlorine, heavy metals, acid aerosols, ammonia, aldehydes, hydrogen chloride, hydrogen sulfide, and particulates (such as cement dust) can be phytotoxic air pollutants. They are released by a variety of industrial sources, accidents, and agricultural applications. They are less widespread and either less concentrated or less phytotoxic than the air pollutants of primary concern to crop loss assessment efforts.

The increased use of fossil fuels and the release of many stable compounds (for example, chlorofluorocarbons) with subsequent reduction in stratospheric O_3 have raised global concerns associated with possible climate modification. Two specific results of these activities are the rising levels of atmospheric carbon dioxide (CO_2) and the increase of solar ultraviolet radiation (UV-B radiation). In general, increasing CO_2 is generally viewed as a positive input to crop productivity. Studies with UV-B have shown mixed results but generally some loss of productivity might be expected.

Air Pollutants of Lesser Concern. Although many phytotoxic air pollutants have been identified, only those that have caused serious damage around point sources or are ubiquitous across the nation and perceived as major problems have received serious attention. Research on the effects of HF on vegetation has decreased as control technologies have been improved. Decreased research also reflects a sound data base and a perception that the fluoride problem is well understood. Minimal research has been done on compounds such as chlorine, hydrogen sulfide, ammonia, and pesticides or on heavy metals (as air pollutants), suspended and deposited particulate matter, or combustion products from automobiles. Most of these substances are associated with point sources, acci-dental releases, or occasional releases or are not considered to be very phyto-toxic.

Several pollutants deserve a few comments and in selected areas around point sources may produce severe problems. Others are widespread around the country but by themselves do not appear to be damaging to crop growth and productivity. Fluoride is one of the most phytotoxic of the first group and de-serves some recognition. Hydrogen chloride and hydrogen sulfide are less toxic, but they can damage plants and they are associated with point sources or acci-dental releases. PAN is much more phytotoxic than O_3 and has been identified in all major metropolitan areas west of Denver. Nitrogen dioxide has been studied principally in combination with O_3 and SO_2 because it is not particularly phy-totoxic at current ambient concentrations. In addition, it is a primary agent in ozone formation. Some mention is made of the rising concern for increases in atmospheric CO_2 and UV-B radiation.

Fluoride. Research on the effects of gaseous fluoride on plants has used HF as the principal test chemical. Little research is being done in the United States now even though worldwide interest continues because of losses around point sources. Although results are reviewed in many articles, the proceedings of a 1983 symposium (Shupe et al. 1983) are comprehensive and should be accepted as a thorough compendium on research covering the effects of HF on vegetation. To highlight more recent work, a few examples can be cited of the research on HF.

Movement of fluoride through the cuticle does not appear to be an important mode of entry (Chamel and Garrec 1977). Fluoride-induced injury to plants may require light (MacLean et al. 1982); the nature of the light activation step is not known. Lesion growth of *Xanthomonas campestris* (bacteria) on red kidney bean was decreased by 5 days of continuous exposure to 1 to 3 micrograms per cubic meter (μg/m3) of HF (Laurence and Reynolds 1984). The fluoride was propor-tional to the content in unpolished rice (Sakuri et al. 1983). Yield and seed number in soybean were reduced when plants were exposed to about 2.2 ppb of HF (Ghiasseddin et al. 1981). Continuous exposure of grapevine for 64 days to 0.31 ppb HF resulted in some foliar necrosis, no effect on yield, and no signifi-cant fluoride accumulation in the berries (Murray 1983). Long-term chronic exposures of radish to HF resulted in fluoride accumulation in radish leaves and translocation to the roots but no accumulation in the edible tubercle (Kisman et al. 1983). A linear relationship between HF dose and foliar uptake in sweet cherry occurred at concentrations below 19.4 ppb (Facteau et al. 1978).

Hydrogen chloride. Solid rocket fuel forms large quantities of hydrogen chloride (HCl) gas when it combusts. This stimulated research on short-term exposures of plants to fairly high concentrations of HCl. Phytotoxicity to most of the tested species showed a threshold of 5–10 ppm after a 15–60-minute ex-posure period (Heck et al. 1980). Permeability changes were found in bean leaves exposed to HCl (Heath and Endress 1979). Phytotoxicity was similar to that reported for NO_2.

Hydrogen sulfide. Moderate to severe injury was reported in several crop plants after continuous exposure to either 0.3 or 3.0 ppm H_2S (Thompson and Kats 1978); growth stimulation was found for several species at 0.03 ppm.

Peroxyacetyl nitrate. Peroxyacetyl nitrate is a member of the peroxyacyl nitrate family of nitrogenous organic compounds produced in the photochemical complex. This family of compounds, especially PAN, was studied at length in the 1960s because they are found in the Los Angeles area at phytotoxic levels; they are more phytotoxic than O_3. A review of known worldwide concentrations of PAN that would be expected to cause plant injury suggested that only in southern California should PAN be considered a major problem (Temple and Taylor 1983). However, PAN may contribute to plant effects and crop yield reductions in other areas of the country since it is a component of the photochemical complex.

On tomato, PAN symptoms may be confused with those resulting from a mixture of O_3 and SO_2. Yield was reduced in lettuce and swiss chard after a long-term intermittent exposure to PAN at 0.04 ppm (Sigal and Taylor 1979). The plant water potential of a sensitive, but not a resistant, bean cultivar was affected by PAN (Starkey et al., 1981). The inbred parents of a susceptible and resistant petunia hybrid were used to study the inheritance of PAN resistance (DeVos et al. 1980). A genotype by environmental interaction was found; resistance appeared to be partially dominant in some cases, but when severe injury occurred, genes for susceptibility showed almost complete dominance.

Nitrogen dioxide. Several reports conclude that NO_2 can be used as a nitrogen source by plants. Uptake of $^{15}NO_2$ and use of the ^{15}N by wheat were reported at NO_2 concentrations above 1.0 ppm (Prasad and Rao 1980). The first published study using ambient levels (0.097, 0.152, or 0.325 ppm NO_2 for 3 hours) of $^{15}NO_2$ reported a linear relation between exposure concentration and uptake for snapbean; essentially all the nitrogen was metabolized (Rogers et al. 1979). Although several studies have reported greater NO_2 uptake at night, Yoneyama et al. (1979) found that night absorption was only about 14 percent of that absorbed during the day.

Uptake rates of NO_2 in potato correlated well with exposure concentrations, and the NO_2 did not affect leaf conductance (Sinn et al. 1984). Herbicide-treated soybean plants released NO_2 at night (Klepper 1979); such releases can be stimulated by sulfite additions (Tingey 1979). Daylight emissions of NO_2 were not found, owing to the photosynthetic reduction of nitrite, which apparently is the toxic component of NO_2 exposure.

Both nitrogen oxide and NO_2, associated with the production of high levels of greenhouse CO_2, substantially decreased growth and productivity of greenhouse crops (Mansfield et al. 1982). This finding raises serious concern for greenhouse operators who add CO_2 in the greenhouse.

Carbon dioxide. Increasing concentrations of CO_2 have been well documented and are expected to double their preindustrial levels in the next 50 years (Strain and Cure 1985). The Department of Energy (Strain and Cure 1985) developed an intensive program to determine the effects of increasing levels of CO_2 on plant productivity with effects on crop production as an early area of interest. A primary thrust was to develop models that could be used to predict impacts as CO_2 concentrations continued to rise. Research to date (Strain and Cure 1985) supports the contention that CO_2 is a limiting factor in crop production. Increasing levels of CO_2 increase photosynthate production and productivity of crops studied. Additionally, there is an increase in water-use efficiency that could help counteract the expected increase in global warming. This is an area of continuing interest. Future research should concentrate on interactions with other stresses.

UV-B radiation. The expected increase in UV-B radiation (280–320 nm) associated with the decrease in stratospheric O_3 has been an area of increasing concern since the mid–1970s. The potential effects on crop productivity have been a concern, and considerable research has been supported in this area. Teramura (1983) developed a good review of the subject in 1983 and, although some research has continued, there has been no substantive change in our understanding of UV-B effects on crop productivity. A primary problem with research to date is the lack of supportive field research. Thus most of the available information comes from studies in controlled environments. Depending upon the species studied, UV-B radiation has been shown to affect stomatal resistance, chlorophyll, soluble proteins, lipid and carbohydrate pools, leaf area, biomass, carbon partitioning, and crop yield. About 50 percent of the species studied were not affected while 30 percent were tolerant and 20 percent were sensitive at moderate exposures to increases in UV-B. A brief review of thirty-four experiments involving 20 different species studied in the field under supplemental UV-B radiation showed no measurable effects on yield in 25, a positive effect (8 percent) in 1 case, and negative effects (19–75 percent) in 8 cases.

Air Pollutants of National Importance. The air pollutants of greatest national concern are O_3, SO_2, NO_2, and the transformed products of SO_2 and NO_2 ($SO_4{}^{2-}$, $NO_3{}^-$) that are largely responsible for acidic precipitation. The effects of acidic deposition are briefly highlighted in the following section and NO_2 alone was just discussed. The effects of O_3, SO_2, and mixtures of these two will be discussed at some length in the remainder of this chapter.

In the United States, O_3 has the greatest effect on agricultural productivity because it is found at damaging concentrations in most sections of the country. Sulfur dioxide is associated mainly with many point sources and is of importance due to the presence of high stacks, which makes it more of a regional multiple-source problem. Although SO_2 at low concentrations over weeks or months can injure sensitive vegetation, its primary impact may occur when it is associated with O_3 or NO_2. The significance of NO_2 in terrestrial ecosystems is as an ingredient in photochemical reactions (forming O_3) and when present in association with SO_2 or O_3. Ozone and SO_2 have been the two gaseous pollutants most intensively studied during recent years. These three pollutants (O_3, SO_2, NO_2) are critical components in the formation of acidic precipitation.

Effects of Acidic Deposition on Crop Systems. The concern for effects of acidic precipitation (often incorrectly called acid rain but encompassing all forms of precipitation) on crops was highlighted in the United States in an international symposium (Dochinger and Seliga 1976). Although the term *acidic deposition* is often used, it includes both wet and dry deposition of acidifying substances such as SO_2 and NO_2. The term *acidic precipitation* is used to separate the transformed products ($SO_4{}^{2-}$ and $NO_3{}^-$) from gaseous SO_2 and NO_2; acid rain denotes only that specific form of acidic precipitation. Vegetation growing where acidic fogs or clouds are common may be affected.

Acidic deposition has been a significant air pollution concern over the last decade and has attracted the attention of environmentalists. The pollution control strategy of using tall stacks to reduce ground-level concentrations of sulfur and nitrogen oxides near fossil fuel combustion sources is a contributing factor in the long-range transport of acidic substances. Sulfur and nitrogen oxides are oxidized to sulfate and nitrate during long-distance transport and may be deposited hundreds of kilometers downwind in precipitation. Dry deposition processes are also significant in the eastern United States, where local sources play an important role in regional air quality. Atmospheric processes follow no political boundaries, and frequently downwind receptors have been across state, provincial, and national boundaries. It is this latter factor that has served to make acidic deposition a major policy issue.

Acidic precipitation may directly or indirectly affect the growth, reproduction, quality, and/or yield of agricultural crops (Heck et al. 1984a). Direct effects include changes in leaf surface morphology, foliar nutrient leaching, uptake of additional S or N, and changes in metabolic function or reproductive processes; with perennial plants, the effect may be cumulative across growing seasons. Indirect effects include altered physiocochemical characteristics of soils (water-holding capacity), nutrient availability, availability of toxic elements, and susceptibility of plants to biotic and other stress.

Although effects on foliage (Heck et al. 1984a) have not been reported at ambient rain pH levels, lower pH values do show leaf spotting, acceleration of epicuticular wax weathering, and changes in foliar leaching rates. Some research has shown inhibition of certain pathogens at lower rain pH values but a possibility of increased plant susceptibility to the pathogens. A decrease in spore formation in fungi has been related to simulated rain application. Plant dry weight has been reduced when acid rain has been applied simultaneously with O_3. These reports have been on a few species; thus we do not know the extent of these types of effects.

Research efforts in the early 1980s developed around field studies utilizing rain exclusion systems. The first such system was designed and tested by Evans et al. (1981). The system permitted growth of plants under field conditions with the exclusion of ambient rain and the addition of simulated rain treatments. Subsequent systems were designed to permit the study of the gaseous pollutants in combination with simulated rain. Johnston et al. (1986) developed rain exclusion covers over open-top chambers ("Open-Top Field Chambers" [p. 252]), and Kuja et al. (1986) inserted an air exclusion system ("Open-Air Exposure Systems" [p. 251]) for control of gaseous release. These systems have been used in a number of extensive field studies.

Several reviews of large-scale field studies have been published (Evans 1984, Evans et al. 1986a, 1986b). Evans (1984) gave an excellent review of the effects of acidic rain on botanical systems with an emphasis on crop growth and yield. He reviewed specific research relating to effects on soybean from several locations. The research from his laboratory provided evidence of effects on soybean while research from other locations showed no effects. Evans (1984) suggested that the experimental designs were the primary reason for the differences found. Later, in a document that summarized field research on soybeans at Brookhaven, Evans et al. (1986a) reported significant acidic rain effects on Amsoy 71 over five consecutive growing periods. However, they found no effects on other soybean cultivars studied. Evans et al. (1986b) also reported a significant decrease in total protein content in the Amsoy cultivar.

In similar field research, DuBay and Heagle (1987) reported no effects on growth or yield from rain acidities as low as pH 2.7 for the cultivar Forrest. Banwart et al. (1987) found no effect on two field corn cultivars at present acidity levels, although a significant reduction in yield was found in one cultivar at a pH of 3.0. Pell et al. (1987) detected no effects on two potato cultivars at pH treatments as low as 2.8 but the coefficients of variation (CVs) were very high and they felt additional research would be justified.

Although there is continued concern that acidic precipitation may affect plant systems, the evidence is very limited. It seems safe to suggest that we have not succeeded in developing an experimental approach that will permit the identification of small impacts. Likewise, the presence of increasing nitrogen and sulfur with decreasing pH (increasing H^+) probably confounds experimental results; this has not been adequately dealt with by scientists.

Although the National Acid Precipitation Assessment Program has published its Interim Assessment and reported no effects on crop yield (Irving 1987) in the United States, it will be several years before sufficient information is available to do an actual assessment of acidic precipitation on crop productivity. The Interim Assessment document was premature in its judgment and will hinder effective crop research so that a reasonable assessment may not be possible in 1990.

The Air Pollution System

An understanding of the air pollution system (Heck 1982) permits a conceptualization of how its physical aspects affect our ability to interpret impacts on crop systems (see figure 6.1). Table 6.2 lists major factors that must be determined in assessing the effects of gaseous pollutants. This section highlights the sources, emissions, transport, transformations, monitoring, and deposition processes that must be understood for a national assessment of gaseous pollutant affects on crops. Plant uptake, discussed in "Physiological and Biochemical

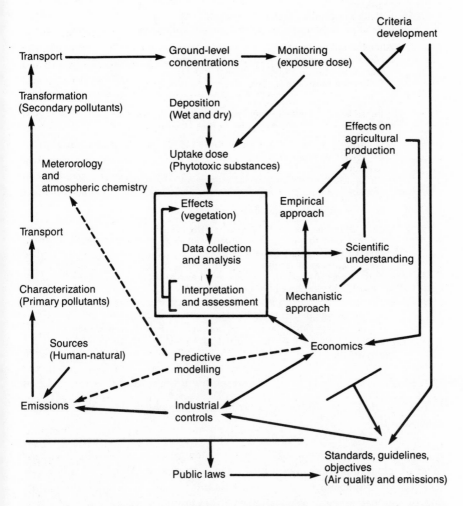

Figure 6.1. Model of the air pollution system and its major components. Solid lines suggest direct linkages; dotted lines are areas of uncertainty associated with predictive concepts. Lines with an arrow to criteria, standards, and public laws show that all aspects of the system are included in developing these three pivotal parts. The central effects box refers to all effects, although this chapter deals specifically with crops. Source: Heck 1982.

Table 6.2. *Major factors to be determined in an assessment of the effects of gaseous pollutants.*

Sources
　Natural versus anthropogenic
　Significant sources (coal, oil, natural gas, nonfossil fuels, natural
　　processes)
　Nature of source (stationary, mobile)

Emissions
　Nature of primary gaseous emissions
　Relative importance of natural versus anthropogenic emissions

Atmospheric transport
　Modes of gas movement in the atmosphere
　Nature of transport (local, regional, global)
　Residence times (up to days, weeks, months, longer)

Atmospheric transformations
　Physical and chemical transformation (type, degree)
　Participation of natural sources in these transformations

Monitoring
　Rural versus urban
　Instrumentation
　Biologically important ways to average air-quality data

Deposition
　Gaseous absorption and adsorption

Biological consequences for plants in terrestrial habitats
　Toxicity (low to high, acute to chronic, affecting few to many plants
　Likely to accumulate in organisms
　Detoxification and/or repair capability (low to high)
　Type of effect (on health, development, growth, reproduction, pheno-
　　logy, behavior, heredity)

Source: Adapted from W. W. Heck 1984. Defining gaseous pollu-
tion problems in North Carolina. In *Gaseous Air Pollutants and Plant
Metabolism,* ed. M. J. Koziol and F. R. Whatley, pp. 35–48. London:
Butterworth.

Effects" (p. 257), is the linkage process between the physical and the biological aspects of the system.

A knowledge of both natural and anthropogenic sources and their emissions is critical to the assessment of crop losses. Such information permits an assessment of maximum atmospheric loading from any source(s) and for any given locality. Anthropogenic sources are both mobile (usually associated with pollutants of regional concern) and stationary (from point sources).

A crop loss assessment requires a knowledge of transport and transformation processes. Transformations must be understood to determine the distance a pollutant can be transported and the rate of loss or formation of different pollutants in the air masses. The relationships among long-distance transport, transformations to secondary pollutants, and their relative concentrations at receptor sites have been calculated with the use of meteorology and atmospheric chemistry. Rates of transformation are associated with trace components, air movement, and stagnation periods. Both O_3 and SO_2 can be transported long distances and cause problems in rural as well as urban areas.

Knowledge of the concentrations of O_3 or SO_2 to which plants are exposed is critical to an assessment of crop production. Measurement of concentration from 1 to 5 meters above the ground can be used to determine flux to the plant (uptake). The atmospheric concentration is the exposure concentration and, with the exposure duration, defines the exposure dose. The exposure dose is the air quality unit that links the atmospheric scientist, the plant scientist, and the control official. It is the air quality measure necessary to set a standard, can be used to determine gas flux (uptake), and is essential for a regional assessment.

Air pollutants are removed from the atmosphere by wet and dry deposition processes. Dry deposition processes for O_3 and SO_2 are by sorption on plant leaves, including both adsorption on surfaces and absorption through the cuticle or stomata. Stomatal absorption rates are related to flow resistances associated with wind speed and internal plant factors. The absorption (uptake) rate of O_3 or SO_2 into the leaf is used to calculate the uptake dose. The exposure and uptake doses are related by physical factors associated with gas entry into the leaf. The portion of the uptake dose that alters leaf physiology and induces leaf pathology is the stress-causing dose (effective dose). Both the uptake and effective dose are essential to the understanding of plant response, but the exposure dose is the most functional dose concept for the air pollution system because it must be used to develop control strategies.

Evaluation of Information Sources and Other Key Reviews

This chapter makes extensive use of information in selected reviews published since 1982 that present excellent and in-depth coverage of the literature on effects of air pollutants on crops. The primary source is Heck et al. (1986a). Secondary sources include the recent criteria documents released from the U.S. Environmental Protection Agency (1986) for O_3 and the U.S. Environmental Protection Agency (1982b) for SO_2. Information relating to assessment of O_3 impacts on crops is obtained primarily from overview papers from the National Crop Loss Assessment Network (NCLAN) (Heck et al. 1982c, 1983a, 1984b, 1984c).

Other excellent reviews (Heck and Brandt 1977, Heck et al. 1977, P. R. Miller 1977, McCune and Weidensaul 1978, U.S. EPA 1978; U.S. EPA 1982b), compilations of papers (Altshuller and Linthurst 1984, Dochinger and Seliga 1976, Guderian 1985, Heck et al. 1979, Koziol and Whatley 1984, Mansfield 1976, Shriner et al. 1980, Shupe et al. 1983, Treshow 1984, Unsworth and Ormrod 1982, Winner et al. 1986), and books (Guderian 1977, Ormrod 1978, U.S. EPA 1976) permit the inquiring reader to gain a fuller appreciation of our knowledge of effects of air pollutants on crops (see table 6.3). A selected group of references presents an excellent historical perspective (Heck and Brandt 1977, Heck et al.

Table 6.3. *Recommended books and review articles.*

Subject	References
General reviews	Heck and Brandt 1977, Heck et al. 1986a, Ormrod 1978, Strain and Cure 1985, Teramura 1983, U.S. EPA 1976
Books	Altshuller and Linthurst 1984, Dochinger and Seliga 1976, Guderian 1977, Guderian 1985, Mansfield 1976, Ormrod 1978, Shupe et al. 1983, Treshow 1984, Unsworth and Ormrod 1982, U.S. EPA 1976, Winner et al. 1986
Criteria documents	U.S. EPA 1978, 1982a, 1982b, 1986
Materials on:	
Ozone	Heck and Brandt 1977; Heck et al. 1977, 1984a, 1986a; Mansfield 1976; Unsworth and Ormrod 1982; U.S. EPA 1978, 1986
Sulfur dioxide	Heck 1984, Heck and Brandt 1977, Heck et al. 1986a, Jager and Klein 1980, Linzon 1978, McCune and Weidensaul 1978, Roberts 1984, Unsworth and Ormrod 1982, U.S. EPA 1982b, Winner et al. 1986
Nitrogen oxides	Heck and Brandt 1977, Heck et al. 1986a, Nat. Acad. Sci. 1977, Unsworth and Ormrod 1982, U.S. EPA 1982a
Fluoride	Heck and Brandt 1977, Heck et al. 1986a, Mansfield 1976, Shupe et al. 1983, Unsworth and Ormrod 1982
Physiology/metabolic function	Heck and Brandt 1977; Heck et al. 1977, 1984a, 1986a; Koziol and Whatley 1984; Lechowicz 1987; Mansfield 1976; Shupe et al. 1983; Unsworth and Black 1981; Unsworth and Ormrod 1982; U.S. EPA 1986
Environmental-biological factors	Alstad et al. 1982; Heck and Brandt 1977; Heck et al. 1977, 1984a, 1986a; Koziol and Whatley 1984; Laurence 1981; Shupe et al. 1983; Unsworth and Ormrod 1982; U.S. EPA 1986
Acidic precipitation	Altshuller and Linthurst 1984; Dochinger and Seliga 1976; Evans 1984; Heck et al. 1984a, 1986a; Irving 1987; Shriner et al. 1980
Genetic variability	Heck and Brandt 1977; Heck et al. 1977, 1984a, 1986a; Mansfield 1976; Reinert et al. 1982; Shupe et al. 1983; Unsworth and Ormrod 1982; U.S. EPA 1986
Mixtures	Heck et al. 1977, 1984a, 1986a; Lefohn and Ormrod 1984; Reinert 1984; Shupe et al. 1983; U.S. EPA 1986
Growth and reproduction/yield	Heck and Brandt 1977; Heck et al. 1977, 1982a, 1982c, 1983a, 1984a, 1984b, 1984c, 1986a; Roberts 1984; Shupe et al. 1983; Unsworth and Ormrod 1982; U.S. EPA 1982b, 1986
Methodology	Guderian 1977; Heck et al. 1979, 1984a, 1986a; U.S. EPA 1986

1977, P. R. Miller 1977; McCune and Weidensaul 1978, National Academy of Sciences 1977, Shupe et al. 1983, U.S. EPA 1976). In-depth coverage is provided in publications of the National Academy of Sciences (Heck et al. 1977, P. R. Miller, 1977; McCune and Weidensaul 1978, National Academy of Sciences 1977) and in a number of the books that address effects on crops, several of which are compilation of papers presented at symposia or special meetings. The best of these include Heck et al. 1979, Koziol and Whatley 1984, Shriner et al. 1980, Shupe et al. 1983, and Unsworth and Ormrod 1982.

The most in-depth program to determine the effects of air pollutants on crops is the NCLAN program, which was initiated in 1980 and ended in 1987 with a major international conference entitled Assessment of Crop Loss from Air Pollutants. This conference had 23 invited speakers who presented the most comprehensive and up-to-date coverage of our current knowledge of air pollution effects on crops, including an economic assessment and the role of crop loss assessment in policy decisions. Their papers have been published in a special volume by Elsevier (Heck et al. 1988a). In addition, 29 contributed papers and 31 posters presented some of the most current research results in the field. These papers and summaries of the posters have been published in a special issue of *Environmental Pollution* (Heck et al. 1988b). Although NCLAN gave primary focus to O_3, the organizational approaches and methods used could be guides for future efforts on other pollutants or for additional O_3 research.

Assessment of the Impact of Air Pollutants on Crops

An assessment of impact, by implication, includes an economic aspect as a final step. Yet an in-depth assessment of effects can be done without the final economic step. The NCLAN program featured an economic analysis that utilized an assessment of crop losses as a basis for the economic assessment. In this chapter the economic aspects of a crop loss assessment are not included, although a brief summary of economic impacts is provided in a later section. Since assessment implies an economic end point, the significant crop response is the plant part intended for human use—the yield component. Air pollution/plant specialists use the terms *injury* and *damage* to separate general adverse plant response to air pollutants (injury) from a response that affects the intended use of the plant (damage), a distinction that is followed in this chapter. Thus, in assessing the impact of air pollution, one is concerned with *damage* to crops.

Needs of an Assessment Program

Local, regional, or national assessments of crop losses require three basic types of information (Heck et al. 1984b, Shriner et al.1984): a crop census (what crops are grown and their yields within a geographic unit such as a county or state); an air quality data base for use in estimating crop exposure, which covers the same geographic unit as the crop census and can be used to develop the same exposure statistic (pollutant concentration over some duration of exposure) as used for the crop response functions; and a response function relating crop yield to the exposure statistic. Although an assessment effort should use as small a geographical area as possible for greater accuracy, the primary interest is to address state, regional, and national impacts. The first approach to an assessment effort is to determine crop losses as a function of pollutant concentration for all crops where a reasonable data base exists.

The U.S. Department of Agriculture provides data on crop production on a county level every 5 years. In addition, crop production and yield data are reported each year on a state basis. These yield and acreage figures are averaged across cultivars to produce a county- or state-wide census for each crop. Therefore an assessment must utilize dose-response models that combine data across cultivars of each crop for which data are available. If the combination model adequately represents the crop response to the pollutant, the introduction of new cultivars may cause only minor changes in the assessment predictions.

The second critical component of an assessment effort is the availability of an air quality data base at the same level of detail as the crop census (county). Although monitoring data are available within the United States for SO_2 and NO_2, the data are fragmentary and occur more around point sources and in urban areas. Additional SO_2 data should be available from industrial monitoring of operations, but we do not yet know how to extrapolate such data for regional and national assessment efforts. However, a large data base for O_3 is available from the U.S. Environmental Protection Agency (EPA) from their Storage and Retrieval of Aerometric Data (SAROAD) program. This information is stored using average hourly O_3 concentrations from both urban and rural sites throughout the country.

Although most of the SAROAD data come from urban monitoring sites, several facts suggest that extrapolation over large distances will give reasonable estimates of rural concentrations: O_3 is a secondary pollutant and can be transported over considerable distances, O_3 precursors are transported over great distances and are emitted in both rural and urban areas, and the available rural monitoring data suggest that rural concentrations are similar to urban concentrations outside city centers. It is known that city-center O_3 concentrations are lower than at many rural sites due to titration by nitrogen oxide. Thus it seems reasonable to use some means of extrapolating O_3 data across large areas of the country where a reasonable number of monitoring sites exist. An interpolation technique called kriging (Heck et al. 1983b) was developed and initially applied on a national basis. Kriging is a statistical technique that estimates gridded values from irregularly spaced, generally clustered monitoring data (Heck et al. 1983b, 1985, Lefohn et al. 1987). This technique was used to interpolate the SAROAD data and provide estimates of county-level O_3 concentrations (Heck et al. 1984b; Heck et al. 1983b, 1985; Lefohn et al. 1987; Shriner et al. 1984).

In addition to the availability of a large data base, an exposure statistic (concentration and averaging time-exposure dose) must be developed that adequately explains the crop response to the pollutant of interest. The acceptance of an exposure statistic for use both as an air quality standard and in crop response models would be the most useful. However, if one statistic could be used as a surrogate for the other, different exposure statistics could be used for the standard and the model; this was the original intent for the current secondary standard (0.12 ppm hourly maximum).

The third element of an assessment is the development of pollutant dose-crop response models for each crop and cultivar. The functional form for the model should permit a reasonable biological interpretation of the results, and the exposure statistic should not be site specific. It should also be possible to use the model to test the homogeneity of data across cultivars and across sites for a given species. In the NCLAN program, a model was developed by using a seasonal mean O_3 statistic from one site to estimate yield from data obtained from a second site, or for a different year at the same site (Heck et al. 1984c). This could not be done using seasonal peak statistics for model development. The dose statistic is a critical aspect of any dose response model.

It should be noted that an in-depth assessment effort should include as much known plant response data as possible. Although much of the information on plant response highlighted in "Effects of Ozone and Sulfur Dioxide on Crops" (p 255) is not directly applicable to crop loss assessment, the information is supportive of any assessment effort. The response data included in the aforementioned section helps the scientist understand the physiological and genetic basis for crop (cultivar) sensitivity to the pollutant(s) of concern. The better these processes are understood, the more mechanistic the models can become and the better predictions the scientist can make on pollutant effects. Additional information is needed to permit a reasonable assessment of the importance of climatic and biotic factors on crop response to pollutants. It is essential to understand that plant response data obtained under both controlled environment and greenhouse conditions have been used in assessment efforts with reasonably valid results. However, field data are necessary to verify results from nonfield studies.

In summary, a regional or national assessment of pollutant impacts on agricultural crops includes the development of functional relationships (models) or pollutant dose and crop response for important crops within the area to be assessed, the development of estimates of air quality (pollutant statistics) in rural areas, and the merging of these with the crop census on yield and acreage. The result is a relative estimate of crop damage from exposure to the pollutant, expressed in yield units, that allows an economist to predict the economic losses on a regional or national basis.

Approaches to the Development of Functional Relationships

The primary goals of crop assessment research are to measure the effects of pollutants on productivity and to develop principles and techniques for exposure facilities and assessment approaches. Early assessments related loss of productivity to foliar injury using data from growth chamber, greenhouse, and closed-top field chamber studies supported with information from field surveys.

The studies used plants grown under conditions that differed from those in the field; whether these conditions affected plant response to the pollutants is not known. Results of field surveys were questionable because the relationships of pollutant dose, injury, and yield were poorly defined; also, the cause of injury was often difficult to determine. Regardless of the weaknesses in these assessment efforts, the results furnished a solid basis for early assessments and helped substantiate more current assessments based primarily on field data.

Pollutant dose–crop yield response information should be obtained from plants exposed to a range of pollutant doses that mimic ambient concentrations and grown under ambient climatic conditions common for the species/cultivar of interest. These exposure conditions will help prevent confounding of results from different environmental conditions or pollutant regimes that can affect growth and yield. A number of field approaches for the development of crop-pollutant functional relationships have been used. They include concentration gradients in ambient air, protective chemicals, comparison of sensitive and resistant cultivars within a species, open-air exposures of several types, and field exposure chambers (Heagle and Heck 1980).

Concentration Gradients. Ambient pollution gradients involve epidemiological/statistical approaches to functional relationships and should not be used to prove cause-effect relationships. The technique is useful in certain cases, provided that both careful measurement of pollutant concentration and any environmental factors that can affect plant growth are included. The gradient approach is valid where SO_2 and NO_2 doses vary within fairly short distances (point sources with stacks); it is not valid for SO_2 from tall stacks or for NO_2 or O_3 because the distances required to obtain meaningful dose differences impose environmental variations that will confound the results.

The gradient approach was used in California to correlate O_3 dose with yield of tomato (Oshima et al. 1975) and alfalfa (Oshima et al. 1976); the results indicated that the O_3 dose was more important than temperature or humidity in explaining the differences found. Although the investigators identified locations with major differences in O_3 dose and the plants were container grown, the environmental differences across the gradient could have confounded the results. For assessment efforts on a regional or national basis, the gradient method is not feasible. A method to control pollutant concentration around the test crop while providing relatively normal environmental conditions for plant growth is essential for developing functional relationships for predictive purposes.

Protective Chemicals. Various chemicals that protect plants from O_3 have been tested to determine their value in estimating crop yield losses at existing ambient levels of O_3 (Carnahan et al. 1978, Foster et al. 1983, Heck et al. 1977, Hofstra et al. 1978, Smith et al. 1987, Weidensaul 1980). The method allows for the exposure of many plants under similar environmental conditions without the use of chambers. Major disadvantages include the following: the chemical might affect plant growth and yield, dose response studies with O_3 are not possible, and the amount of protection from O_3 for different crop species is not known. Thus, the approach is not feasible for developing functional relationships necessary for assessment efforts that include a predictive capability. However, the method could be used to supplement other field information, to verify predictive capabilities of exposure dose-crop response functions, and to assess current-year impacts at different places if background information is obtained for each chemical on each crop.

For each crop, growth and yield responses should be measured under charcoal-filtered and unfiltered air conditions both with and without a chemical treatment. This would permit an evaluation of the direct effects of the chemical on the crop and the ability of the chemical to protect the crop from O_3 injury. If the results justified the effort, the chemicals could be used to add support to an assessment effort. To date, the type of supportive research necessary to utilize chemicals in assessment research has not been done. Problems in using protective chemicals in assessment efforts increase when two or more phytotoxic pollutants are present in the ambient air.

Sensitive and Resistant Cultivars. Two cultivars of a crop, similar except for their relative sensitivity to the pollutant of interest, could be used to assess the impact of a given pollutant in the field (Heagle and Heck 1980). Sensitive and resistant cultivars have been used to show that yield losses can occur from ambient levels of O_3. Both the advantages and the major disadvantages are similar to those listed for the protective chemicals. Suitable sensitive and resistant cultivars have not been identified for most species. Confounding of results would relate to the possible differential response of the two cultivars to variable environmental conditions. The concept may have value if background information similar to that suggested for the protective chemicals were developed. Correlations between yield of sensitive and resistant cultivars at different pollutant doses could result in a useful indicator network. Again, the presence of two or more phytotoxic pollutants in the ambient air could confound results.

Open-Air Exposure Systems. Several such systems are able to control pollutant concentrations around plants growing under field or near field conditions. In one technique the pollutant is released into the air from holes in pipes. The first such system (Lee et al. 1978) released SO_2 through horizontal pipes at canopy height simultaneously from the four sides of a field plot area. Improved SO_2 control and the ability to compensate for variable wind speeds were accomplished using computer control of SO_2 release from a variable upwind side of the experimental field (Greenwood et al. 1982). Similar systems have used horizontal release pipes at intervals within a field, a single horizontal release pipe at one side of the field, and vertical release pipes at regular intervals throughout an experimental area. A detailed description of a computer-controlled open-air system has been provided (McLeod and Fackrell 1983).

All open-air systems have the advantage of growing plants under normal field practices. However, control of pollutant levels is difficult because there are changes in wind direction and velocity, because the experimental plot should have a number of continuous monitors, and because a control treatment cannot be provided for ubiquitous pollutants such as O_3.

A variation of the full open-release system is called the linear gradient system (Shinn et al. 1977), which uses plastic tubes along crop rows to exclude or add pollutants. Varying the number of release points along the tubes allows a gradient of pollutant concentration to be developed. The approach is a hybrid between the true open-release system and the open-top chamber system (discussed next). It has some attributes of both systems but the disadvantages are similar to those stated for the other open release systems. Olszyk et al. (1986b) designed an air exclusion system for low growing plants that performed essentially as well as the open-top chambers described in the next section. There is a need to continue studying and experimenting with exposure systems to assure that experimental methods are characterized and stay close to natural systems.

Open-Top Field Chambers. This technique uses cylindrical open-top chambers for field exposures (Heagle et al. 1973, 1979; Mandl et al. 1973). These chambers provide controllable pollutant concentrations. The uniformity of concentration depends on the height of the chambers, the presence of a top attachment (frustum, which decreases ingress of ambient air), and wind speed. The most widely used chamber is 3 m diameter by 2.4 m tall without a frustum. Several reports discuss the effectiveness of these chambers and the chamber effects on microclimate (Heagle et al. 1973, 1979; Unsworth et al. 1984a, 1984b). The microclimate changes reported include a constant wind speed (2.5 κm hr⁻¹), reduced daytime wind speed, a 10 to 15 percent decrease in light intensity, and a maximum 1–3° C rise in daytime temperature. The chambers may, but have generally not, affected the yield of the species tested (Heagle and Heck 1980; Heck et al. 1982c) and are now in use at many field sites within the United States.

Approaches to the Development of an Acceptable Pollutant Statistic

A single exposure statistic for O_3 or SO_2 to relate plant response to pollutant concentration over a growing season has not been accepted. It is known that the severity of plant response is affected more by pollutant concentration than by exposure duration. Thus, a similar seasonal mean concentration should produce a greater effect if the mean includes a number of high pollutant concentrations rather than many medium-high concentrations. Although this concept has not been verified in field studies, it has validity based on greenhouse and control chamber exposures. It is possible that no single exposure statistic for a given pollutant will be adequate for all crops under all environmental conditions.

Several investigators have identified a threshold value, summed all hourly values over that threshold for the exposure duration, and used that summation as the exposure statistic (Oshima et al. 1976). This approach does provide a range of doses to define the response curve, but the response of the test plants to the seasonal exposure cannot be interpreted. One approach that has gained many advocates is to use the frequency distribution of pollutant concentrations and to define the exposure statistic as the percentage of time the critical pollutant concentration is exceeded. This permits assessment of single- or multiple-peak concentrations but it is not possible to identify when the peaks occur during the growing season. The NCLAN program used a seasonal 7 daily mean O_3 exposure statistic and included a frequency distribution of the daily 7-hr and 24-hr average ambient concentrations in time sequence through the growing season (Heck et al. 1982c). This time-sequenced frequency distribution permits a discussion of the importance of peak O_3 concentrations at different growth stages. Later, the NCLAN scientists utilized a seasonal 12-hr daily mean O_3 statistic with the associated frequency distribution (Heagle et al. 1987, Heck et al. 1986b).

This section develops a recommended approach to the formulation of acceptable exposure statistics for both O_3 and SO_2. Although it is important to understand that the plant actually responds to the effective dose and not to the exposure dose, in an assessment effort we must use some measure of the exposure dose since that is the concentration over time that can be routinely measured.

An Acceptable Ozone Exposure Statistic. Yield loss is caused by the cumulative effects of chronic daily O_3 exposures over the growth period. In general, chronic O_3-induced foliar injury is difficult to distinguish from normal foliar senescence associated with leaf age and crop maturity. Early senescence was observed for crops tested in open-top field chambers without charcoal filtration. Early senescence decreases photosynthetic activity in plant leaves, thereby decreasing plant growth rate, total photosynthate production, and final yield. Usually periods of relatively high O_3 concentration (peak values of 0.10–0.15 ppm) are interspersed with periods of lower concentration. Although the effects of a high O_3 concentration may be severe, a 1- to 2-day exposure with peaks above 0.10 ppm would not affect the yield of most plants unless additional episodes occurred. The cumulative and possibly interactive nature of the plant response to daily O_3 exposure was shown for the sensitive Bel-W$_3$ tobacco cultivar (Heagle and Heck 1974). Over eight different seven-day periods, plants exposed to ambient O_3 for 7 consecutive days developed more than twice as much foliar injury as the sum of the injury in seven groups of plants that were exposed for 1 day each for 7 days. Thus, preexposure to O_3 appeared to sensitize the tobacco cultivar to subsequent exposure to O_3.

Results to date suggest that crop response and the fluctuation of O_3 concentration should be considered in formulating an exposure statistic. Although a rationale could be made for accepting any of a number of exposure statistics, a simple statistic may be best for implementing the ozone standard. Whatever is used, a mean value that considers the fluctuation in O_3 concentration may be best.

The NCLAN program identified four mean statistics that appeared usable in defining the O_3 statistic for interpreting the yield response functions (Heck et al. 1984c). These were the seasonal mean of the daily 7-hr (0900–1600 ST) mean O_3 concentration, the seasonal mean of the daily maximum 1-hr means, the peak (maximum) daily 7-hr mean O_3 concentration over the season, and the peak 1-hr mean O_3 concentration over the season. These four statistics were compared utilizing several NCLAN data sets. The seasonal mean values were better predictors than the maximum peak values and were recommended as the best dose statistics to use.

From results to date, the use of some seasonal mean O_3 value that uses a fixed number of maximum hourly values occurring during the light hours is recommended. This value should appear with the same daily mean and the 24-hr mean in time sequence across the growing season. Although other exposure statistics could be used, these give an adequate presentation of the O_3 to which plants are exposed and no other statistic has been shown to be superior.

An Acceptable Sulfur Dioxide Exposure Statistic. Natural sources of SO_2 do not appear to be a significant factor, thus monitored levels of SO_2 are of anthropogenic origin. The absence of monitoring data in rural areas away from point sources is more acute than for O_3. Where data are available, concentrations away from point sources rarely exceed 0.01 ppm (U.S. EPA 1982b). Thus agricultural areas of concern are in the vicinity of single or multiple large sources (power plants and smelters), where concentrations in excess of 0.10 ppm are common. The lack of regional monitoring sites makes generalizations difficult, but available data suggest that a kriging approach to estimates of regional SO_2 concentrations would not be valid. Rather, if the required monitoring data around major point sources could be used and modeled to show the extent and pattern of expected SO_2 concentrations around the source or sources, this information could be utilized in a modeling approach to assess SO_2 impacts on crop production; this model would incorporate a number of meteorological variables.

In evaluating SO_2 effects on crops, one must consider the concentration/time/frequency relationships for different situations (for example, urban-industrial multiple sources, point sources with low and high stacks, and topographical differences associated with sources). Experimental research has involved distance from source, open-field release, field chamber designs, and greenhouse exposure chambers. Experimental exposure regimes have included natural regimes around sources, simulated natural regimes using open-field release or computer-programmed controlled releases, and constant additions (giving a square wave in greenhouse chambers and fluctuations around the mean in open-top chambers). The development of a dose metric or a statistic for use in the interpretation of the biological response has been similar to those developed for O_3, with some type of mean concentration being the most common. However, summations of ppm-hr, frequency distributions, averages over time of occurrence, and long-term averages have also been used. For ambient exposures, mean SO_2 concentrations across the time of occurrence for all concentrations above 0.10 and 0.01 ppm should be considered. This information should be used in conjunction with a time-related frequency distribution figure. For controlled exposures, averages for elapsed times should be used with length of exposure and number of exposures included.

It may be impossible to develop a single biologically meaningful dose metric that will apply in all circumstances. There is certainly a need for further innovative efforts in the analysis of SO_2 monitoring data for use in experimental designs and crop loss assessment evaluations.

Effects of Ozone and Sulfur Dioxide on Crops

This section presents concise summary information on the effects of O_3 and SO_2 on crop growth, physiology, and productivity. Most of the results included are from research done since 1970, although selected earlier results have been included. The extensive literature attests to the importance of O_3 as a regional/national problem. From the early 1950s to the 1970s little research on the effect of SO_2 on crops was done, as it was generally accepted that these effects were understood and that the importance of SO_2 on a national level was decreasing because of control measures. New interest in SO_2 was stimulated by increased power generation and the subsequent increased SO_2 emissions, and by the results of long-term assessments around point sources. The results convinced the scientific community that earlier perceptions of SO_2 thresholds were flawed and would have to be reexamined. This section provides background information on plant response, including some review of field research that supports efforts to assess pollutant dose and crop yield responses for assessment purposes. The final section presents functional relationships for O_3 that can be derived from the NCLAN data base.

Symptomatology

The effects of O_3 and SO_2 on crops is either visible or subtle. Visible effects are morphological, pigmented, chlorotic, or necrotic foliar patterns resulting from major physiological disturbances in plant cells. Subtle effects are measurable growth or physiological changes without visible injury that may affect yield or reproductive or genetic crop systems.

Visibile injury is either acute or chronic. Acute symptoms are associated with short exposures (measured in hours) to relatively high concentrations of O_3 or SO_2 and usually appear within 24 to 48 hr after exposure. Chronic symptoms are associated with long-term or intermittent exposures (measured in days) to lower concentrations.

Acute injury from both O_3 and SO_2 results in cell plasmolysis, then death and collapse of the tissue. Injury is characteristic for a pollutant: random necrotic foliar lesions often bleached white are typical of SO_2 injury, whereas O_3 injury often appears as flecks (small, bleached necrotic areas) or stipple (small pigmented areas) on upper foliar surfaces (see table 6.4). Often, the first symptom is a water-soaked or bruised area or areas on foliar surfaces. The affected area may recover or dry out and produce a necrotic pattern characteristic of O_3 or SO_2. Chlorosis may be associated with acute exposure to O_3 or SO_2.

Chronic injury may be mild or severe but does not immediately result in cell death. Initial disruption of cellular activity may be followed by chlorosis or other color or pigment changes that may eventually lead to cell death. Subtler effects may induce early senescence with or without leaf abscission. Chronic injury patterns are generally not characteristic of O_3 or SO_2 exposure and are easily

Table 6.4.   *Acute foliar injury patterns from ozone and sulfur dioxide exposure.*

Ozone	
Leaf markings	divided into categories by plant type:
Broadleaf	upper surface, pigmented, red-brown spot (stipple); bleached tan to white areas (fleck); small irregular bifacial collapsed (necrotic) areas that may coalesce to form irregular necrotic blotches; chlorosis and premature senescence may occur
Grasses	scattered bifacial necrotic areas (fleck); sometimes larger lesions or necrotic streaking may occur
Similar markings	red spider mite and certain insects may cause an upper-surface fleck; some leaf spot fungi may give similar patterns; severe ozone-induced lesions may resemble those caused by sulfur dioxide; chronic injury may resemble normal leaf senescence
Sensitive plants	oat, petunia, pinto bean, potato, radish, soybean, tobacco, tomato
Resistant plants	beet, geranium, gladiolus, mint, pepper, rice
Sulfur dioxide	
Leaf markings	divided into categories by plant type:
Broadleaf	irregular, bifacial, marginal, and interveinal necrotic areas bleached white to tan or brown; chlorosis may be associated with necrotic areas, or a general chlorosis of older leaves may develop; diffuse to stippled colors ranging from white to reddish-brown have been observed
Grasses	irregular, bifacial, necrotic streaking that is bleached light tan to white between larger veins; chlorosis usually is not pronounced
Similar markings	white spot of alfalfa, leafhopper injury, rose chafer injury, various mosaic viruses, cherry leaf spot, and other fungal diseases producing blotchy markings; Victoria blight on oats, bacterial blight of barley and other grains; terminal bleaches in cereals
Sensitive plants	alfalfa, barley, cotton, squash, wheat
Resistant plants	cantaloupe, celery, corn

Source: W. W. Heck and C. S. Brandt, 1977. Effects on vegetation: Native, crops, forests. In *Air Pollution*, ed. A. C. Stern, 3d ed. New York: Academic Press.

confused with symptoms caused by diseases, insects, other stresses, and normal leaf senescence.

Descriptions of injury to crops from O_3 and SO_2 are found in many reviews and original papers. The most complete descriptions for both pollutants are found in the 1970 pictorial atlas (Jacobson and Hill 1970). Descriptions are found in other review articles and criteria documents (Heck and Brandt 1977; Thomas 1961; U.S. EPA 1982b, 1986). The most complete discussion of SO_2 effects is in the German atlas (van Haut and Stratmann 1970).

Acute foliar injury can be symptomatic of both O_3 and SO_2 exposure and thus, for trained observers, is a good field indicator of cause-effect relationships. However, since chronic foliar injury may be mimicked by other biotic and abiotic environmental stresses or may resemble normal senescence, it is not a good field diagnostic tool. From a practical standpoint, visible foliar injury is the only conclusive way to identify O_3 or SO_2 injury in the field. Research has shown relationships between visible injury and growth and yield for many plant species (Heck et al. 1977). Such relationships show high correlations where experiments are designed for that purpose but do not necessarily correlate well if the experiments are designed for other purposes.

Physiological and Biochemical Effects

Changes in chlorophyll content and release of "stress" ethylene are two objective indicators for quantifying crop response to O_3 or SO_2. An inverse relationship between increasing O_3 concentration and foliar chlorophyll (Reich et al. 1986) and a positive correlation between loss of chlorophyll and visible injury are commonly found in controlled exposures (Knudson et al. 1977). Similarly, ethylene release from plants under controlled conditions correlates with O_3 concentration (Tingey 1976). Chlorophyll reduction and the release of stress ethylene are biochemical responses of crops to different stresses. Under controlled conditions correlations between these responses and O_3 or SO_2 should be high. Under field conditions, however, using these responses as a specific test for O_3 or SO_2 is not possible because of the presence of many other stress factors.

Ozone. Ozone enters leaves through stomata. The flux of O_3 into leaves may be measured more effectively using leaf resistance techniques than micrometeorlogical techniques since the latter do not separate flux to soils and other surfaces from flux to the leaves (Leuning et al. 1979). The effect of O_3 on stomatal closure was more important than stomatal number in determining sensitivity between a resistant and a sensitive bean cultivar (Butler and Tibbitts 1979). Partial stomatal closure was reported for bean exposed to a mixture of O_3 plus SO_2 (Kobriger et al. 1984). No relationship was found between O_3 sensitivity, stomata number, or gas exchange rate for two sweet corn cultivars (Harris and Heath 1981) or two soybean cultivars (Taylor et al. 1982). Reich et al. (1985) reported no interaction among O_3, leaf age and water stress in control of leaf conductance in soybean, and a linear decline in conductance with increasing ozone. Olszyk and Tingey (1986) found a fairly complex relationship between stomatal conductance in garden pea and various combinations of O_3 and SO_2 with O_3 being more effective than SO_2 in inducing stomatal closure. No single factor appears related to stomatal control of plant response to O_3, but any stress causing stomatal closure reduces O_3 uptake and thus protects the plant. Stomatal closure is a resistance mechanism referred to as avoidance, as opposed to tolerance (the plant's ability to withstand some level of O_3 within the cell system).

The fate of O_3 after entry into plant leaves is not known. Ozone gas- and liquid-phase pathways into the leaf may be different from those of water vapor and CO_2 (Taylor et al. 1982). Some fraction of O_3 may pass through the cell membrane (Mudd et al. 1984), O_3 may react with protein or lipid membrane components (Heath 1980), or free radical products of O_3 activity within the substomatal cavity may react with membrane components (Grimes et al. 1983). Regardless of the exact mechanism of action, the cell membrane is probably the site of initial O_3 reaction as evidenced by lysis and fluorescein diacetate staining to potato (Illman and Pell 1985), by the leakage of potassium and other electrolytes (Keitel and Arndt 1983, Heath and Frederick 1979), ultrastructural changes (Athanassious 1980), evidence of membrane repair (Sutton and Ting 1977), and the induction of senescence (Pauls and Thompson 1981). Rhoads and Brennan (1978) reported that chloroplast electron transport was inhibited more in the leaves of O_3-sensitive Bel-W_3 tobacco than in the more tolerant Bel-B, but that isolated chloroplasts from a study of the two tobacco cultivars suggest that resistance is related to the structure or physiology of the cell membrane that affects O_3 movement into the cell interior.

Changes in metabolite levels occur after O_3 exposure and in some cases the results differ between susceptible and resistant cultivars. Metabolite changes related to O_3 exposures include changes in ribulose–1,5-bisphosphate carboxylase, total glycoalkaloids, isoflavonoid production, total carbohydrate, mineral content, nitrate reduction, phospholipid composition, photosynthetic pigments, lipids, and peroxidase (Heck et al. 1986a). Increased peroxidase activity did not help identify O_3-sensitive tobacco selections (Petolino et al. 1983) but may serve as a marker in sweet corn (Podleckis et al. 1984). Ozone lowered foliar concentrations of Ca, Mg, Fe, and Mn and increased the concentrations of K, P, and Mo in the pods (Tingey et al. 1986).

Mechanisms of O_3 action are not known but may be related to results from a number of experiments. Ozone depression of nitrate reduction in soybean was reversed by sucrose additions, suggesting that O_3 decreased available photosynthate and did not directly inactivate the nitrate reductase (Purvis 1978). Ozone-induced loss of superoxide dismutase and L-ascorbate in spinach leaves resulted in higher levels of active oxygen radicals that subsequently led to loss of chlorophyll-a and carotenoids (Sakaki et al. 1983). Ozone resistance has been associated with high concentrations of metabolites, compact cell arrangements, higher concentrations of reducing sugars, lower osmotic potential, lower transpiration rates, higher glutathione reductase activity, and increased levels of superoxide dismutase (Heck et al. 1986a).

Research supports the conclusion that primary crop effects are associated with direct effects of O_3 (or oxygen radicals) on plasma membrane integrity. Secondary effects may be associated with a variety of oxygen radicals that overpower the plant's defensive mechanisms. These two phenomena may be viewed as separate mechanisms of O_3 response but they may phase together. The membrane response may induce acute symptoms; both primary and secondary responses may initiate chronic symptoms that resemble normal leaf senescence. In summary, both stomatal control (avoidance) of gas exchange and plant physiological factors (tolerance) control plant response to O_3.

Sulfur Dixiode. Unsworth and Black (1981) presented an excellent overview of available information on stomatal response to SO_2. Low concentrations of SO_2 can injure epidermal and guard cells, leading to increased entry of SO_2 into some crop species. In other species, stomatal conductance is reduced and less SO_2 enters the plant. These contrasting effects may be related to the susceptibility of the crop cultivar (sensitive versus resistant cultivars). Alternatively, stomatal closure may be related to high SO_2 concentrations and stomatal opening to low concentrations (Unsworth and Black 1981). Temple et al. (1985a) reported a decrease in stomatal conductance in pinto bean with increasing SO_2 to 0.5 ppm; conductances returned to normal a day after termination of exposure. A study using sunflower (Omasa et al. 1985) exposed to 1.5 ppm of SO_2 compared conductance and width of stomatal aperture; these related well, showing reduced conductance and closure over time with the SO_2 concentration used; stomatal closure was more rapid in uninjured areas of the leaf.

Estimates of SO_2 fluxes into leaves vary with species, conditions of exposure, SO_2 concentration, and possible entry routes (stomatal, cuticular) into the leaf. Based on meteorological and plant canopy conditions affecting SO_2 flux (deposition) to plant tissues, the use of an average stomatal resistance was suggested (Wesley and Hicks 1977). Sulfur $^{35}SO_2$ has been used to determine uptake rates and the distribution of ^{35}S in plant tissues (Garsed and Read 1977); the ^{35}S was translocated through the plant and was found in organic compounds.

Sulfur dioxide reduced net photosynthesis, increased dark respiration (which could reduce net photosynthesis), and increased transpiration. Results were generally the same for all species studied at many SO_2 concentrations for short- and long-term exposures. Rates of these processes returned to preexposure levels soon after termination of exposures that lasted up to several days (Heck et al. 1986b). In sunflower, the rate of reduction of net photosynthesis caused by SO_2 was related to leaf age (Furukawa and Totsuka 1983). A slight increase in both net photosynthesis and transpiration at low SO_2 concentrations was reported for short time periods, followed by a decrease in both processes (Taylor and Selvidge 1984); higher SO_2 concentrations induced immediate decreases in both processes.

Increasing interest in understanding SO_2 effects on crops has led to investigations of enzyme systems. These studies show the extent to which changes in enzyme activity are affected by SO_2 concentration, plant species, environmental factors, and plant age. The enzymes studied include ribulose- bisphosphate carboxylase, glutamate-pyruvate transaminase, glutamate- oxaloacetate transaminase, peroxidases, alanine aminotransferase, aspartate aminotransferase, glutamate dehyrogenase, glucose–6-phosphate dehydrogenase, isocitrate dehyrogenase, malate dehyrogenase, adenosine 5′-phosphosulfate sulfotransferase, glycolate oxidase, glyceraldehyde–3-phosphate dehydrogenase, ribulose–5-phosphate kinase, and fructose–1,6-bisphosphatase (Heck et al. 1986a). In some cases, enzyme activity is increased by exposure of the plants to low levels of SO_2 and decreased by higher SO_2 concentrations.

Plant metabolism is affected by SO_2. Phosphorus metabolism is stimulated by SO_2 (Plesnicar 1983), and carbohydrate levels are increased by low concentrations and decreased by higher SO_2 concentrations (Koziol and Jordan 1978). Photosynthesis is suppressed and photophosphorylation is inhibited after exposure of C_3 and C_4 plants to 2 or 4 ppm of SO_2 for 1 minute; the significance of these findings is unclear. Griffith and Campbell (1987) reported that SO_2 concentrations that do not cause visible symptoms can interfere with carbon metabolism and transport in snapbean. An interesting note on differential phloem loading, as measured by carbon-11, was shown for C_3 and C_4 plants when exposed to high concentrations (ppm levels) of SO_2 for short periods; an immediate reduction in phloem loading was noted in C_3 but not in C_4 plants (Menchin and Gould 1986).

The mechanism of crop response to SO_2 has been studied by using sensitive and resistant cultivars of the same species or by comparing sensitive and resistant species. Sensitivity differences in four cultivars of cucumber were related to SO_2 uptake (stomatal activity), but leaves of different sensitivity on the same plant involved a biochemical or developmental resistance mechanism (Bressan et al. 1978) related to the formation and loss of hydrogen sulfide (H_2S); resistant leaves lost H_2S more rapidly than did sensitive leaves (Sekiya et al. 1982). Evidence for the photodetoxification of SO_2 was associated with increased injury to plants on which stomata remained open during dark exposures to SO_2; a strong case was presented for sulfite as the primary phytotoxicant for acute plant injury (Olszyk and Tingey 1984). Plant physiological and biochemical processes are probably more important controllers of plant resistance to SO_2 (tolerance) than is control of gas entry via the stomata (avoidance).

Factors Affecting Plant Response

Many biological and physical factors are known to affect the response of plants to O_3 and SO_2; these include genetic (cultivar and species differences, effects on reproductive structures, inheritance of sensitivity); biological (plant diseases, insects); environmental [climatic (temperature, light, humidity) and edaphic (nutrition, soil moisture)]; and chemical (herbicides, insecticides, special additives) factors. This section includes a discussion of our current under-standing of these factors, and tables highlight results of different investigations.

Genetic Factors. Research of a genetic nature has concentrated on the deter-mination of relative sensitivities of species and their genotypes, and on the effects on pollen and pollen germination of both O_3 and SO_2. Little attention has been given to understanding the mechanism of inheritance, although some infor-mation is provided by Reinert et al. (1982a). Research has dealt with the effects on pollen, meiosis, and adaptation; inheritance of resistance; and cultivar screens.

Ozone. The sensitivity of in vitro germinating pollen to O_3 correlated with the O_3 sensitivity of the pollen parent (Feder 1981); O_3-resistant and O_3-sensitive pollen was identified from tomato, petunia, and tobacco cultivars and used as a bioassay for atmospheric O_3. Significant meiotic damage to chromo-somes during meiosis was found when *Vicia faba* (field bean) pollen was ex-posed to 2 ppm of O_3 for 4 hr (Janakiraman and Harvey 1976); pollen chromoso-mal damage may not occur under field conditions but physiological injury may.

A summary of selected studies on the heritability of O_3 resistance is given in table 6.5; the studies suggest that O_3 resistance is heritable. Heritability may involve several major genes, with partial control by other genes. Heritability of resistance may be both dominant and recessive, even in the same species (bean), suggesting that heritability of O_3 is poorly understood (Guri 1983). Damicone and Manning (1987) evaluated 35 soybean genotypes (parental and F1, F2, and backcross populations) and found that foliar injury to O_3 was qualitatively inher-ited; they suggested that breeding for O_3 tolerance in soybean was possible. Dragoescu et al. (1987) developed an autotetraploid model for use in analyzing tolerence in potato to O_3; they found that O_3 tolerance was determined by genes with a dominant tendency and that breeding programs could be used to develop tolerant genotypes. Environmental factors present during growth and exposure of the plants may affect O_3 heritability in bean (Hucl and Beversdorf 1982). The heritability of O_3 resistance will receive increasing attention as the agricultural community becomes aware of the extent of crop losses associated with O_3.

Reinert et al. (1982a) also compiled an excellent summary of cultivar screen-ing efforts through 1980. Table 6.6 highlights screening efforts since 1975. Re-sults suggest that such screens, when dealing with extremes of sensitivity, main-tain the same relative separations when longer-term exposures are performed. Species have been tested under both field and controlled conditions. Results indicate that generalizations based on a single exposure screening effort should be verified through seasonal exposure to chronic O_3 doses, using yield as the response measure. In general, extremes in sensitivity appear to keep their rela-tive ranking for crop species for both acute screens and chronic studies.

Considerable research has gone into screening bean germplasm for sen-sitivity to O_3 (table 6.6) because bean, as a species (*Phaseolus vulgaris*), is an important world food source; bean is a good experimental plant unit and can be

Table 6.5. *Genetic control of ozone sensitivity in selected studies.*

Test plant	Plant selections	Ozone exposure characteristics	Response
Tobacco	7 parents, 21 hybrids	0.50 ppm, 4hr (greenhouse)	Heterotic response of hybrids was to increase O_3 susceptibility
Bean, snap	2 sensitive, 2 resistant; F_1, F_2 generations	1.33 ppm, 1 hr (controlled)	Heritability of resistance to O_3 was 0.83; O_3 resistance was recessive and regulated by a few major genes
Bean, snap	2 sensitive, 2 resistant, F_1, F_2 generations	0.28–0.32 ppm, 8 hr (controlled)	Ozone resistance was dominant; two interacting genes control trait; many genes have minor control
Bean, snap	1 sensitive 3 resistant; F_1, F_2 generations	0.41 ppm; 6 hr/day, 2 days (controlled)	Heritability of resistance was 0.66–0.88 (control chambers) and 0.16–0.21 (in the field)
Potato	14 cultivars F_1, hybrids	0.40 ppm, 3 hr (controlled)	Dominance for resistance fit model best, general (70%) and specific (20%); combining abilities were significant
Tall fescue	6 parents, F_1 hybrids	0.50 ppm, 3 hr (controlled)	General combining ability and reciprocal effects were significant; additive genetic variance suggests that breeding for O_3 resistance is possible
Petunias	7 inbred lines, F_1 generation	Ambient at max of 0.3 ppm, 4 hr; controlled at 0.25–0.50 ppm, 11 hr	Additive gene action for ambient, partial dominance for sensitivity
Sweet corn	Inbred lines; F_1, F_2 generations	Field exposures	High degree of inheritance for resistance (r) and susceptibility (s); sXr gave mostly susceptible F_1; rXr gave r F_1 and F_2; incomplete dominance for susceptibility

Source: W. W. Heck, A. S. Heagle, and D. S. Shriner, 1986. Effects on vegetation: Native, crops, forest. In *Air Pollution,* ed. A. S. Stern, vol. 6. New York: Academic Press; individual references are found in the original table.

Table 6.6. *Relative cultivar sensitivity of selected species to ozone.*

Test plant	Number of cultivars or genotypes	Ozone exposure characteristics	Results
Bean, snap	387 127	Field ambient 0.35 ppm, 3 hr (controlled)	270 resistant, 31 sensitive Good separation of cv was obtained [injury index of 3.8–8.2 (of 10.0) for a 20-cv sample]; correlation with field results was low ($R^2 = 0.20$) but significant
Bean, snap	>2000 Plant introductions	0.60 ppm, 2 hr (controlled)	Identified 54 resistant and 67 highly sensitive introductions
Turfgrass species	8 species	0.10 ppm; 3.5 hr/day, 5 days (controlled)	Warm season (2 species) were most tolerant; cool season (6 species) most sensitive
Potato	59	Field, 2 locations in Ohio	Broad range of resistance: later maturing were more resistant; relative resistance was similar across tests
Soybean	4	Acute: 0.15, 0.30, 0.45, 0.60, 0.75 ppm; 1.5 hr Chronic: 0.03, p.06, 0.09, 0.12, 0.15 ppm; 6 hr/day, 10 days (greenhouse)	Relative resistance affected by concentration, duration, and measure of response
Soybean	4	0.025 and 0.101 ppm; 7 hr/day, 3 months (field, open-top chambers)	Ransom was not predictable but Bragg, Forrest, and Davis kept same relative sensitivity
Soybean	35	Field ambient; 124 hr above 0.06 ppm for 1982; 372 hr above 0.06 ppm for 1983	Nineteen genotypes were evaluated as sensitive; 16 were tolerant. (Damicone and Manning 1987)
Sweet corn	20	0.25 ppm, 3 hr (controlled)	Three genotypes were sensitive and four were tolerant; peroxidase markers may make a good screening technique (Podleckis et al. 1984)
Tobacco	7	0.10, 0.15, 0.20, 0.25 ppm; 2 hr (controlled)	Relative sensitivities were similar in controlled exposures and field studies
Tomato	6	0.40 pm, 1 or 2 hr, 3 ages, 2 ex-	Relative sensitivity changed with age; plants

(continued)

Table 6.6.   (Continued)

Test plant	Number of cultivars or genotypes	Ozone exposure characteristics	Results
		posures (green-house)	were most sensitive at 4 weeks; relative rankings were different, depending upon whether response was growth or injury
Watermelon	14	Field ambient; above 0.05 ppm, 9 hr/day, during growing season	Relative injury varied at two study sites; all cvs were injured; a range of injury was noted with Sugar Baby being the most sensitive (Decoteau et al. 1986b)

Source: See table 6.5.

grown under a variety of conditions; the gene pool is broad and includes sensitive and tolerant selections; since bean is a legume, nitrogen fixation can be studied; and commercial bean cultivars may be as sensitive to O_3 as the cultivars of any crop species. Heck et al. (1988c) identified two sensitive and two tolerant bean cultivars from a massive screening effort and exposed them in field chambers to different O_3 concentrations; results of regression analysis showed that the two sensitive cultivars gave homogeneous response functions, as did the two resistant cultivars. However, the sensitive and tolerant cultivars were different and maintained the same relative sensitivity shown in the screening results.

Sulfur dioxide. Sulfur dioxide inhibited pollen germination and germ tube growth on artificial media, and moist pollen was more sensitive than dry (Varshney and Varshney 1981). Pollen germination was also inhibited in wild geranium, resulting in reduced seed production (Dubay and Murdy 1983a) and in *Lepidium virginicum* with no effect on seed set (Dubay and Murdy 1983b); pollen tube growth in the style was not affected in either species.

The relative sensitivity of several plant cultivars or species to SO_2 is shown in table 6.7. Relatively high SO_2 exposures were used in most of these studies; thus the resultant cultivar separations may not be representative of results from chronic exposures. One study utilized chronic SO_2 exposures over a number of days to determine the relative sensitivity of six species (Markowski et al. 1975). Although barley was generally the most sensitive, no consistency was found across species when different response parameters were used. Corn was more resistant than pea due partly to increased SO_2 uptake by pea, but biochemical mechanisms were involved (Klein et al. 1978). Pande (1985) exposed five barley cultivars to different SO_2 concentrations over 3 weeks and found stomatal re-

Table 6.7. *Relative susceptibility of selected plant species and cultivars to sulfur dioxide.*

Test plant	Number of cultivars (cv) or species	Sulfur dioxide exposure characteristics	Results
Poinsetta	17	3.0 ppm, 3 hr (controlled)	Bracts were generally less sensitive than true leaves; all cv were injured, most about 50%
Coleus	15	1.5, 2.3, 3.0 ppm; 4 hr (controlled)	Nine cv were sensitive; found wide range in tolerance
Turfgrass species	6	2.0 ppm, 6 hr (controlled)	Injury was 4–67% with good separation; relative sensitivity stayed the same in two tests
Marigold	39	1.0 ppm, 4 hr; 2.0 ppm, 2 hr (greenhouse)	Injury was 0–42% with only one cv showing no injury (1.0 ppm, 4 hr); generally cv were up to 10 times more sensitive at 2 ppm for 2 hr
Tomato	26	1.0 ppm, 3 hr; 2.0 ppm, 2 hr (greenhouse)	Injury 17–71% with good cv separation
Selected species	6	0.3 or 0.5 ppm; 5 hr/day, 12 or 16 days	No consistent separation; studied visible injury, biomass, chlorophyll, height
Barley	12	0.35 ppm, 9 hr (controlled)	Injury 2–12% with cv separation
Barley	5	0.04, 0.08, 0.12 ppm; 3 weeks	Relative sensitivity varied with SO_2 concentration but cvs maintained their same relative sensitivity (Pande 1985)

Source: See table 6.5.

sistance was highest in the most tolerant cultivar (Midias) and was lowest in the most sensitive cultivar (Koru), suggesting stomatal control of sensitivity.

Perennial ryegrass populations from areas of low and high SO_2 concentration were compared, and results support the hypothesis that populations growing in areas of high SO_2 have evolved greater tolerance to SO_2 (Horsman et al. 1979). Additional support was found in five grass species growing in areas differing in SO_2 concentration (Ayazloo et al. 1982); the authors suggested that evolution of tolerance could occur within 17–25 years. Additional research using lawn grasses showed tolerance developing in 3 or 4 years from sowing near an SO_2 source; the tolerance was through a rapid selection process and was lost as the SO_2 concentrations were reduced (Wilson and Bell 1985). It is probable that the more sensitive selections were not killed, just depressed, during the higher SO_2 years and recovered as the SO_2 decreased.

Morphological and physiological studies suggested that acute injury differences may be avoidance (stomatal control) and that chronic injury differences may have a biochemical mechanism (Ayazloo et al. 1982). Sulfur dioxide flux does not explain differential sensitivity in *Geranium carolinianum* that is genetically controlled and quantitatively inherited; a physiological-biochemical basis is suggested (Taylor and Tingey 1981). In earlier work on infraspecific variation of SO_2 responses in *G. carolinianum*, Taylor (1978) concluded that inheritance of SO_2 susceptibility is quantitative and that neither tolerance nor susceptibility is dominant.

Biological Factors. Interrelationships between biological stresses (insects, diseases) and plant response to air pollutants must be understood as part of a crop loss assessment effort. Available information (Koziol and Whatley 1984, Unsworth and Ormrod 1982) regarding effects of O_3 and SO_2 on plant parasites suggests that obligate fungal parasitism is generally inhibited by O_3, and that some facultative parasites may benefit; effects are probably indirect through the host. Foliar parasites may provide localized protection of the host from O_3 injury.

Alfalfa resistance to pea aphid was not affected when plants were exposed to O_3 (Elden et al. 1978), and tobacco inoculated with a root-knot nematode was as susceptible to ambient O_3 as the noninoculated plants (Bisessar and Palmer 1984). The tomato pinworm developed faster on tomato plants injured by O_3 and survived better (Trumble et al. 1987); effects were related to more readily available N in O_3-injured plants. Mexican bean beetle larva were fed preferentially on soybean foliage exposed to chronic concentrations of SO_2; they grew larger, had lower mortality rates, and produced 50 percent more progeny after one generation than larva fed on foliage not exposed to SO_2 (Hughes et al. 1983). Hughes et al. (1985) found maximum beetle weight gains when soybean was exposed to 0.30 ppm SO_2 for 24 hr with a threshold effect between 0.025 and 0.050 ppm SO_2. Glutathione (GSH) levels increased in soybean leaves in the same way as the insect response and may be a good predictor of insect feeding strength on soybean (Chiment et al. 1986). Similar field (Chappelka et al. 1988) and greenhouse (Endress and Post 1985) studies with O_3 found preferential feeding of Mexican bean beetles on soybean foliage with increasing but chronic concentrations of O_3. It is not known whether the beetle could be a more destructive pest on soybean in areas of chronic O_3 and/or SO_2 pollution. A similar stimulation of growth of the black bean aphid on field bean was mediated through the host plant after exposure to SO_2 or NO_2 (Dohmen et al. 1984). Numerous observations suggest that plants exposed to air pollutants may be attacked more readily by insect species.

Review articles by Heagle (1982) and Laurence (1981) cover much of the research on plant-parasite interactions. Highlights of studies similar to those reported are given in table 6.8. The degree of interaction between O_3 and parasites depends on the growth and exposure conditions, and the concentration and timing of the exposures in relation to inoculation time. Generally, mycorrhizal root infection was inhibited by O_3 exposure. Sulfur dioxide inhibited infection of corn by the southern corn leaf blight fungus and of wheat by the stem rust fungus (Laurence et al. 1979). The titer of two viruses was increased in host plants exposed to chronic doses of SO_2 (Laurence et al. 1981).

Environmental Factors. Discussions of the relationships between climatic or edaphic factors and plant response to pollutant exposure are found in several reviews for O_3 (Heck et al. 1977, U.S. EPA 1986) and for SO_2 (U.S. EPA 1982b).

Pinto bean primary leaves were less sensitive when exposed to O_3 at 24°C than at 15 or 32°C; results were not related to stomatal conductance (Miller and Davis 1981). Studies on the effects of temperature, light, and humidity on the response of pinto bean and Bel-W$_3$ tobacco to O_3 were designed to determine possible interactions among O_3, growth conditions, exposure conditions, and species (Dunning and Heck 1977). Three-way interactions were found for temperature and light with O_3, and two-way interactions for humidity with O_3. Ozone uptake was tripled or quadrupled with increasing relative humidity (McLaughlin and Taylor 1981); the humidity effect on O_3 uptake was controlled by internal leaf resistance to O_3 and not by stomatal regulation. The results reinforce the concept that the pollutant-environment-species system is both dynamic and complex.

Generalizations on the modifications of plant response to O_3 by climatic factors are difficult because there usually are known exceptions. However, evidence suggests that lower light intensity during growth, higher light intensity during exposure, higher growth temperatures, and higher growth and exposure humidities increase the sensitivity of many species to O_3 (Heck et al. 1977). Available information suggests that increased O_3 sensitivity, although related to stomatal conductance, reflects changed physiological conditions within the plant. Environmental conditions in the field can affect stomatal control of plant response the same day and will start changing physiological control of plant response on the following day. A 2–4-day period is usually necessary before physiological conditions that affect plant response to O_3 will completely reflect the changed environmental conditions. An in-depth discussion with tabular presentations on climatic and edaphic factors affecting plant response to O_3 is found in Heck et al. (1977).

Although environmental interactions have generally not been discussed in this chapter, it seems pertinent to highlight selective research that addresses the effects of soil moisture on the response of plants to O_3. Although similar results should be expected with SO_2, the research has not been reported. Tingey et al. (1982), using a uniform water stress, reported that O_3 sensitivity in bean decreased with increasing plant water stress; protection occurred in 1 day at the

Table 6.8. *Interactions of plant parasites, ozone, and sulfur dioxide.*

Test plant	Parasite	Exposure characteristics	Results
Ozone			
Soybean	*Pseudomonas* sp (bacterium)	0.25, 0.30, 0.35, 0.40 ppm; 2, 3, 4 hr (greenhouse)	Fewer O_3 symptoms if inoculated 24 hr prior; enhanced symptoms if inoculated 4 hr prior; O_3 treated did not show a bacterial response
Wild strawberry	*Xanthomonas fragariae* (bacterium)	0.20 ppm; 3 hr (greenhouse)	Ozone inhibited infection of strawberry by the bacterium
Onion	*Botrytis* sp. (fungus)	Ambient O_3 (field studies)	Increased *Botrytis* infection in O_3 chambers
Onion	*Botrytis cinerea*	0.16 ppm, 4 hr; 0.12 ppm, 5 hr/day, 4 days (controlled chambers) Innoculated after exposure	Increased lesions on oldest nonsenescing leaves of O_3 exposed plants. (Rist and Lorbeer 1984)
Corn	*Helminthosporium maydis* (fungus)	0.06, 0.12, 0.18 ppm; 6 hr/day, various times before or after inoculation (controlled)	Effecs on lesion length and sporulation depended on O_3 concentration and timing of exposure
Tomato	*Glomus fasciculatus* (endomycorrhizal fungus)	0.15, 0.30 ppm; 3 hr, 1 or 2 times/week, 9 weeks (greenhouse)	Infection reduced by 46 and 63% in the 0.15 and 0.30 ppm treatments, respectively
Soybean	*Glomus geosporum* (endomycorrhizal fungus)	0.025, 0.049, 0.079 ppm; 7 hr/day, 139 days (field)	Soybean less sensitive to O_3 in presence of fungus; greater P uptake by mycorrhizal plants
Tobacco	Tobacco streak (virus)	0.30 ppm; 3 hr, 1 or 2 times (greenhouse)	Virus plus O_3 gave a synergistic response for increased injury and decreased biomass
Soybean	Tobacco ringspot (virus)	0.35 or 0.40 ppm, 4 hr (greenhouse)	Virus increased soybean resistance to O_3
Tomato	Tobacco mosaic virus (TMV) or cucumber mosaic virus (CMV)	0.15, 0.30, 0.45, 0.60, 0.90 ppm; 3 hr (greenhouse)	Amount of injury from O_3 depended upon the virus, tomato cultivar, O_3 concentration, and virus incubation time; TMV plants were more O_3 sensitive than CMV plants; O_3

Table 6.8. (Continued)

Test plant	Parasite	Exposure characteristics	Results
			exposure 21 days after inoculation caused less leaf injury, possibly due to leaf age
Wheat	*Puccinia recondita* (Brown rust)	0.085 ppm, 3 days; 0.105 ppm; 7 hr/day, 7 days; innoculated after exposure	Ozone reduced number of uredospore pustules (Dohmen 1987)
Sulfur dioxide			
Bean, kidney	*Xanthomonas phaseoli* (bacterium)	0.20 ppm; 6 hr/day, 10 days	Sulfur dioxide had no effects on growth and development of the bacteria
Ectomycorrhizal fungi (4 isolates)	(Direct effects on mycelial mats)	0.05, 0.50 ppm; 1 hr (for both SO_2 and O_3)	Reduced mycelial respiration; SO_2 had greater effect than O_3

Source: See table 6.5.

highest stress, and recovery occurred within 6 days. The sensitivity change was associated with small changes in leaf water potential and thus with a change in leaf stomatal conductances. Bean treated with a chemical to induce stomatal opening (fusicoccin) in water-stressed plants was as sensitive to O_3 as nonwater-stressed plants (Tingey and Hogsett 1985). Results suggest that O_3 protection in water-stressed plants is a result of stomatal control and not biochemical control. Heggestad et al. (1985) reported on O_3 by soil moisture stress interaction in soybean grown in open-top chambers when comparing the charcoal filtered and nonfiltered treatments. They found that plants were more sensitive to ambient concentrations of O_3 under a small soil moisture stress than when adequate moisture was available. In a similarly designed field study with cotton, Temple et al. (1985b) found an interaction between O_3 and soil moisture stress in a very dry year (soil moisture protected plants from O_3) but not in a cooler and moister year. In the first case, soil moisture stress gave some protection to cotton from O_3. In a separate 2-year study with soybean, Flagler et al. (1987) reported no interaction between O_3 and soil moisture stress even though the two years were quite different. They found that both O_3 and water stress affected N partitioning among plant parts and that the status of soil water had a pronounced effect on plant response to O_3. These studies were part of the NCLAN program and demonstrated that no simple explanation is adequate to address the effects of soil moisture on plant response to O_3. Stomatal control must play some role but it is reasonable to argue that physiological changes due to moisture stress can affect plant response to O_3. Because of the importance of soil water stress in agricultural areas, it is imperative that we try to understand this interaction and include it in any assessment effort.

There is interest in preconditioning plants by treatment with noninjurious concentrations of O_3. Bean seedlings exposed to 0.02 ppm O_3 for 6 hr per day were more susceptible to subsequent acute O_3 exposures than seedlings not pretreated (Runeckles and Rosen 1977). Similar results were reported for tobacco preexposed to 0.03 ppm O_3 for 12 hr prior to an acute exposure (Steinberger and Naveh 1982). Finally, soybean preexposed to different chronic concentrations of O_3 for 6 hr per day were more sensitive to an acute O_3 exposure when the chronic exposure did not produce visible symptoms; when visible injury was caused by the chronic exposure, subsequent injuries from acute O_3 episodes caused additive or less-than-additive injury (Johnston and Heagle 1982). These results suggest that metabolites necessary for the repair mechanisms may be utilized by the exposure to low concentrations of O_3, resulting in less protection during the subsequent acute exposure. Low SO_2 concentrations prior to exposure to short-term acute concentrations did not increase plant sensitivity to the acute SO_2 doses (Laurence et al. 1985).

Environmental factors that influence plant response to SO_2 include light, temperature, humidity, CO2, freezing, soil moisture, and soil nutrition; several of these have been studied in combination. A summary of representative studies is shown in table 6.9. Perhaps the most in-depth studies were performed on oat and included interactions among growth temperature, exposure temperature, and SO_2 and among exposure temperature, relative humidity, and SO_2 (Heck and Dunning 1978); root growth was more affected than top growth at all SO_2 concentrations. Plants are generally more sensitive to SO_2 as light intensity, wind speed, temperature, and humidity increase; elevated CO2 levels protect plants; freezing may increase plant sensitivity whereas low soil moisture tends to make plants more resistant. Krizek et al. (1986) found that soil moisture stress and abscisic acid treatment caused poinsettia cultivars to respond in a similar fashion to SO_2; lower stomatal conductance and transpiration rates were reported in both cases. It seems that stomatal control plays a primary protective role but changes in plant physiology must affect the plant response.

Chemical Protectants. Interest in chemical protectants for protecting sensitive plants from O_3 or SO_2 injury has lessened in recent years for two reasons; mixed results were obtained and one or two seasonal chemical treatments are not sufficient. Results of selected experiments are shown in table 6.10. The antitoxidant chemical ethylenediurea (EDU) is not included here; it is discussed in Heck et al. (1986a).

Cadmium (Cd) enhanced O_3 injury in plants cultured in increasing concentrations of nutrient Cd; this enhancement was especially apparent at low concentrations of both O_3 and Cd (Czuba and Ormrod 1981). Benomyl, a fungicide, suppressed O_3 injury in several plants including three turfgrass spe-

Table 6.9. *Effects of environmental factors on the response of plants to sulfur dioxide.*

Test plant	Factors[a]	Sulfur dioxide exposure characteristics	Results
Timothy	Light: 100 or 400 $\mu E/m^2/s$; temp.: 30°C day by 19°C night vs. 26°C day by 12°C night	0.11–0.12 ppm, 44 days (controlled)	Lower temp. and light gave greater reduction in stem growth (light, temp., and SO_2 appeared to be interactive); roots were more sensitive to SO_2 regardless of the environmental conditions; net assimilation rate reduced 14.7%
Oat, 2 cultivars	Four growth and exposure temp.; four growth temp. and two exposure RH	0.5, 1.0, 2.0, 4.0 ppm or 0.75, 1.5, 3.0 ppm, 1.5 hr, 2 times (controlled)	Coker 227 more sensitive than Carolee; root dry wt. reduced more than top in all treatments; plants more sensitive at higher growth temp.; exposure temp. did not affect response; plants more sensitive at higher RH; interactions found for SO_2 with growth temp. and exposure RH
Bean, snap	Exposure temp.: 13, 21, 32°C; RH: 40, 60, 80%	0.9 ppm; 1 or 2 hr (controlled)	Higher temp. and RH gave most injury; stomatal conductance increased with temp. and RH
Bean, snap	RH: 35, 78%	0.23, 0.35, 0.47 ppm; 1 hr (controlled)	Two- to threefold increase in SO_2 absorption at higher RH
Wild geranium	RH: 50, 70, 80, 90%	0.2, 0.4, 0.8, 1.2, 1.5 ppm; 4.7 hr (controlled)	Decreased seed set at RH of 80%; 0.2 ppm at 90% RH gave 32% reduction in seed set; no visible injury
Geranium	RH and CO_2	1 ppm; 1 hr (controlled)	Injury to leaves greater at high RH and low CO_2; SO_2 in dry air hastened stomatal closure
Six species 3 C_3; 3 C_4	CO_2: 300, 600, or 1200 ppm	0.25 ppm; 8 hr/day, 5 days/week for 2d and 4th weeks of elevated CO_2 (greenhouse)	C_3 plants more sensitive to SO_2 at ambient CO_2 levels; C_4 were more sensitive at elevated CO_2; effect related to effect of CO_2 on stomatal conductance
Soybean	CO_2: 300, 450, 600, and 1200 ppm	0.18–1.1 ppm, 2 hr (greenhouse)	Photosynthesis decreased with increasing SO_2; effect reduced with increasing CO_2

(continued)

Table 6.9. (Continued)

Test plant	Factors[a]	Sulfur dioxide exposure characteristics	Results
Ryegrass	Wind speed: 10 or 25 m/minute	1.1 ppm, 4 weeks (controlled)	High wind speeds increased effect of SO_2; reduced leaf area, all dry wts and root/shoot ratios
Ryegrass	Freezing	0.1 ppm, 3 weeks (controlled)	Greater sensitivity to freezing temp., sensitivity enhanced by added nutrients; effect may be important in field selection of SO_2-resistant genotypes
Plants: 4 species	Soil moisture [adequate (40% FC) and drought (above wilting point)]	0.5 ppm; 5 hr/day, 6 day/week (total length not specified) (field)	Plants in higher soil moisture showed more depression of photosynthesis; corn showed no leaf injury; barley and sunflower did; homozygous inbred corn lines more sensitive than heterozygous lines
Bean, pinto	Soil moisture at $-\frac{1}{3}$, -1, -3, -5 atm	2.2 ppm, 3 hr (controlled)	Severe injury at $-\frac{1}{3}$ and -1 atm; little at -3 and -5 atm; injury correlated with soil moisture, stomatal conductance, and water potential; SO_2 decreased stomatal conductance at $-\frac{1}{3}$ and -1 atm
Grassland (native)	Nutrients (S and N)	0.02, 0.05, and 0.10 ppm; growing season (field)	Sulfur and nitrogen addition gave increased biomass; high SO_2 plus nitrogen and sulfur addition decreased aboveground biomass; SO_2 at all levels tended to decrease biomass

Source: See table 6.5.

[a]Abbreviations: μE (microeinstein), temp. (temperature), wt (weight), RH (relative humidity), FC (field capacity), atm (atmosphere), m (meter).

cies; carboxin sprays did not (Papple and Ormrod 1977). Two sensitive tobacco cultivars that were protected from O_3 injury by two of three herbicides when grown and exposed under controlled conditions were protected for only several weeks after transplant when exposed in the field to ambient levels of O_3 (Reilly and Moore 1982). Exposure of greenhouse-grown sorghum seedlings to O_3 interfered with the herbicide antidote CGA–43089 in protecting the seedlings against

Table 6.10. *Chemical protectants for crops against sulfur dioxide.*

Test plant	Treatment[a]	Sulfur dioxide exposure characteristics	Results
Petunia, sensitive and resistant cultivars	EDU at 250–5000 ppm (foliar spray, soil drench)	0.75–3.75 ppm, 3 hr (controlled)	No protection against SO_2 concentrations that gave foliar injury
Chrysanthemum, 2 cultivars	Ancymidol, 0.16 and 0.48 mg, soil drench	0.2, 0.5, or 2.5 ppm; 2, 8, or 24 hr (controlled)	Reduced SO_2 injury in both cultivars; related to stomatal conductance
Pea	Cadmium; 5, 10, 15 ppm in soils	0.067 ppm	Cd and SO_2 affect peroxidase; no strong interactions are shown
Species, five	Cadmium, lead, copper, manganese, and silicon, mixture of the oxides in a 1:30:10:15:35 ratio; 1 dusting week, 4 weeks	0.08 ppm; 13 hr/day, 28 days	Sulfur dioxide did not affect uptake or translocation of heavy metals, did affect yield loss from heavy metals; foliar injury was increased
Barley	Copper, 50 or 100 ppm as soil drench	1 ppm; 6 or 7 hr/day, 2 days	Cu reduced stomatal conductance and plant injury
Cowpea	Ca(OH)$_2$; 0.5% aqueous spray, 4 times at 20-day intervals	About 0.25 ppm; 1.5 hr/day, 40 days	Reductions in chlorophyll and biomass from SO_2 were less for plants receiving the Ca(OH)$_2$ spray
Wheat	Urea spray (3%), 3 times	1 ppm; 2 hr/day, 80 days (field)	Effects were reduced by urea spray
Bean, snap	Paclobutrazol (gibberellin inhibitor) 0.02 mg/pot	1.0 ppm, 3 hr 1.5 ppm, 2 hr 2.0 ppm, 1 hr (controlled)	Gave complete protection which was partially reversed by a gibberellin spray (Lee et al. 1985)

Source: See table 6.5.

the herbicide metolachlor (Hatzios 1983); results suggest that O_3 may alter the effectiveness of agricultural chemicals in an O_3-polluted atmosphere.

Rice seedlings pretreated with abscisic acid (a natural plant hormone) gave protection against O_3 injury (Jeong et al. 1981), while natural concentrations of abscisic acid were not related to the severity of O_3 injury in pea (Kobriger et al. 1984).

Certain agricultural chemicals tended to protect plants from SO_2 injury, whereas heavy metals, except copper, did not affect plant response. Generalizations are premature since relatively few research results are available.

Pollutant Interactions. The effects of pollutant mixtures on terrestrial eco-systems have been reviewed in several documents (Lefont and Ormrod 1984, Reinert 1984). Generally, plants are more severely affected by mixtures of pollut-ants than by the individual pollutants. In many studies mixtures tend to give a greater-than-additive (synergistic) response when the concentrations are below those causing visible effects from pollutants singly; concentrations around the injury threshold tend to produce an additive response; and concentrations above threshold tend to cause a less-than-additive (antagonistic) response. The term *synergistic*, although statistically appropriate, is not necessarily biologi :ally appropriate since response functions may not be linearly related to pollutant concentration. The mixture of O_3 with SO_2 has received the most study, fol-lowed by mixtures of SO_2 with NO_2, and mixtures involving all three pollutants.

It should be noted that an additive response to two or more pollutants is equal to the sum of the individual effects. The term *interactive response* includes the concepts of antagonism and synergism. These concepts are clear and simple and can be handled with statistical processes. Except for the concept of synerg-ism, the statistical and biological interpretations generally agree. The terms *additive, antagonism,* and *synergism* are used in this chapter.

Table 6.11 is a summary of the results from selected studies. Only a rela-tively few from a number of responses are shown. Our knowledge of the effects of pollutant mixtures on terrestrial ecosystems is fragmentary. However, available information suggests that mixtures of pollutants, at concentrations similar to those expected in ambient air, could produce synergistic responses in a number of species (cultivars, selections). Where multiple responses have been studied, similar trends are reported.

Most laboratories are not equipped to perform the experimental designs necessary to understand the response of plants to mixtures. Some studies were not designed to account for pollutant interactions. In other studies the results were presented without adequate consideration of possible interactions. Statisti-cal means of handling and interpreting the results of mixture studies are high-lighted in three papers (Larsen et al. 1983, Ormrod et al. 1984, Reinert et al. 1982b).

Few studies address possible biological mechanisms for interactions. Most reports address stomatal conductance that affects pollutant entry (avoidance). Although a reasonable physiological approach, it does not address biochemical mechanisms. An interesting biochemical explanation for the synergistic action of SO_2 and NO_2 on several grass species involves the inability of the plant to detoxify nitrite in the presence of SO_2 (Wellburn et al. 1981).

Growth, Biomass, and Yield Effects

The effects of O_3 and SO_2 on the biomass and yield of selected plant species from greenhouse and controlled-environment research are highlighted in this section. The effects of O_3 and SO_2 on partitioning of photosynthate continues to be a major focus of research to explain observed biomass or yield changes.

Table 6.11. *Plant response to pollutant mixtures.*

Mixture	Test plant	Exposure information	Plant response to the mixture
$O_3 + SO_2$	Bean, snap	0.20 ppm each gas; 7 hr/day, 4 days; varied salinity	Stomatal conductance: synergistic (variable); foliar injury: antagonistic; growth: additive; effect changed with salinity
	Bean, snap	0.15 ppm each gas; 6 hr/day, 5 days, consecutive	Foliar injury: antagonistic; stomatal closure: synergistic, variable response
	Bean, field	0.05–0.30 ppm O_3; 0.04 ppm SO_2; 4 hr	Net photosynthesis: additive or antagonistic, depending on O_3 concentration
	Tomato	0.2 ppm O_3, 0.2 or 0.8 ppm SO_2; 3 or 4 hr, 15 times; infection with nematodes	Growth and yield additive at 0.2 ppm SO_2, antagonistic at 0.8 ppm SO_2
	Tomato	0.005 to 0.468 ppm SO_2; 0.015 ppm or 0.056 ppm O_3; 5 hr/day, 5 da/wk, 57 days; field study with open-top chambers	Ripe fruit decreased 16% by O_3 (low SO_2 treatment), 18% by SO_2 (low O_3 treatment), 32% in high O_3– high SO_2 treatment; additive response (Heggestad et al. 1986)
	Lettuce, radish	0.4 ppm O_3, 0.8 ppm SO_2; 6 hr	Use of covariates increased precision for lettuce and radish; lettuce growth and injury effects antagonistic; radish was additive
	Potato	Four O_3 concentrations, filtering of ambient O_3; 0.1 ppm SO_2; for 6 hr/day, 255 hr	Reductions in various growth and yield parameters were additive
	Soybean, 2 cultivars	0.25–1.0 ppm O_3, 0.50–1.5 ppm SO_2; 0.75, 1.5, or 3 hr	Foliar injury and reduced shoot fresh weight: additive, antagonistic, or synergistic, depending upon concentration and time
	Soybean	0.06 or 0.08 ppm O_3, 0.06 or 0.11 ppm SO_2; 5 hr/day, 16 days; in open-field facility	Both O_3 and SO_2 caused decreases in a number of yield measures; mixture responses were additive
	Soybean	0.20 ppm O_3, 0.70 ppm SO_2; various combinations of individual and both gases over 2 hr of exposure	No effect on carbon exchange rate (CER) by individual gases; O_3 or SO_2 for 1 hr followed by mixture for 1 hr gave a 38 and 59% reduction in CER; mixture

(continued)

Table 6.11. (*Continued*)

Mixture	Test plant	Exposure information	Plant response to the mixture
		(controlled chambers)	gave 57% reduction; suggests not a stomatal response; synergistic (Chevone and Yang 1985)
	Soybean	0.04 to 0.08 mean ppm O_3, 0.00 to 0.11 mean ppm SO_2; 5 hr/day, 16 days, during pod fill; linear gradient system in field	Ozone caused 26% seed yield reduction, SO_2 a 6% reduction with no significant interactions (Reich and Amundson 1984)
$O_3 + SO_2$	Grape, 2 cultivars	0.20 and 0.40 ppm O_3, 0.15 and 0.30 ppm SO_2; 4 hr	Foliar injury and reduced shoot length were antagonistic to synergistic, depending on concentration; leaf abscission was synergistic
$O_3 + NO_2$	Four species: bean, mint, radish, wheat	0.08 to 0.10 ppm NO_2 for 3 hr prior to 0.08 to 0.10 ppm O_3 for 6 hr; greenhouse exposures	NO_2 increased growth in radish and bean; O_3 reduced growth in wheat and bean; apparent synergism from mixture in radish and wheat, additive to antagonistic in bean and no effect in mint (Runeckles and Palmer 1987)
$SO_2 + NO_2$	Bean, snap	0.1 ppm or each gas; 5 days	Transpiration: increased by individual gases, decreased by combination
	Tomato	0.11 and 0.05 ppm each gas; continuous, 14 and 28 days	No effects of individual gases; mixture caused decrease in leaf fresh weight and area (14 days), root fresh weight, and dry weight (28 days)
	Grass, *Poa pratensis*	0.10 ppm of each gas; 104 hr/week, long term	Reduced growth of roots and shoots: synergistic in late winter, but not later; shoot recovery during summer; effects on flowering were additve
	Alfalfa, 12 strains	0.08 ppm SO_2, 0.120 ppm mean NO_2 (singly and as mixture); 15 days; greenhouse-type chambers	Strains showed form +8 to −35% effect from SO_2; from +13 to −27% effect from NO_2; from −8 to −50% effect from mixture; mixture gave from additive to synergistic responses on different strains (Lorenzini et al. 1985)
	Soybean	0.2, 0.4, 0.6 ppm of each gas; 2 hr	Photosynthesis: synergistic stomatal conductance: syn-

(continued)

Table 6.11. (*Continued*)

Mixture	Test plant	Exposure information	Plant response to the mixture
			ergistic; respiration: additive
	Soybean	0.13–0.42 ppm SO_2, 0.06–0.40 ppm NO_2; 3 hr/exposure, 10 days during pod fill (2-year study, ambient air with O_3)	Chlorophyll reduction, synergistic; yield reduced from 9 to 25% in 2 years, synergistic
	Soybean	0.13 to 0.42 ppm mean SO_2; 0.06 to 0.40 ppm mean NO_2; 3 hr/exposure, 10 times during pod fill; two-year study in field open-release plots	Yield not affected by NO_2 in either year; 6% reduction from SO_2 in second year; mixtures reduced yield from 9 to 25%, response was synergistic (Irving and Miller 1984)
	Soybean	0, 0.2, 0.3 ppm SO_2; 0, 0.1, 0.2 ppm NO_2; 3 hr/day, every other day, 15 exposures; greenhouse chambers	Leaf wt was increased 2% and 9% by SO_2 and NO_2, respectively; low combination reduced leaf wt by 7%, high combination by 16%; root wts were −10% and +2% fo SO_2 and NO_2, and −31 and −38 for low and high combinations. Probably synergistic (Klarer et al. 1984)
O_3 + PAN	Petunia; bean, kidney	0.10–0.40 ppm O_3; 0.01–0.10 ppm PAN; 4 hr	Foliar injury, antagonism to synergism depending on concentration of pollutants
SO_2 + HF	Corn, sweet	0.09 ppm SO_2, 0.55 ppb HF; 32 days (continuously)	No effect of HF or SO_2, mixture reduced fresh and dry weight of stalk and reduced yield, reduced accumulation of foliar F; interaction not tested
NO_2 + HF	Corn, sweet	0.6 and 1.2 ppm NO_2, 6 hr/day, 4 to 5 days/week; 0.6 and 1.9 ppb HF (continuous)	Foliar injury, additive or antagonistic; interaction shown for increased stomatal resistance
O_3 + SO_2 + NO_2	Radish	0.1, 0.2, 0.4 ppm of each gas; 3 hr	Reduced yield: no three-way interaction but an NO_2 + O_3 and an NO_2 + SO_2 interfaction; interactions appear synergistic
	Radish, marigold	0.3 ppm each gas; 3 hr, 9 times	Reduced yield: primarily additive responses; NO_2 +

(continued)

Table 6.11   (Continued)

Mixture	Test plant	Exposure information	Plant response to the mixture
	Turfgrass, 6 species, 18 cultivars	0.15 ppm O_3 for 6 hr/day; 0.15 ppm SO_2 and 0.15 ppm NO_2 continuously; 10 days	$SO_2 + O_3 + SO_2$ interactions antagonistic Foliar injury and reduced leaf area, primarily additive with some antagonism
	Sunflower	0.2 ppm each gas, 2 hr	Reduced net photosynthesis: $NO_2 + O_3$ and $SO_2 + O_3$, synergistic response; three-way similar to two-way responses

Source: See table 6.5.

Selected field studies are included for both O_3 and SO_2. Field studies associated with the National Crop Loss Assessment Network are reported and discussed in the following section because they were designed for use in assessment efforts. Results from the NCLAN studies reflect effects of ambient O_3 concentrations on crop growth and productivity.

The use of the NCLAN studies for assessment activities does not detract from the importance of growth chamber, greenhouse, and field pot studies in developing an understanding of the effects of O_3 on plant growth and development. Early assessments using available data have underestimated the effects of O_3 found in field studies, but the estimates have been reasonable. Additionally, the understanding of effects and methodology under nonfield conditions permitted the rapid development of useful field experimental methods and designs.

Ozone. The effects of O_3 on growth and yield (see table 6.12) are briefly summarized here: O_3 usually affected crop growth and productivity; cultivar differences were usually observed when studied; root growth was affected more than shoot growth when carbon allocation was studied; changes in growth rate occurred when exposures were during early vegetative growth, but normal growth resumed shortly after the exposure ended; changes in the quality of usable product occurred in some experiments; the results suggest that growth and yield reductions occur at ambient O_3 concentrations.

Ozone-sensitive species may be lost from plant communities with increasing O_3 stress. It is important to understand the effects of O_3 on ecosystem dynamics. In agroecosystems the loss of clover from a clover/ grass community has been studied with respect to increasing O_3 concentrations. In a crimson clover–annual ryegrass mixture, the clover decreased and ryegrass increased under O_3 stress (0.03 or 0.09 ppm O_3 for 8 hr a day for 6 weeks) (Bennett and Runeckles

Table 6.12. *Growth and yield of selected crop species in response to ozone exposure.*

Test plant	Ozone exposure characteristics	Results
Bean, lima (8 genotypes)	Comparison of charcoal filtered (<0.02 ppm mean daily max) and nonfiltered (<0.06 ppm mean daily max) greenhouses	Yield reductions of 3.4 to 68.5% were found across the 8 genotypes (Meredith et al. 1986)
Bean, snap	0.30, 0.60 ppm; 1.5 hr, 2 times; 6 growth stages; harvest 7 days after exposure or at fresh harvest (controlled)	Reduced relative and absolute growth rates, pod production, nodulation, and fixed nitrogen; magnitude varied with O_3 concentration and growth stage; partial recovery from exposure at early growth stage
Bean, snap; a sensitive and resistant cultivar	Variable concentrations and durations (controlled)	Short-term changes in allocation to root but no yield effect on resistant cultivar; significant allocation changes persistent for at least 1 week in sensitive cultivar
Tomato, Tiny Tim	0.08–0.10 ppm; 5 hr/day, 5 days/week, 5 weeks (greenhouse)	86% reduction in fruit number, 91% reduction in fruit weight
Tomato, 6718 VF	Ambient O_3 gradient in southern California (field)	Multiple regression models predicted O_3 was responsible for 85% of reeduced fruit size along the gradient; model predicted 50% yield reduction at O_3 dose of 20 ppm-hour for O_3 at concentrations >0.10 ppm
Potato, 2 cultivars	0.2 ppm; 3 hr at several growth stages (greenhouse)	Decreases in tuber and total solids and increases in reducing sugar in O_3-exposed plants
Pepper	0.12, 0.20 ppm; 3 hr, 3 times/week (greenhouse)	Decrease in root, stem, and leaf dry matter; 16 and 54% loss in total fruit dry wt, 77% loss in mature fruit (0.2 ppm)
Parsley, Banquet	0.20 ppm; 4 hr, 2 times/week, 8 weeks (greenhouse)	23% decrease in plant dry wt, 43% decrease in root dry wt; greatest effects caused by initial exposure
Alfalfa	Charcoal filtered vs. nonfiltered; over growing season (pots); compared linear-gradient with open-top chambers; 50 to 80% O_3 exclusion	In 1983: 18% reduction in fresh wt for linear gradient, 22% reduction for open-top chambers; in 1984: the values were 12 and 1%, respectively (Olszyk et al. 1986a)
Soybean, Corsoy	0.022 and 0.112 ppm; 7 hr/day, seasonal (field)	Yield reduced 39% at higher O_3 treatment; no effect on seed protein; 12.6% reduction in seed oil
Soybean	0.02 to 0.097 ppm; 341	Yield was +15%, −34%, and −40%

(continued)

Table 6.12 (Continued)

Test plant	Ozone exposure characteristics	Results
	hr, intermittent over 113 days; greenhouse	at 0.046, 0.070 and 0.097 ppm in comparison to 0.02 ppm. The increase at 0.046 ppm was not significant (Endress and Grunwald 1985)
Cotton, Acala SJ-2	0.20 ppm; 6 hr, 2 times/week, 1 group started at 8 day, 1 at 42 days from seed (greenhouse)	Vegetative biomass and boll production reduced; greatest reduction in boll and root wts; 48% reduction in boil number in both treatment
Clover, Ladino	0.3, 0.6 ppm; 2 hr, 2 times, 7 days apart; 4 stages of development (controlled)	Shoot and root growth reduced; response affected by O_3 concentration and exposure stage, nodulation decreased by exposure at first three stages
Clover, Ladino	0.03, 0.05, 0.08 ppm; 7 hr/day, 6 months (field: pots)	Total forage and forage regrowth reduced for clover and clover-fescue mixture in relation to O_3 concentration; fescue unaffected; O_3 response unaffected by N fertilization
Clover, Ladino	0.05, 0.10, 0.15 ppm; 4 hr/day, 6 days, 32 days after seeding; $^{14}CO_2$ exposures at 48 hr after last O_3 (controlled)	Max. root reduction, 42%; max. shoot reduction, 24%; O_3 reduced net photosynthesis; carbon allocated favored roots at 0.10 ppm O_3 and shoots at 0.15 ppm
Rice, 3 cvs	0.05 to 0.20 ppm, 5 hr/pday, 5 days/wk, 15 wks; in pots in open-top chambers	Yield at highest O_3 compared to lowest was reduced from 12 to 29% in the three cultivars (Kats et al. 1985)
Wheat (7 cvs)	0.35 ppm, 3 hr/day, 5 days/wk, over growth periods (controlled)	Yield reductions of 43 to 57% were reported across the 7 cvs (Decoteau et al. 1986a)
Wheat (6 cvs)	Charcoal filtered (<0.025 ppm daily max) vs nonfiltered with added (0.1 ppm) O_3 for 4 hr/day (<0.13 ppm daily max), 5 days, during anthesis (field)	Yield reductions of 14 to 36% reported for the 6 cvs (Mulchi et al. 1986)

Source: See table 6.5.

1977). A fescue-ladino clover mixture (Kochlar et al. 1980) exposed for 2 hr at 0.30 ppm O_3 gave similar results. Leaf leachates from O_3-exposed fescue inhibited clover growth and nodulation in the latter study. In a series of studies (Blum et al. 1983a, Blum et al. 1983b, Letchworth and Blum 1977; Montes et al. 1982, Montes et al. 1983), Blum and associates reported a decrease in clover and an increase in fescue in a clover/fescue mixture that correlated with increasing O_3 concentration; a loss of total biomass occurred at elevated O_3 concentrations. A 2-year field study, as part of the NCLAN program, has confirmed these results (Heagle et al. 1989).

Sulfur Dioxide. The effects of SO_2 on carbon translocation and partitioning and on plant growth and yield are shown in table 6.13. Generally, assimilates move to developing leaves rather than to roots under low SO_2 stress, although the reverse has been reported (Coughenour et al. 1979). Root growth was generally reduced more than shoot growth and occurred at relatively low SO_2 concentrations. The results support the contention that plants are sensitive to low SO_2 concentrations (<0.10 ppm) when exposed continuously. Improved experimental designs and facilities are needed to confirm these reports.

The effects of low SO_2 concentrations on grasses have been extensively studied in both the United States (Lauenroth and Preston 1984) and the United Kingdom (Bell 1982). In the Colstrip study (Lauenroth and Preston 1984) assimilate partitioning, translocation, and forage quality of native grasses were studied and considerable variability in response was shown; the trends appeared consistent. The most extensive and controversial studies on the effects of low SO_2 concentrations on growth and productivity in forage grasses have been done in the United Kingdom. Bell (1982) summarized research results from the United Kingdom with an emphasis on ryegrass (*Lolium perenne*). His summary of results from continuous 24-hour exposures to low SO_2 concentrations indicated that effects on the exposed plants were from +11 to −68% of the effects on the control plants (the greatest effect was at the lowest concentration). The experiments reported were complicated by differences in experimental design (no replication in some instances) and by differences in experimental facilities. Generally, results suggest that forage grass (specifically ryegrass) productivity is decreased as a result of exposure to ambient SO_2 concentrations that occur in the United Kingdom.

Sulfur dioxide concentrations in agricultural areas away from point sources rarely exceed 0.01 ppm for extended periods of time (U.S. EPA 1982b). Daily 4- or 7-hour exposures of cotton and tomato (Heck et al. 1983b) or soybeans (Heagle et al. 1983) in open-top field chambers demonstrated that SO_2 concentrations, likely to occur regionally in the United States, probably do not cause decreased yield. A similar conclusion was reached in a review by Roberts (1984) on the effects of SO_2 on plant productivity; results included those from open-air studies. However, emissions of SO_2 near point sources can cause decreased yield in sensitive crop species.

Table 6.13. *Carbon partitioning, growth, and yield of selected crop species in response to sulfur dioxide exposures.*

Response measured	Test plant	Sulfur dioxide exposure characteristics	Results
Carbon partitioning, translocation (use of $^{14}CO_2$)	Native grasses	0.10 ppm; continuous (field)	Proportion of total assimilated carbon in roots varied with SO_2 treatment, species, and time of year; generally SO_2 caused increased translocation to roots
	Timothy	0.11 ppm for 3 weeks; 0.06 ppm for 6 weeks; continuous (controlled)	Reduced assimilate movement to roots and increased movement to developing regions of the shoot
Yield/biomass	Six species	Constant and stochastic concentration, 0–0.20 ppm (greenhouse)	Constant SO_2 concentration underestimated effects compared to the time series treatments; excellent discussion of time series concept; interpretation of results difficult
	Tobacco and cucumber	0.02 ppm; 4 weeks, continuous (controlled)	Reduced plant dry wt fractions; cucumber more sensitive than tobacco; roots affected more than shoots; no replication
	Barley	0.010, 0.023, 0.038, and 0.058 ppm mean over growing seasin (field, open release system)	Controls not defined but three lowest SO_2 concentrations gave about 5, 20 and 12% yield increase, respectively; the high SO_2 reduced yield about 18% (McLeod et al. 1986)
	Barley	0.04 to 0.20 ppm mean concentration across growing season; a two-year field study; open release of SO_2	Yield reduction found in both years; when average SO_2 during fumigation was used in regression analysis found a 2.2% loss per 0.01 ppm of SO_2 (Baker et al. 1986)
	Alfalfa	0.036 ppm, continuous (greenhouse)	Slight increase in biomass with no added sulfate, no effect with added sulfate to growth medium

(continued)

Table 6.13 *(Continued)*

Response measured	Test plant	Sulfur dioxide exposure characteristics	Results
	Sunflower	0.05, 0.1 ppm; 5 weeks, continuous (controlled)	Variable results; higher concentrations gave 20–25% reduction in net assimilation rate; both concentrations accelerated senescence
	Rice (3 cvs)	0.05 to 0.20 ppm, 24 hr/day, 5 days/ week, 15 weeks; in pots in open-top chambers	Yield at highest SO_2 compared to lowest was reduced from 11 to 29% in the three cultivars (Katz et al. 1985)
	Lolium perenne	0.012 to 0.029 ppm winter mean at 4 selected sites with differing SO_2 concentrations (field correlational)	Two sites with lowest mean SO_2 showed same total dry wt yields; two highest sites showed a −19 and −54% reduction, respectively. The site with highest SO_2 also suffered winter injury (Ashenden 1987)
	Lolium perenne	0.02, 0.16 ppm continuous (greenhouse)	No effects in any growth parameters measured; some visual injury at 0.16 ppm and a reduction in specific leaf area
	Lolium perenne	Variable, means <0.08 ppm for growing season (field)	Growth in filtered and unfiltered air not different in highly polluted area; added SO_2 gave variable results with a maximum growth reduction of 12%
Plant quality	Western wheatgrass	0.02, 0.038, 0.065 ppm: seasonal means; continuous (field)	Increase in sulfur; no effect on forage digestibility
	Tomato	0.12 ppm; 72 hr/ week, 5 or 10 weeks (controlled)	Slight decrease in ascorbic acid of ripe fruit; no effect on fruit yield, soluble or total solids

Source: See table 6.5.

Dose response studies using an open-air SO_2 release system have simulated exposures of soybeans to SO_2 near point sources. Sulfur dioxide released at canopy height produced a SO_2 concentration gradient downwind from the release point (Miller et al. 1980). Soybean yield was decreased by periodic SO_2 exposures after flowering to doses of approximately 10–15 ppm-hr (Miller et al. 1980; Sprugel et al. 1980). The 10 to 15 ppm-hr dose statistics were products of mean exposure durations of 2.5–4.2 hr, mean concentrations of 0.12 to 0.31 ppm, and 19–25 exposures. Doses in the 5 ppm-hr range were either stimulatory or inhibitory. Maximum peak-to-mean-SO_2-concentration ratios were about 2.5. Figure 6.2 shows results of these studies.

Crop Yield Responses to O_3 under Field Conditions: Results from the NCLAN Program

The National Crop Loss Assessment Network was initiated by the Environmental Protection Agency to develop pollutant dose–crop yield response functions for use in assessing the economic effects of several gaseous pollutants on crop production. The pollutant of primary interest to the program was O_3 since it is generally perceived to be the pollutant of most concern to agricultural production. The NCLAN program was designed to develop information that could be used to strengthen the scientific basis for the secondary National Ambient Air Quality Standard for O_3 and to help EPA determine whether the current standard

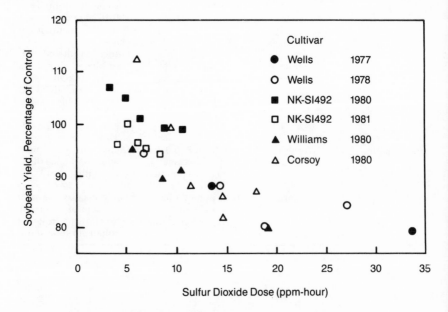

Figure 6.2. Effects of periodic SO_2 exposures in open air on the yield of soybeans. Source: Heck et al. 1986a.

was adequate (0.12 ppm, 1-hr mean, occurring no more than once a year). The scientific basis for the current standard comes principally from information on vegetation effects presented in chapters 11 and 12 of the 1978 O_3 criteria document (U.S. EPA 1978). A review of those data suggested a need for functional relationships between O_3 dose and crop response where the crops were grown and exposed under field conditions.

The NCLAN program was intiated in 1980 as a cooperative field research program involving EPA, the U.S. Department of Agriculture's Agricultural Research Service, Argonne National Laboratory, Boyce Thompson Institute, the University of California at Riverside, and Lawrence Livermore National Laboratory. The program ended in 1987 after seven years of intensive field research and a final year of data analysis and writing to complete the documentation of results from all research sites. In addition to sponsoring the international conference Assessment of Crop Loss from Air Pollutants (Heck et al. 1988a, 1988b), the program published four assessment papers (Heck et al. 1982c, 1983a, 1984b, 1984c) to highlight its early accomplishments, Annual Reports for 1980–85 (Heck et al. 1981, 1982d, 1983b, 1984d, 1985, 1986b), and over 100 journal articles.

The program had three primary objectives: (1) define the relationships between yields of major agricultural crops and varying exposure to O_3, SO_2, NO_2, and their mixtures; (2) assess the primary economic consequences resulting from the exposure of agricultural crops to O_3, SO_2, NO_2, and their mixtures; and (3) advance the understanding of the cause and effect relationships that determine crop response to pollutant exposure. The NCLAN program stressed O_3 because funding became limited, because the O_3 standard was due for review, and because the perception persisted that O_3 was the pollutant of greatest national concern.

Research Approach: NCLAN. Although program cooperators realized that a real national assessment of the impacts of O_3 was not possible with funds available, a first effort required regional representation so that some understanding of regional differences could be developed. It was expected that differences in regional growing conditions would affect plant response to O_3, and thus research programs with field sites in the Southeast (2 sites), Northeast (1 site), North Central (1 site), and Southwest (2 sites) were identified for participation in the program. Comparison of results across years at the same site and across sites was facilitated by using standardized procedures when possible, following an acceptable quality assurance program, and by developing functional relationships that could compare relative yield values and not actual yield values across years and sites.

Crops were grown using standard agronomic practices for the crop and region, and efforts were made to minimize perturbation of the plant environment. Irrigation was used to reduce plot-to-plot variation in soil moisture within an experimental design and to minimize the possible confounding of an interaction between O_3 and soil moisture within and across experiments. Although the irrigation practices have been the source of some criticism, that criticism is largely misplaced in terms of the goals of the program. Also, irrigation practices are standard in the Southwest and are common in the Southeast. In the last years of the program research was designed to assess the importance of soil moisture on crop response to O_3.

All crops were planted in the field using acceptable row and in-row plant spacing. After emergence, the fields were assessed for crop uniformity and test plots were identified. In most studies the fields were blocked for replications. Treatments were then randomly assigned to the test plots within the blocks and the fields were set up for experimental exposures. The basic protocol was designed to include exposures during the log phase of growth and all reproductive phases. Open-top chambers were erected over each plot [except for the ambient air (AA) plots], and monitoring and dispensing systems were installed (Heagle et al. 1973, 1979) to control O_3 concentrations around the crop plants. Plants were harvested across treatments when the plants growing in charcoal-filtered air reached an acceptable harvest maturity; O_3 exposures were terminated several days prior to harvest.

Four to six O_3 chamber treatments were included: a charcoal-filtered (CF) air (control) treatment, a non-charcoal-filtered (NF) air treatment, and two to four NF treatments with added O_3. The NF treatments were slightly below the ambient O_3 concentrations due to losses on particular filters and chamber structures. Ozone was introduced as a constant addition to ambient O_3 for 7 hr per day (0900–1600 standard time) for the first 4 years of the program, and as a proportional addition to ambient O_3 for 12 hr per day (0800–2000 standard time) for most experimental designs from 1984 to 1986. The latter protocol was accepted by the program because the proportional addition permitted a more realistic rise and fall of O_3 throughout the day, and high O_3 concentrations often occurred well into the evening. Two experiments were run to determine possible differences between constant and proportional additions, and between the 7- and 12-hr exposure regimes. The 7-hr constant and 7-hr proportional addition gave similar results for yield in soybean (Heagle et al. 1986) and tobacco (Heagle et al. 1987). However, differences were found between the 7- and 12-hr proportional additions in tobacco unless the 7-hr addition used a 12-hr averaging statistic.

Experimental designs in 1980 focused on O_3 but most designs from 1981 were factorial [O_3 as the primary factor with one from a number of other factors (cultivars, soil moisture, sulfur dioxide, exposure dynamics) included in the design]. In all cases the yield response was the one of primary concern, but many other responses were measured that helped explain the effects of O_3 and other factors on crop response.

The most economically important crops were first selected for study, including corn, cotton, peanut, sorghum, soybean, and wheat. Later alfalfa, barley, several clovers, and tobacco were included. Several horticultural crops (kidney bean, lettuce, tomato, turnip) were also included. In general, the most commonly used cultivars in a region were selected.

Issues Addressed by NCLAN. One program concern was the importance of the chamber in affecting plant response to O_3. Most experimental designs included both the AA and NF treatment plots where the O_3 concentrations were approximately equal (the NF plots were slightly lower due to O_3 losses on particulate filter and plenum walls). Possible chamber effects were tested by comparing the mean yields in the AA and NF treatments (Heck et al. 1982c) or by the deviation of the AA plot yield from the O_3 dose–crop yield response curve estimated for the chambered plots (Heck et al. 1983a, 1984b). The chamber effect was evaluated with a *t*-test (Heck et al. 1983a) and was significant for peanut, turnip, and for three of four wheat cultivars but not for soybean, corn, cotton, kidney bean, or lettuce. Heagle et al. (1988) addressed yield results from the entire NCLAN program, including a nonstatistical comparison of the NF and AA treatments. They reported NF/AA ratios of 0.77 to 1.47 for 24 of the experimental designs. Eighteen of the ratios were between 0.86 and 1.14, suggesting that, in general, chamber effects were not a major problem. A more in-depth assessment of chamber effects was not attempted but reasons for effects were discussed. In the first assessment (Heck et al. 1982c), the AA plots were included in the dose response functions when no significant chamber effect was found. In subsequent papers, the AA plots were not included in the response functions regardless of their significance level because AA treatments were intended only to test for chamber effects. However, the AA plots were always included in the analyses for the estimation of experimental error. Actually the presence of a chamber effect is not in itself important because yield losses are shown on a relative basis. Possible interactions between O_3 and the general chamber environment cannot be tested because the experimental designs were not set up to test this possible interaction; the interaction needs to be tested.

Efforts were made to reduce treatment variability by making the field conditions as uniform as possible across treatments. The uniform use of fertilizers, irrigation, and pesticides was stressed. Also, topsoil depth, nutrition level, and amount of organic material were used to block treatments across the field and in the selection of the plots. Yields from "comparison" plots were used in covariate analyses to adjust treatment means in several early experiments, but these covariates were not needed.

The relative importance of both genetic and environmental uncertainties in a national assessment is not understood. They may not be critical in a first-cut assessment such as NCLAN. No associated stress factors (such as SO_2 and soil moisture) were included in the NCLAN yield functions for the final assessment effort although the final economic assessment made some corrections for soil moisture based on a modeling study by King (1988). However, other environmental factors were indirectly considered when the homogeneity or heterogeneity of combined data sets was determined. The homogeneity found across many data sets within a crop suggests that there may be fewer interactions between O_3 and other factors on a relative yield loss basis than might be expected. However, the small number of data sets and the variability within data sets make it impossible to draw definitive conclusions on the significance of genetic and environmental factors on the response of crops to O_3. It is critical that the importance of these factors be understood before firm national assessments can be made.

The NCLAN program addressed the selection of a natural background O_3 concentration to use as a reference to estimate crop loss. Natural O_3 results from stratospheric transfer and from photochemical reactions involving biogenic hydrocarbons and naturally released nitrogen oxides. It is necessary to know the concentration of natural O_3 in order to determine the effects of anthropogenic O_3 on crop yields. One NCLAN assessment paper (Heck et al. 1984b) highlighted these issues and concluded, from the information presented, that a seasonal 7-hr/day mean O_3 concentration of 0.025 ppm was a reasonable estimate for natural O_3. More precise estimates would be of value, but under the current anthropogenic loading of O_3 precursors it may be impossible to obtain a more definitive estimate. This estimate approximates the CF treatment values from the NCLAN data sets and does not require extrapolation of derived dose-response information to a possible lower natural O_3 concentration. A seasonal 12 hr/day mean O_3 of 0.020 ppm seems to be a reasonable estimate for calculating yield losses based on a 12-hr exposure metric. It should be noted that lack of definitive information does not preclude the use of other estimates of natural O_3 in any assessment effort.

The NCLAN program utilized a shared-time monitoring system (Heagle et al. 1979) in which a single monitor was used to monitor up to 20 points for 2–3 minutes at each point (two to three measurements per hour). Air samples were continuously drawn through long lines to a sampling manifold and diverted to the monitor as noted. The efficiency of the individual monitoring tubes was determined on a regular basis to be between 85 and 100 percent. Correction factors were used to calculate actual O_3 concentrations. Several monitoring studies done to determine the adequacy of the shared-time approach (Heck et al. 1986b) found that the approach gave as good results as using a continuous monitor for all long-term means, good results for calculating 7-hr means, but not such good results for calculating individual hourly averages.

The NCLAN program adopted the seasonal 7 hr/day mean because the 0900 to 1600 time period included the time when plants are generally the most physiologically active and when O_3 concentrations are usually the highest. The 12 hr/day seasonal mean was used for the experiments where O_3 was dispensed over that time. The seasonal 7 hr/day statistic was compared with the seasonal 1 hr/day and with two short-term statistics (Heck et al. 1984c), and the seasonal statistics were found to be the best. It is interesting to note that Larsen and Heck (1984) presented a method to calculate an effective O_3 mean for use in assessing crop losses. The NCLAN publications also include the daily 7- and 24-hr ambient O_3 means through the growing season (Heck et al. 1983a, 1985) as a way to help interpret the biological responses.

Summary Results from the NCLAN Program. The crops and cultivars for which NCLAN has developed O_3 dose–yield response data are shown in table 6.14, with the sites and years for each experiment. Experiments involving O_3 interactions with genetic and environmental factors are shown in table 6.15. No attempt has been made in this chapter to summarize all the results reported by the various investigators, but many such reports are in the proceedings of the international conference (Heck et al. 1988a, 1988b). Heagle et al. (1988) has an excellent summary of yield data and predicted losses at different seasonal (7 and 12 hr) mean O_3 concentrations.

Data from the early experiments were analyzed using linear equations (Heck et al. 1982c) because they fit the data well. However, it was clear that nonlinear models could better reflect yield losses associated with O_3, and these were used in subsequent assessment papers (Heck et al. 1983a, 1984c). A number of three-parameter models were tested (Rawlings and Cure 1985) using data from the first three years of NCLAN and, except for the quadratic, they did not differ in terms of goodness of fit. From the models tested, the Weibull model was chosen for several reasons: The flexible form covers the range of responses observed; the form is biologically realistic; the parameters are easily interpreted; it provides direct estimates of proportional yields; and tests for homogeneity of proportional yield responses over data sets are easy to accomplish. The Weibull model (Rawlings and Cure 1985) is given as

$$y = \alpha \exp\{-(x / \sigma)^c\} + \epsilon,$$

where y is the observed yield and x is the O_3 concentration (ppm). The three estimated parameters are α, the hypothetical maximum yield at zero O_3 concentration; σ, the O_3 concentration when y is 0.37 α; and c, a dimensionless shape parameter that gives the model flexibility. The term ϵ is the random variation associated with each experimental unit. Factors that affect yield will affect α. An analysis of variance was used to obtain the estimate of experimental error for hypothesis testing when fitting the Weibull model to each data set. Predicted relative yield losses (percent) at four seasonal 7 hr/day mean O_3 concentrations are shown for major field crops tested through 1982 (table 6.16). Predicted relative yield losses (percent) at three seasonal 12 hr/day mean O_3 concentrations are shown for selected field crops tested from 1983 through 1986 (table 6.17). Figure 6.3 shows results drawn from summaries of NCLAN data through 1982; later results are very similar and are shown in Heagle et al. (1988). Additionally, graphic results are shown for nine species in the economic paper by Adams et al. (1988). The results are representative of data from the NCLAN program collected from 1980 through 1986.

Table 6.14. *Crops and cultivars for which NCLAN developed ozone dose response data.*

Crop/cultivar	Crop coverage sites and years grown
Alfalfa	
WL-312	Argonne, Ill. (1984)
WL-514	Shafter, Calif. (1984, 1985)
Barley	
CM-72	Shafter, Calif. (1983)
Poco	Shafter, Calif. (1982)
Clover	
Ladino clover/fescue	Raleigh, N.C. (1984, 1985)
Red clover/timothy	Ithaca, N.Y. (1984, 1985)
Corn	
PAG 397	Argonne, Ill. (1981)
Pioneer 3780	Argonne, Ill. (1981, 1985)
FR20A × FR634	Argonne, Ill. (1985)
FR20A × FR35	
FR23 × LH 74	
Cotton	
Acala SJ2	Shafter, Calif. (1981, 1982, 1985)
Stoneville 213	Raleigh, N.C. (1982)
McNair 235	Raleigh, N.C. (1985)
Grain sorghum	
A28+	Argonne, Ill. (1982)
Kidney bean	
California Light Red	Ithaca, N.Y. (1980, 1982)
Lettuce	
Empire	Shafter, Calif. (1983)
Peanut	
NC-6	Raleigh, N.C. (1980)
Soybean	
Davis	Raleigh, N.C. (1981, 1982, 1983, 1984)
Young	Raleigh, N.C. (1986)
Hodgson	Ithaca, N.Y. (1981)
Williams 79	Beltsville, Md. (1981, 1982, 1983)
	Argonne, Ill. (1983)
Corsoy 79	Argonne, Ill. (1980, 1983, 1985, 1986)
	Beltsville, Md. (1983)
Essex	Beltsville, Md. (1981)
Forrest	Beltsville, Md. (1982)
Amsoy 71	Argonne, Ill. (1983)
Pella	Argonne, Ill. (1983)
Tobacco	
McNair 994	Raleigh, N.C. (1983)
Tomato	
Jet Star	Beltsville, Md. (1980)
Murietta	Tracy, Calif. (1981, 1982)
Wheat	
Abe	Argonne, Ill. (1982, 1983)
Arthur 71	Argonne, Ill. (1982, 1983)
Roland	Argonne, Ill. (1982)
Vona	Ithaca, N.Y. (1982, 1983)

Source: W. W. Heck et al. 1986b. *National Crop Loss Assessment Network (NCLAN) 1985 Annual Report*, EPA/600/3-86/041, Environmental Research Laboratory, Office of Research and Development. Corvallis, Oreg.: U.S. Environmental Protection Agency.

Table 6.15. *NCLAN field experiments involving interactions of genetic and environmental factors and O_3 response.*

Crop/cultivar	Year	Site
Cultivar sensitivity ranking		
Corn (14 hybrids)	1981	Argonne, Ill.
Soybean		
Williams 79, Pella,	1983	Argonne, Ill.
Corsoy 79, Amsoy 71		
Interactions: O_3 and soil moisture		
Alfalfa, WL-514	1984, 1985	Shafter, Calif.
Barley, CM 72	1983	Shafter, Calif.
Cotton		
Acala SJ2	1981, 1982, 1985	Shafter, Calif.
McNair 235	1985	Raleigh, N.C.
Ladino clover/fescue	1984, 1985	Raleigh, N.C.
Soybean		
Williams 79, Forrest	1982	Beltsville, Md.
Williams 79, Corsoy 79	1983	Beltsville, Md.
Corsoy 79	1985, 1986	Argonne, Ill.
Davis	1983, 1984	Raleigh, N.C.
Young	1986	Raleigh, N.C.
Interactions: O_3 and SO_2		
Alfalfa, WL-312	1984	Argonne, Ill.
Corn	1985	Argonne, Ill.
Pioneer 3780		
FR20A × FR634		
FR20A × FR35		
FR23 × LH74		
Cotton		
Stoneville 213	1982	Raleigh, N.C.
Red clover/timothy	1984, 1985	Ithaca, N.Y.
Soybean		
Davis	1981	Raleigh, N.C.
Williams 79	1981, 1982	Beltsville, Md.
Essex	1981	Beltsville, Md.
Forrest	1982	Beltsville, Md.
Amsoy 71	1983	Argonne, Ill.
Corsoy 79	1983	Argonne, Ill.
Tomato		
Jet-Star	1980	Beltsville, Md.
Murrieto	1981, 1982	Tracy, Calif.
Wheat		
Vona	1983	Ithaca, N.Y.
Ozone exposure dynamics		
Soybean		
Davis	1982	Raleigh, N.C.
Tobacco		
McNair 944	1983	Raleigh, N.C.

Source: W. W. Heck et al. 1986b. *National Crop Loss Assessment Network (NCLAN) 1985 Annual Report*, EPA/600/3-86/041, Environmental Research Laboratory, Office of Research and Development. Coravillis, Oreg.: U.S. Environmental Protection Agency.

Table 6.16. *Predicted relative yield losses (percent) at four seasonal 7 hr/day mean O_3 concentrations using the Weibull function.*

Species/cultivars[a]	Concentration (ppm)			
	0.04	0.05	0.06[b]	0.09[b]
Barley				
Poco	0.1	0.2	0.5	2.9
Bean, kidney				
Calif. Light Red (FP)	11.0	18.1	24.8	42.6
(PP)	1.7	4.4	9.1	38.4
Corn				
PAG 397	0.2	0.7	1.5	8.1
Pioneer 3780	1.2	2.6	4.8	16.7
Common response	0.6	1.5	3.0	12.5
Cotton				
Acala SJ-2 81 (I)	5.9	10.0	14.0	25.9
Acala SJ-2 81 (D)	1.4	2.7	4.2	10.1
Acala SJ-2 82 (I)	11.3	20.9	31.4	62.4
Acala SJ-2 82 (D)	4.5	10.4	19.1	57.4
Stoneville 213	4.8	9.9	16.4	42.2
Common response[c]	4.0	6.9	10.0	20.0
Peanut				
NC-6	6.4	12.3	19.4	44.5
Sorghum				
DeKalb 28	0.8	1.5	2.5	6.5
Soybean				
Corsoy[d]	5.6	10.4	15.9	34.8
Davis 81	11.5	18.1	24.1	39.0
Davis 82 (CA)	5.1	9.8	15.4	35.6
Davis 82 (PA)	1.8	4.8	10.2	43.7
Essex	4.7	8.2	12.0	b
Williams 81	6.3	10.4	14.4	b
Forrest (I)	1.3	2.8	5.0	15.3
Williams 82 (I)	7.1	11.5	15.7	27.2
Williams 82 (D)	5.6	9.9	14.5	29.1
Hodgson (FP)	6.7	10.9	14.9	25.9
Common response[e]	10.3	16.6	22.4	b
Tomato	7.3	12.1	17.0	30.7
	0.7	1.7	3.6	
Murrieta 81	0.7	1.7	3.6	16.0

(continued)

Integration of Response Functions/Ozone Statistics/Crop Inventories to Assess National Yield Effects and Economic Losses

Ozone Functions for Use in Crop Loss Assessments

The ozone dose–crop yield response functions were discussed in the preceding section. The NCLAN program determined that the Weibull function had a number of redeeming features and accepted the relative response portion of the Weibull function for use in assessment efforts. The Weibull permitted the assess-

Table 6.16 (Continued)

Species/cultivars[a]	Concentration (ppm)			
	0.04	0.05	0.06[b]	0.09[b]
Murrieta 82	8.2	17.7	b	b
Wheat, winter				
Abe	3.1	6.2	10.2	26.7
Arthur	3.7	7.2	11.4	27.4
Roland	9.4	16.4	23.7	45.4
Vona	24.6	37.6	48.3	70.7
Common response[f]	3.5	6.9	11.1	27.4

Source: W. W. Heck et al. 1984c. *Journal of Air Pollution Control Association.* 34: 810.

Note: Yield losses are calculated relative to a seasonal 7 hr/day O_3 mean of 0.025 ppm.

[a]Identifying information is given where necessary, including the year for the experimental design.

[b]The b indicates that the highest treatment concentration was below the column heading value. The data should not be extrapolated beyond the treatment concentration.

[c]This combines the I and D Acala SJ-2 data sets from 1981. When either of the Shafter 1982 or Raleigh data sets were incorporated, the combined model became heteogeneous. The two Shafter 1982 data sets could be combined but would not combine with the Raleigh data set. The Raleigh data using Stoneville 213 may better represent southern cotton cultivars.

[d]The Corsoy soybean yield data were adjusted for the effects of a virus infection using percentage of infected plants as a covariate.

[e]A number of combinations of the 11 soybean data sets werd tried. The 11 data sets gave a heterogeneous response ($F = 3.69$ with 24 and 112 dF). A combination model including all but the Davis 1981 and the Davis 1982 (PA) data sets gave a response with an F value of 1.177 with 18 and 83 dF that was just significant at 0.05; these nine data sets were used for the combination model shown.

[f]This is a combined model for the Abe and Arthur 71 data sets. When either Roland or Vona was added, the model became heterogeneous.

ment either to utilize a homogeneous response function for a species (a single or several cultivars) developed from experimental designs across years or sites, or to use a heterogeneous response function if such a function did not seem to push the data too far (if no apparent inconsistencies were observed). In the assessment of yield effects (tables 6.16 and 6.17, figure 6.3), the combined data sets shown in table 6.16 were homogeneous; in the 1988 economic assessment, heterogeneous data sets were used, where necessary, to describe the effects on a single species. For a general estimate of crop losses, heterogeneous data sets are probably useful.

Table 6.17. *Predicted relative yield losses (percent) at three seasonal 12 hr/day mean O_3 concentrations using the Weibull function.*

Species cultivars/years	Concentration (ppm)		
	0.04	0.06	0.08
Alfalfa (1984, 1985)			
WL-514	5.0	11.5	19.0
Corn (1985)			
Pioneer 3780 and LH74 X FR23	2	6	15
Cotton (1985)			
McNair 235	7	21	40
Fescue/ladino clover mix (1984, 1985)			
Kentuck 31 Regal	5.5	14.5	27.0
Soybean (1986)			
Young	6	17	34
Tobacco (1983)			
McNair 944	7	17	28

Source A. S. Heagle et al. 1988. In Assessment of Crop Loss from Air Pollutants, ed. W. W. Heck, D. T. Tingey, and O. C. Taylor. London: Elsevier Science Publishers.

Note: Yield losses are calculated relative to a seasonal 12 hr/day O_3 mean of 0.020 ppm.

However, as more detailed information becomes available, homogeneous data sets will permit better estimates of loss.

Ozone dose–crop yield response functions can take on a variety of forms depending upon the dose statistic used. There is no preferred function but it should utilize an exposure statistic that adequately describes the biological response to O_3. The NCLAN program used 7 and 12 hr/day seasonal mean O_3 values because they adequately described the yield responses of the crops tested. These functions were used by Adams et al. (1989) in their final economic assessment for NCLAN.

Interpolation of an Ozone Statistic across a Region or across the Country

Defining any O_3 statistic across a specific region requires a detailed monitoring program at as many sites as possible. The NCLAN program utilized data from EPA's Storage and Retrieval of Aerometric Data system for interpolation processes. Limitations of the SAROAD system are twofold. First, most monitoring sites are urban, with few in rural crop-growing areas. Second, many areas of the United States have only a few monitoring sites. Even with these limitations, monitoring data for O_3 are much more extensive than for any other pollutants of concern. Several factors enhance our ability to interpolate O_3 data across broad areas. First, O_3 precursors are transported over great distances. Second, O_3 is

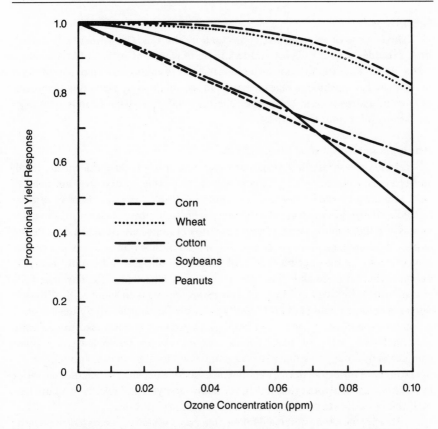

Figure 6.3. Effects of chronic O_3 exposures on the proportional yield loss of five crop species as predicted by the Weilbull model, using results from open-top chambers. The O_3 concentration is the seasonal 7 hr/day mean. Source: Heck et al. 1986a.

more stable as air masses move into rural areas because concentrations of reactive chemical species are reduced (Heck et al. 1984b).

There is no universally accepted way to interpolate O_3 monitoring data. Many have felt that it is not reasonable to perform any type of interpolation. However, NCLAN (Heck et al. 1983b, 1984b) used the kriging spatial interpolation process to develop county-level seasonal (7 and 12 hr/day) mean O_3 concentrations for use in crop loss assessment. This technique was fully reviewed by an outside group (Heck et al. 1985, Lefohn et al. 1987) and found to be a reasonable approach to estimate county-level O_3 concentrations. Detailed results of the kriging operation are found in Heck et al. 1985 and in Lefohn et al. 1987. It should be noted that this technique does not handle mountainous terrain well and is weak where monitoring points are too far apart, as in much of the western United States.

Crop Data

The Census of Agriculture, conducted by the U.S. Department of Agriculture, provides a county-level yield statistic for crops of interest (Adams et al. 1989, Heck et al. 1982b, Shriner et al. 1984). It involves an extensive national inventory of crops and acreage grown. Data are obtained by analyzing responses to questionnaires mailed out approximately every five years. County estimates are adjusted for nonrespondents.

Analysis of Crop Yield Reductions

Ozone dose–crop yield response functions, crop yields at the county level, and seasonal 7 hr/day mean O_3 concentrations at the county level are used to calculate crop losses related to reductions in yield. The impacts of O_3 are reflected in the yield data found in the Census of Agriculture. The county-level O_3 values and the response function for the crop of interest are used to calculate the expected percentage yield reduction of the crop in the county. The yield reduction is based on comparing the yield found at the observed county-level O_3 values with the expected yield at a 0.025 ppm (seasonal 7 hr/day mean O_3 concentration in clean air) O_3 concentration. Increases in yield with different percentage improvements in air quality can then be calculated. Values for each crop and each county where a given crop is grown are then used to calculate national yield losses for each crop of interest. Simple dollar losses are often calculated by using the value of a unit of the commodity and multiplying this by the number of units lost to O_3. This can be used for each crop to calculate a dollar loss value, but the approach is not an economic analysis and results in an unclear picture of real losses associated with O_3 as an air pollutant.

This approach was first used for an analysis with four species (corn, peanut, soybean, and wheat) by Heck et al. (1982b) as part of a larger assessment by Shriner et al. (1984). The assessment was done to show the value of the selected approach in documenting regional and national losses of crops due to O_3. Nine NCLAN type data sets (one corn, one peanut, three soybean, and four wheat) obtained from two NCLAN and from seven pre-NCLAN studies at North Carolina State University were utilized in the assessment effort. County O_3 and crop inventory data for 1978 were used. Kriging of the 1978 O_3 data was first attempted and used on a national basis by James Reagan of the U.S. EPA (Heck et al. 1983b) to predict county-wide yield losses. Several ozone concentration–crop yield response functions were tested and the simple linear form was chosen. The data sets were used with the 7 hr/day seasonal mean O_3 statistic in calculating yield reductions. Results were calculated for county units, and tables and maps were developed to summarize and show patterns of the O_3 effects on soybean, corn, wheat, and peanuts. The assessment estimated that approximately $3 billion of productivity in the four crops would be gained if current maximum 7 hr/day seasonal O_3 concentrations were reduced to 0.025 ppm. Soybean represented 64 percent of the impact, corn 17 percent, wheat 12 percent, and peanut 7 percent. The methodology is sound but the assessment of dollar losses cannot be considered an economic analysis.

The same basic approach was utilized by Adams et al. (1984b, 1988) in developing yield reduction/increase estimates for their interim (1984) and final (1988) NCLAN economic assessment efforts. In the early assessment (Adams et al. 1984b) the Weibull model was used for the response function. Two O_3 data sets were utilized: the kriged 1980 O_3 data on a county basis and the 1978–82 five-year average O_3 data base kriged to the county level. Crop inventory data were used and losses were based on 1980 dollars. Six crops (barley, corn, cotton, sorghum, soybean, and wheat) were used in this assessment. The second assessment (Adams et al. 1988) utilized a similar approach, incorporated three additional crops (alfalfa, hay, and rice) and used 1982 as the base year. Results from the two assessments were in good agreement.

Economic Effects of Ozone on Agriculture

Since 1980 crop response data from many experimental designs have been employed as a basis for a number of regional and national economic assessments. A summary of regional and national assessments, as developed by Adams et al. (1984b), is shown in tables 6.18 and 6.19. Similar information with additional

Table 6.18. *Summary of recent regional air pollution control benefits estimates.*

Region	Reference	Annual benefits or loss estimate ($ million)	Comments
Southern California	Adams et al. 1982	43–45	Estimated as economic surplus in 1976 dollars for 14 annual crops. Employs mathematial programming model to evaluate benefits of reducing current ambient levels to seasonal 7-hr average of 40 ppb
South coast air basin (California)	Leung et al. 1982	93–103 (300)[a]	Losses estimated as economic surplus in 1975 dollars for citrus, avocados, and selected annual crops. Employs econometric procedures to compare "clean air case" (no oxidant pollution) with ambient levels
Ohio River basin	Page et al. 1982	278[b] (6,960)[c]	Losses estimated as producer losses for corn, soybean, and wheat in 1976 dollars. Region includes Illinois, Indiana, Ohio, Kentucky, West Virginia, and Pennsylvania

(continued)

Table 6.18 (Continued)

Region	Reference	Annual benefits or loss estimate ($ million)	Comments
Minnesota	Benson et al. 1982	30.5[d]	Losses estimated in 1980 dollars for corn, alfalfa, and wheat under alternative O_3 assumptions. Farm level dollar losses obtained from econometric model of national commodity markets
Corn belt	Adams and McCarl 1985	668	Uses a sectoral model of U.S. agriculture to record economic effects of changes in yields of corn, soybeans, and wheat due to alternative oxidant standards. Benefits include effects on both consumer and producer of a more stringent federal standard (80 ppb)
Illinois	Mjelde et al. 1984	55–200[e]	Uses profit functions to measure effect of O_3 on producers' profits. Aggregate effect over corn, soybean, and wheat assuming a 25 percent reduction in ambient O_3
California	Howitt et al. 1984	37	Benefits measured as an increase in economic surplus from a reduction in ambient O_3 to 40 ppb seasonal 7-hr average. Analysis based on a mathematical programming model of California annual crops
California	Rowe et al. 1984	46–117	Benefits measured as increase in economic surplus arising from meeting three alternative California oxidant standards. Uses mathematical programming model of major California annual and perennial crops

Source: R. M. Adams, S. A. Hamilton, and B. A. McCarl 1984b. The Economic Effects of Ozone on Agriculture. EPA-600/6-84-090, Environmental Research Laboratory. Corvallis, Oreg.: U.S. Environmental Protection Agency.

[a]Estimate of direct and indirect losses for entire state.

[b]Estimated annual equivalent loss due to oxidants.

[c]Present value of losses due to oxidants for 25-year period (1976–2000).

[d]Worst case O_3 situation, which ignores production effects outside Minnesota. If other regions are included in analysis, worsening of O_3 *increases* total gross returns to Minnesota producers by $67 million due to inelastic nature of commodity demand.

[e]Range of economic benefits due to a 25 percent reduction in O_3 from ambient levels over a four-year period (1978–81).

Table 6.19. *Summary of recent estimates of national economic consequences of pollution.*

Study	Crops	Annual benefits of control ($ million)	Comments
Stanford Research Institute 1981	Corn, soybean, alfalfa, and 13 other annual crops	1,800	Updated version of Benedict-SRI model. Loss measured in 1980 dollars for 531 counties
Shriner et al. 1984	Corn soybean, wheat, peanuts	3,000	Effects estimated in 1978 dollars, measured at producer level. Control assumes a background or "clean-air" oxidant level of 25 ppb O_3. Uses NCLAN response information for 1980
Adams and Crocker 1984	Corn, soybean, cotton	2,200	Benefits measured in 1980 dollars using economic surplus. Benefit represents difference between current production and production if an ambient ozone level of 40 ppb had been achieved. Uses NCLAN response information for 1980
Adams et al. 1984a	Corn, soybeans, wheat, cotton	2,400	Benefits measured as economic surplus in 1980 dollars. Benefits arise from the increase in production due to a reduction in O_3 from 53 ppb to approximately 40 ppb. Response information from 1980, 1981, and 1982 NCLAN data
Kopp et al. 1984	Corn, soybeans, wheat, cotton, peanuts	1,200	Benefits measured as economic surplus in 1978 dollars. Benefits due to reduction in federal standard from 120 ppb to 80 ppb hourly maximum. Uses NCLAN response information from 1980 and 1981

Source: See table 6.18.

summary information is found in the O_3 criteria document (U.S. EPA 1986) in tables 6–31 and 6–32. Regional estimates ranged from about $30 million (Minnesota) to about $670 million (corn belt). National estimates were based on different groups of crops but ranged from $1.2 to $3.0 billion; corn and soybean were included in all estimates.

Adams et al. (1984b) used early NCLAN data to develop an economic model with which they estimated the effects of reduced and increased levels of O_3 on "producer" and "consumer" surplus to give an estimate of total surplus. The economic model was validated using the 1980 performance of the agricultural sector. The economic estimates showed benefits of reduced O_3 to both producers and consumers of agricultural produce. They reported that a 25 percent reduction in O_3 would result in a societal benefit of $1.6–$1.9 billion. Conversely, a 25 percent increase in O_3 would result in a negative benefit to society of $1.9–$2.3 billion. Ozone changes of 10–40 percent showed proportionally greater or lesser benefits, as expected. A final assessment effort using all NCLAN data (Adams et al. 1988) and including some farm program provisions in the analysis gave results similar to the 1984 assessment. The latest assessment estimates are strengthened from the use of more accurate exposure-response and aerometric data, and greater spatial resolution and crop coverage than the 1984 assessment. The 1988 results confirm the earlier findings that O_3 causes a substantial economic cost to society; increases in the yields of eight crops associated with a 25-percent reduction in O_3 (1981–83 averages) would result in a $1.9 billion benefit (in 1982 dollars) while a 40 percent reduction would result in almost a $3.0 billion benefit.

A summary of economic estimates currently available suggests that current seasonal O_3 concentrations are causing in excess of $3 billion annual loss in crop productivity.

Conclusions

This chapter presents an in-depth review of research associated with the effects of O_3 on crops and selected information on the effects of other pollutants, especially SO_2, on plants. The in-depth information presented supports the thesis that O_3 has a major impact on crop production in the United States. Details of assessment methodology are developed along with results of field research that are critical to the prediction of O_3 effects on crop productivity. This section summarizes our current knowledge on the effects of O_3 on crops (statements may be true for other pollutants) and highlights areas of uncertainties in relation to the assessment of effects.

Summary of Current Knowledge on O_3 Effects

- Ozone is responsible for most of the crop yield losses from air pollutants on both a regional and national scale within the United States.

- Annual economic losses are estimated at from \$2–\$5 billion for major agronomic crops within the United States.
- Ozone dose–crop yield response functions are essential for predicting yield losses; nonlinear models give the best fit to available field data.
- Based on available technology the open-top chamber system is the best approach for the development of predictive models.
- The extrapolation of O_3 data on a regional basis, using the interpolation technique of kriging, is a useful and necessary part of an assessment effort. The technique is not suitable with SO_2.
- Foliar symptoms on crops under field conditions often appear as early senescence and may be difficult to assess.
- Although the mechanism of plant response to O_3 is not understood, the cell membrane is probably the site of initial impact.
- Ozone affects photosynthesis and carbon allocation in plants; reduced allocation to roots and reproductive structures is usually found.
- Differences in both species and cultivars within species response to O_3 are found in all crops.
- Interactions between O_3 and both biotic and abiotic factors on plant responses are documented.
- The response of many crop species to O_3 is affected by the presence of other pollutants.
- Most crops show growth, biomass, and yield reduction when grown under current ambient air concentrations of O_3.
- The NCLAN data base has permitted a reasonable first estimate of crop yield losses associated with O_3 as an air pollutant of national importance.

Areas of Uncertainty in Assessing the Effects of O_3 on Crop Production

- The available data base is small and thus is not fully representative of the United States.
- Only ten field crops were studied in the NCLAN program; five had four or more experimental designs.
- Just one to four cultivars were studied in all crops except soybean (nine cultivars).
- Potential chamber effects have not been fully addressed.
- Ozone dose–crop yield response models are empirical and not based on mechanistic considerations.
- The effect of soil moisture on crop yield response to O_3 has been studied but the results are not definitive.
- The effects of other biotic or abiotic stresses on crop response to O_3 are not understood and are not included in the predictive models.
- The importance of peak O_3 values occurring throughout the growing season is not well understood.

- There are insufficient rural monitoring sites to collaborate the kriging interpolative process.
- The estimates of economic loss would probably range from $1 billion to $7 billion or more if all crops were considered; losses will vary from year to year depending on both O_3 concentrations and meteorological conditions.

References

Adams, R. M., and T. D. Crocker. 1984. Economically relevant response estimation and the value of information: Acid deposition. In *Economic Perspectives on Acid Deposition Control*, ed. T. D. Crocker, pp. 35–64. Boston: Butterworth Publishers.

Adams, R. M., and B. A. McCarl. 1985. Assessing the benefits of alternative oxidant standards on agriculture: The role of response information. *J. Environ. Econ. Management* 12:264–276.

Adams, R. M., T. D. Crocker, and R. W. Katz. 1984a. The adequacy of natural science information in economic assessments of pollution control: A Bayesian methodology. *Review of Economic Statistics* 66:568–575.

Adams, R. M., T. D. Crocker, and N. Thanavibulchai. 1982. An economic assessment of air pollution damages to selected annual crops in southern California. *J. Environ. Econ. Management* 9:42–58.

Adams, R. M., J. D. Glyer, and B. A. McCarl. 1989. The NCLAN economic assessment: Approach, findings and implications. In *Assessment of Crop Loss from Air Pollutants*, ed. W. W. Heck, D. T. Tingey, and O. C. Taylor. London: Elsevier Science Publishers.

Adams, R. M., S. A. Hamilton, and B. A. McCarl. 1984b. The Economic Effects of Ozone on Agriculture. Corvallis, Oreg.: U.S. EPA. Environ. Res. Laboratory EPA-600/3-84-090. September 1984. 175 pp.

Alstad, D. N., G. F. Edmunds, and L. H. Weinstein. 1982. Effects of air pollutants on insect populations. *Annual Review of Entomology* 27:369–384.

Altshuller, A. P., and R. A. Linthurst, eds. 1984. The Acidic Deposition Phenomenon and Its Effects: Critical Assessment Review Papers, vol. 2. Effects Sciences, EPA-60018-83-016B. Washington, D.C.: U.S. EPA, Office of Research and Development.

Ashenden, T. W. 1987. Effects of ambient levels of air pollution on grass swards subjected to different defoliation regimes. *Environmental Pollution* 45:29–47.

Athanassious, R. 1980. Ozone effects on radish (*Raphanus sativus* L. cv Cherry Belle): Gradient of ultrastructural changes. *Zeitschrift Pfanzen Physiologic* 97:227–232.

Ayazloo, M., S. G. Garsed, and J. N. B. Bell. 1982. Studies on the tolerance to sulphur dioxide of grass populations in polluted areas II. Morphological and physiological investigation. *New Phytology* 90:109–126.

Baker, C. K., J. J. Colls, A. E. Fullwood, and G. G. R. Seaton. 1986. Depression of growth and yield in winter barley exposed to sulphur dioxide in the field. *New Phytology* 104:233–241.

Banwart, W. L., P. M. Porter, J. J. Hassett, and W. M. Walker. 1987. Simulated acid rain effects on yield response of two corn cultivars. *Agricultural Journal* 79:497–501.

Bell, J. N. B. 1982. Sulphur dioxide and the growth of grasses. In Effects of Gaseous Air Pollution in Agriculture and Horticulture, ed. M. H. Unsworth and D. P. Ormrod, eds., pp. 225–246. London: Butterworth.

Bennett, J. P., and V. C. Runeckles. 1977. Effects of low levels of ozone on growth of crimson clover and annual ryegrass. *Crop Science* 17:443–445.

Benson, E. J., S. Krupa, P. S. Teng, and P. E. Welsch. 1982. Economic Assessment of Air Pollution Damages to Agricultural and Silvicultural Crops in Minnesota. Final Report to Minnesota Pollution Control Agency, St. Paul: University of Minnesota.

Bisessar, S., and K. T. Palmer. 1984. Ozone, antioxidant spray and *Meloidogyne hapla* effects on tobacco. *Atmos. Environ.* 18:1025–1027.

Blum, U., A. S. Heagle, J. C. Burns, and R. A. Linthurst. 1983b. The effects of ozone on fescue-clover forage: Regrowth, yield and quality. *Environmental and Experimental Botany* 23:121–132.

Blum, U., E. Mrozek, Jr., and E. Johnson. 1983a. Investigation of ozone (O_3) effect on ^{14}C distribution in ladino clover. *Environ. Expt. Bot.* 23:369–378.

Bressan, R. A., L. G. Wilson, and P. Filner. 1978. Mechanisms of resistance to SO_2 in the cucurbitaceae. *Plant Physiology* 61:761–767.

Butler, L. K., and T. W. Tibbitts. 1979. Stomatal mechanisms determining genetic resistance to ozone in *Phaseolus vulgaris* L. *Journal of American Society of Horticultural Science* 104:213–216.

Carnahan, J. E., E. L. Jenner, and E. K. W. Wat. 1978. Prevention of ozone injury to plants by a new protectant chemical. *Phytopathology* 68:1225–1229.

Chamel, A., and J. P. Garrec. 1977. Penetration of fluorine through isolated pear leaf cuticles. *Environmental Pollution* 12:307–310.

Chappelka, A. H., M. E. Kraemer, T. Mebrahter, M. Rangappa, and P. S. Benepal. 1988. Effects of ozone on soybean resistance to the Mexican bean beetle (*Epilachna varivestus* Mulsant). *Environmental and Experimental Botany* 28:53–60.

Chevone, B. I., and Y. S. Yang. 1985. CO_2 exchange rates and stomatal diffusive resistance in soybean exposed to O_3 and SO_2. *Canadian Journal of Plant Science* 65:267–274.

Chiment, J. J., R. Alscher, and P. R. Hughes. 1986. Glutathione as an indicator of SO_2-induced stress in soybean. *Environmental and Experimental Botany* 26:147–152.

Coughenour, M. B., J. L. Dodd, D. C. Coleman, and W. K. Lauenroth. 1979. Partitioning of carbon and SO_2-sulfur in a native grassland. *Oecologia* 42:229–240.

Czuba, M., and D. P. Ormrod. 1981. Cadmium concentrations in cress shoots in relation to cadmium enhanced ozone phytotoxicity. *Environmental Pollution (Series A)* 25:67–76.

Damicone, J. P., and W. J. Manning. 1987. Foliar sensitivity of soybeans from early maturity groups to ozone and inheritance of injury response. *Plant Disease* 71:332–336.

Decoteau, D. R., L. Grant, and L. E. Craker. 1986a. Failure of ozone susceptibility tests to predict yield reductions in wheat. *Field Crops Research* 13:185–191.

Decoteau, D. R., J. E. Simon, G. Eason, and R. A. Reinert. 1986b. Ozone-induced injury on field-grown watermelons. *HortScience* 21:1369–1371.

DeVos, N. E., R. R. Hill, R. W. Hepler, E. J. Pell, and R. Craig. 1980. Inheritance of peroxyacetyl nitrate resistance in petunia. *Journal of American Society of Horticultural Science* 105:157–160.

Dochinger, L. S., and T. A. Seliga, eds. 1976. Proc. First Internat. Symp. Acid Precipitation and the Forest Ecosystem. USDA Forest Service, General Technical Report NE-23. Upper Darby, Pa.

Dohmen, G. P. 1987. Secondary effects of air pollution: Ozone decreases brown rust disease potential in wheat. *Environmental Pollution* 43:189–194.

Dohmen, G. P., S. McNeill, and J. N. B. Bell. 1984. Air pollution increases *Aphis fabae* pest potential. *Nature (London)* 307:52–53.

Dragoescu, N., R. R. Hill, Jr., and E. J. Pell. 1987. An autotetraploid model for genetic analysis of ozone tolerance in potato, *Solanum tuberosum* L. *Genome* 29:85–90.

DuBay, D. T., and A. S. Heagle. 1987. The effects of simulated acid rain with and without ambient rain on the growth and yield of field-grown soybeans. *Environmental and Experimental Botany* 27:395–401.

DuBay, D. T., and W. H. Murdy. 1983a. Direct adverse affects of SO_2 on seed set in *Geranium carolinianum* L: A consequence of reduced pollen germination on the stigma. *Bot. Gaz.* 144:376–381.

DuBay, D. T., and W. H. Murdy. 1983b. The impact of sulfur dioxide on plant sexual reproduction: In vivo and in vitro effects compared. *Journal of Environmental Quality* 12:147–149.

Dunning, J. A., and W. W. Heck. 1977. Response of bean and tobacco to ozone: Effect of light intensity, temperature and relative humidity. *Journal of Air Pollution Control Association* 27:882–886.

Elden, T. C., R. K. Howell, and R. E. Webb. 1978. Influence of ozone on pea aphid resistance in selected alfalfa strains. *Journal of Economic Entomology* 71:283–286.

Endress, A. G., and C. Grunwald. 1985. Impact of chronic ozone on soybean growth and biomass partitioning. *Agric., Ecosyst. and Environ.* 13:9–23.

Endress, A. G., and S. L. Post. 1985. Altered feeding preference of Mexican bean beetle *Epilachna varivestus* for ozonated soybean foliage. *Environmental Pollution (Series A)* 39:9–16.

Evans, L. S. 1984. Botanical aspects of acidic precipitation. *Botanical Review* 50:449–490.

Evans, L. S., K. F. Lewin, C. A. Conway, and M. J. Patti. 1981. Seed yields (quantity and quality) of field grown soybeans exposed to simulated acidic rain. *New Phytology* 89:459–470.

Evans, L. S., K. F. Lewin, and G. R. Hendry. 1986a. Yields of Field-Grown Soybean Exposed to Simulated Acidic Rainfalls. BNL 52009. Department of Applied Science. Brookhaven National Laboratory, Upton, Long Island, N.Y.

Evans, L. S., M. J. Sarrantonio, and E. M. Owen. 1986b. Protein contents of seed yields of field-grown soybeans exposed to simulated acidic rain: Assessment of the sensitivities of four cultivars and effects of duration of simulated rainfall. *New Phytology* 103:689–693.

Facteau, T. J., S. Y. Wang, and K. E. Rowe. 1978. Response of sweet cherry leaf tissue to hydrogen fluoride fumigation at different nitrogen levels. *Journal of American Society of Horticultural Science* 103:115–119.

Feder, W. A. 1981. Bioassaying for ozone with pollen systems. *Environmental Health Perspectives* 37:117–123.

Flagler, R. B., R. P. Patterson, A. S. Heagle, and W. W. Heck. 1987. Ozone and soil moisture deficit effects on nitrogen metabolism of soybean. *Crop Science* 27:1177–1184.

Foster, K. W., J. P. Guerard, R. J. Oshima, J. C. Bishop, and H. Timm. 1983. Differential ozone susceptibility of centenial russet and white rose potato. *American Potato Journal* 60:127–139.

Furukawa, A., and T. Totsuka. 1983. Effect of SO_2 on photosynthesis in sunflower leaves: Age-dependent inhibition. *Environ. Cont. Biol.* 21:43–49.

Garsed, S. G., and D. J. Read. 1977. The uptake and metabolism of $^{35}SO_2$ in plants of differing sensitivity to sulphur dioxide. *Environmental Pollution* 13:173–186.

Ghiasseddin, M., J. M. Hughes, J. E. Diem, and J. V. Mason. 1981. Accumulation of fluoride by the soybean (*Glycine max* L. Merrill Var Dare). Part I. Visible injury and related effects on seed yield. *Journal of Plant Nutrition* 3:429–440.

Greenwood, P., Greenhalgh, A., C. H. Baker, and M. H. Unsworth. 1982. A computer-controlled system for exposing field crops to gaseous air pollutants. *Atmos. Environ.* 16:2261–2266.

Griffith, S. M., and V. F. Campbell. 1987. Effect of sulfur dioxide on nitrogen fixation, carbon partitioning, and yield components in snapbean. *Journal of Environmental Quality* 16:77–80.

Grimes, H. D., K. K. Perkins, and W. F. Boss. 1983. Ozone degrades into hydroxyl radical under physiological conditions. *Plant Physiology* 72:1016–1020.

Guderian, R. 1977. Air Pollution, Phytotoxicity of Acidic Gases and Its Significance in Air Pollution Control. Ecological Studies 22, Springer-Verlag, Berlin.

Guderian, R., ed. 1985. Photochemical Oxidants-Formation, Distribution, Control and Effects on Vegetation. Berlin and New York: Springer-Verlag.

Guri, A. S. A. F. 1983. Attempts to elucidate the genetic control of ozone sensitivity in seedlings of Phaseolus vulgaris L. Canadian Journal of Plant Science 63:727–732.

Harris, M. J., and R. L. Heath. 1981. Ozone sensitivity in sweet corn (Zea Mays L.) plants: A possible relationship to water balance. Plant Physiology 68:885–890.

Hatzios, K. K. 1983. Effects of CGA 43089 on responses of sorghum (Sorghum bicolar) to metolachlor combined with ozone or antioxidants. Weed Science 32:280–284.

Heagle, A. S. 1982. Interactions between air pollutants and parasitic diseases. In Effects of Gaseous Air Pollution in Agriculture and Horticulture, ed. M. H. Unsworth and D. P. Armrod, pp. 333–348. London: Buttersworth Scientific.

Heagle, A. S., and W. W. Heck. 1974. Predisposition of tobacco to oxidant air pollution injury by previous exposure to oxidants. Environmental Pollution 7:247–252.

Heagle, A. S., and W. W. Heck. 1980. Field methods to assess crop losses due to oxidant air pollutants. In Crop Loss Assessments. ed. P. S. Teng and S. V. Krupa, pp. 296–305. E. C. Stakman Commemorative Symposium. Misc. Publ. 7, Agricultural Experiment Station, University of Minnesota, St. Paul.

Heagle, A. S., D. E. Body, and W. W. Heck. 1973. An open-top field chamber to assess the impact of air pollution on plants. Journal of Environmental Quality 2:365–368.

Heagle, A. S., W. W. Heck, V. M. Lesser, and J. O. Rawlings. 1987. Effects of daily ozone exposure duration and concentration fluctuation on yield of tobacco. Phytopathology 77:856–862.

Heagle, A. S., W. W. Heck, J. O. Rawlings, and R. B. Philbeck. 1983. Effects of chronic doses of ozone and sulfur dioxide on injury and yield of soybeans in open-top chambers. Crop Science. 23:1184–1191.

Heagle, A. S., V. M. Lesser, J. O. Rawlings, and W. W. Heck. 1986. Response of soybeans to chronic doses of ozone applied as constant or proportional additions to ambient air. Phytopathology 76:51–56.

Heagle, A. S., R. B. Philbeck, H. H. Rogers, and M. B. Letchworth. 1979. Dispensing and monitoring ozone in open-top field chambers for plant effects studies. Phytopathology 69:15–20.

Heagle, A. S., J. Rebbeck, S. R. Shafer, U. Blum, and W. W. Heck. 1989. Effects of long-term ozone exposure and soil moisture deficit on growth of a ladino clover–tall fescue pasture. Phytopathology 79:128–136.

Heagle, A. S., L. W. Kress, P. J. Temple, R. J. Kohut, J. E. Miller, and H. E. Heggestad. 1988. Factors influencing ozone dose-yield response relationships in open-top field chamber studies. In Assessment of Crop Loss from Air Pollutants. ed. W. W. Heck, D. T. Tingey, and O. C. Taylor. London: Elsevier Science Publishers.

Heath, R. L. 1980. Initial events in injury to plants by air pollutants. Annual Review of Plant Physiology 31:395–431.

Heath, R. L., and A. G. Endress. 1979. Permeability changes in pinto bean leaves exposed to gaseous HCl. Zeitschrift Pflanzenphysiologie 92:271–276.

Heath, R. L., and P. E. Frederick. 1979. Ozone alteration of membrane permeability in chlorella. Plant Physiology 64:455–459.

Heck, W. W. 1982. Future directions in air pollution research. In Effects of Gaseous Air Pollution in Agriculture and Horticulture, ed. M. H. Unsworth and D. P. Ormrod, pp. 411–435. London: Butterworth Scientific.

Heck, W. W. 1984. Defining gaseous pollution problems in North America. In Gaseous Air Pollutants and Plant Metabolism, ed. M. J. Koziol and F. R. Whatley, pp. 35–48. London: Butterworth.

Heck, W. W., and C. S. Brandt. 1977. Effects on vegetation: Native, crops, forest. In Air Pollution ed. A. C. Stern, 3rd ed., vol. 2b, pp. 157–229. New York: Academic Press.

Heck, W. W., and J. A. Dunning. 1978. Response of oats to sulfur dioxide; interactions of growth temperature with exposure temperature or humidity. *Journal of Air Pollution Control Association* 28:241–246.

Heck, W. W., R. M. Adams, W. W. Cure, A. S. Heagle, H. E. Heggestad, R. J. Kohut, L. W. Kress, J. O. Rawlings, and O. C. Taylor. 1983a. A reassessment of crop loss from ozone. *Environmental Science and Technology* 17:573–580A.

Heck, W. W., U. Blum, W. F. Boss, A. S. Heagle, R. A. Linthurst, R. A. Reinert, J. F. Reynolds, and H. H. Rogers. 1984a. Perspectives of air pollution research on plants. In *Reviews in Environmental Toxicology I*, ed. Ernest Hodgson, pp. 173–249. Amsterdam: Elsevier Scientific.

Heck, W. W., U. Blum, R. A. Reinert, and A. S. Heagle. 1982a. Effects of air pollution on crop production. In *Strategies of Plant Reproduction*, ed. W. J. Meudt, pp. 333–350. BARC Symposium No. 6. Totowa, N.J.: Allanheld, Osman & Co.

Heck, W. W., W. W. Cure, J. O. Rawlings, L. J. Zaragoza, A. S. Heagle, H. E. Heggestad, R. J. Kohut, L. W. Kress, and P. J. Temple. 1984b. Assessing impacts of ozone on agricultural crops: I. Overview. *Journal of Air Pollution Control Association* 34:729–735.

Heck, W. W., W. W. Cure, J. O. Rawlings, L. J. Zaragoza, A. S. Heagle, H. E. Heggestad, R. J. Kohut, L. W. Kress, and P. J. Temple. 1984c. Assessing impacts of ozone on agricultural crops: II. Crop yield functions and alternative exposure statistics. *Journal of Air Pollution Control Association* 34:810–817.

Heck, W. W., W. W. Cure, D. S. Shriner, R. J. Olson, and A. S. Heagle, 1982b. Ozone impacts on the productivitty of selected crops. In *Effects of Air Pollution on Farm Commodities*, ed. J. S. Jacobson and A. H. Millin, pp. 147–176. Washington, D.C.: The Izaak Walton League.

Heck, W. W., J. A. Dunning, R. A. Reinert, S. A. Prior, M. Rangappa, and P. S. Benepal. 1988c. Differential response of four bean cultivars to chronic doses of ozone. *Journal of American Society of Horticultural Science* 113:46–51.

Heck, W. W., A. S. Heagle, and D. S. Shriner. 1986a. Effects on vegetation: Native, crops, and forest. In *Air Pollution*, ed. A. C. Stern, 3d ed., vol. 6, pp. 247–350. New York: Academic Press.

Heck, W. W., W. M. Knott, E. P. Stahel, J. T. Ambrose, J. N. McCrimmon, M. Engle, L. A. Romanow, A. G. Sawyer, and J. D. Tyson. 1980. "Response of Selected Plants and Insect Species to Stimulated Solid Rocket Exhaust Mixtures and to Exhaust Components from Solid Rocket Fuels." Tech. Memo. 74109, KSC TR 51-1. National Aeronautics and Space Administration, John F. Kennedy Space Center, Cape Canaveral, Fla. 146 pp.

Heck, W. W., S. V. Krupa, and S. N. Linzon, eds. 1979. *Handbook of Methodology for the Assessment of Air Pollution Effects on Vegetation*. Upper Midwest Section, Air Pollution Control Association, Specialty Conference Proceedings, Pittsburgh, Pa. 392 pp.

Heck, W. W., J. B. Mudd, and P. R. Miller. 1977. Plants and microorganisms. In *Ozone and Other Photochemical Oxidants*, p. 437. Washington, D.C.: National Academy of Sciences.

Heck, W. W., O. C. Taylor, R. Adams, G. Bingham, J. Miller, E. Preston, and L. Weinstein. 1982c. Assessment of crop loss from ozone. *Journal of Air Pollution Control Association* 32:353–362.

Heck, W. W., O. C. Taylor, R. M. Adams, G. Bingham, J. E. Miller, E. M. Preston, and L. H. Weinstein. 1982d. *National Crop Loss Assessment Network (NCLAN) 1981 Annual Report*. EPA 600/3-83-049. Corvallis Environmental Research Laboratory, Office of Research and Development. Corvallis, Oreg.: U.S. EPA. 190 pp.

Heck, W. W., O. C. Taylor, R. M. Adams, G. Bingham, J. E. Miller, E. M. Preston, L. H. Weinstein, R. G. Amundson, R. J. Kohut, J. A. Laurence, W. C. Cure, A. S. Heagle, J. T. Gish, H. E. Heggestad, L. W. Kress, G. E. Neely, J. O. Rawlings, and P. Temple. 1983b. *National Crop Loss Assessment Network (NCLAN) 1982 Annual Report*. EPA-600/3-84-049. Corvallis, Oreg.: U.S. EPA.

Heck, W. W., O. C. Taylor, R. M. Adams, G. Bingham, J. E. Miller, and L. H. Weinstein. 1981. *The National Crop Loss Assessment Network (NCLAN) 1980 Annual Report.* 89 pp. EPA 600/3-82-001. Corvallis Environmental Research Laboratory, Office of Research and Development. Corvallis, Oreg.: U.S. EPA.

Heck, W. W., O. C. Taylor, R. M. Adams, J. E. Miller, E. M. Preston, L. H. Weinstein, R. G. Admundson, W. C. Cure, A. S. Heagle, T. J. Gish, H. E. Heggestad, D. A. King, L. W. Kress, R. J. Kohut, J. A. Laurence, J. Miller, G. E. Neely, J. O. Rawlings, and P. Temple. 1984d. *National Crop Loss Assessment Network (NCLAN) 1983 Annual Report.* EPA 600/3-85-061. Corvallis Environmental Research Laboratory, Office of Research and Development. Corvallis, Oreg.: U.S. EPA. 227 pp.

Heck, W. W., O. C. Taylor, R. M. Adams, J. E. Miller, D. T. Tingey, L. H. Weinstein, R. G. Admundson, A. S. Heagle, D. A. King, R. G. Kohut, L. W. Kress, J. A. Laurence, A. S. Lefohn, V. M. Lesser, J. R. Miller, G. E. Neely, P. J. Temple, and J. O. Rawlings. 1985. *National Crop Loss Assessment Network (NCLAN) 1984 Annual Report.* EPA/600/3-86/041, Environmental Research Laboratory, Office of Research and Development. Corvallis, Oreg.: U.S. EPA. 228 pp.

Heck, W. W., O. C. Taylor, R. M. Adams, J. E. Miller, D. T. Tingey, L. H. Weinstein, R. G. Admundson, A. S. Heagle, D. A. King, R. G. Kohut, L. W. Kress, J. A. Laurence, A. S. Lefohn, V. M. Lesser, J. R. Miller, G. E. Neely, P. J. Temple, and J. O. Rawlings. 1986b. *National Crop Loss Assessment Network (NCLAN) 1985 Annual Report.* EPA/600/3-86/041, Environmental Research Laboratory, Office of Research and Development. Corvallis, Oreg.: U.S. EPA.

Heck, W. W., D. T. Tingey, and O. C. Taylor, eds. 1988a. *Assessment of Crop Loss From Air Pollutants.* London: Elsevier Science Publishers.

Heck, W. W., D. T. Tingey, and O. C. Taylor, eds. 1988b. Assessment of crop loss from air pollutants. *Environmental Pollution,* special issue, vol. 53.

Heggestad, H. E., J. H. Bennett, E. H. Lee, and L. W. Douglass. 1986. Effects of increasing doses of sulfur dioxide and ambient ozone on tomatoes: Plant growth, leaf injury, elemental composition, fruit yield and quality. *Phytopathology* 76:1338–1344.

Heggestad, H. E., T. J. Gish, E. H. Lee, J. H. Bennett, and L. W. Douglass. 1985. Interaction of soil moisture stress and ambient ozone on growth and yields of soybeans. *Phytopathology* 75:472–477.

Hofstra, G., D. A. Littlejohns, and R. T. Wukasch. 1978. Efficacy of antitoxidant ethylene-diurea (EDU) compared to carboxin and benomyl in reducing yield losses from ozone in navy bean. *Plant Dis. Rep.* 62:350–352.

Horsman, D. C., T. M. Roberts, and A. D. Bradshaw. 1979. Studies on the effect on sulfur dioxide on perennial ryegrass (*Lolium perenne* L.). II. Evolution of sulfur dioxide tolerance. *Journal of Experimental Botany* 30:495–501.

Howitt, R. E., T. E. Gossard, and R. M. Adams. 1984. Effects of alternative ozone levels and response data on economic assessments: The case of California crops. *Journal of Air Pollution Control Association* 34:1122–1127.

Hucl, P., and W. D. Beversdorf. 1982. The inheritance of ozone insensitivity in selected *Phaseolus vularis* L. populations. *Canadian Journal of Plant Science* 62:861–866.

Hughes, P. R., J. J. Chiment, and A. I. Dickie. 1985. Effect of pollutant dose on the response of Mexican bean beetle (Coleoptera: Coccinellidae) to SO_2-induced changes in soybean. *Environmental Entomology* 14:718–721.

Hughes, P. R., A. I. Dickie, and M. A. Penton. 1983. Increased success of the Mexican bean beetle on field grown soybeans exposed to sulfur dioxide. *Journal of Environmental Quality* 12:565–568.

Illman, B. L., and E. J. Pell. 1985. Characterization of the ozone response of potato leaf protoplasts. *Canadian Journal of Botany* 63:1936–1941.

Irving, P. M. 1987. Effects on agricultural crops. In: *Interim Assessment: The Causes and*

Effects of Acidic Deposition. Vol. 4, chapter 6. Washington, D.C.: National Acid Precipitation Assessment Program.

Irving, P. M., and J. E. Miller. 1984. Synergistic effect on field-grown soybeans from combinations of sulfur dioxide and nitrogen dioxide. *Canadian Journal of Botany* 62:840–846.

Jacobson, J. S., and A. C. Hill, eds. 1970. *Recognition of Air Pollution Injury to Vegetation: A Pictorial Atlas.* Pittsburgh, Pa.: Air Pollution Control Association, 102 pp.

Jäger, H. J., and H. Klein. 1980. Biochemical and physiological effects of sulfur dioxide on plants. *Angewandte Botanik* 54:337–348.

Janakiraman, R., and P. M. Harvey. 1976. Effects of ozone on meiotic chromosomes of *Vicia faba. Canadian Journal of Genetic Cytology* 18:727–730.

Jeong, Y. H., H. Nakamura, and L. Ota. 1981. Physiological studies on photochemical oxidant injury in rice plants II. Effect of abscisic acid (ABA) on ozone injury and ethylene production in rice plants. *Japan Crop Science* 50:560–565.

Johnston, J. W., Jr., and A. S. Heagle. 1982. Response of chronically ozonated soybean plants to an acute ozone exposure. *Phytopathology* 72:387–389.

Johnston, J. W., D. S. Shriner, and C. H. Abner. 1986. Design and performance of an exposure system for measuring the response of crops to acid rain and gaseous pollutants in the field. *Journal of Air Pollution Control Association* 36:894–899.

Kats, G.., P. J. Dawson, A. Bytnerowicz, J. W. Wolf, C. R. Thompson, and D. M. Olszyk. 1985. Effects of ozone or sulfur dioxide on growth and yield of rice. *Agric., Ecosyst. and Environ.* 14:103–117.

Keitel, A., and U. Arndt. 1983. Ozone-induced turgidity losses of tobacco (*Nicotiana tobacum* var. Bel W_3)—an indication to rapid alterations of membrane permeability. *Angewandte Botanik* 57:193–204.

King, D. A. 1988. Modeling the impact of ozone x drought interactions on regional crop yields. *Environmental Pollution* 53:351–364.

Kisman, S., C. Charlot, S. Brun, and J. C. Cabanis. 1983. Migration du fluor chez *Raphanus sativus. Plant and Soil* 74:417–429.

Klarer, C. I., R. A. Reinert, and J. S. Huang. 1984. Effects of sulfur dioxide and nitrogen dioxide on vegetative growth of soybean. *Phytopathology* 74:1104–1106.

Klein, H., H. J. Jager, W. Domes, and C. H. Wong. 1978. Mechanisms contributing to differential sensitivities of plants to SO_2. *Oecologia* 33:203–208.

Klepper, L. 1979. Nitric oxide (NO) and nitrogen dioxide (NO_2) emissions from herbicide-treated soybean plants. *Atmos. Environ.* 13:537–542.

Knudson, L. L., T. W. Tibbitts, and G. E. Edwards. 1977. Measurement of ozone injury by determination of leaf chlorophyll concentration. *Plant Physiology* 60:606–608.

Kobriger, J. M., T. W. Tibbitts, and M. L. Brenner. 1984. Injury, stomatal conductance, and abscisic acid levels of pea plants following ozone plus sulfur dioxide exposures at different times of the day. *Plant Physiology* 76:823–826.

Kochhar, M., U. Blum, and R. A. Reinert. 1980. Effects of O_3 and (or) fescue on ladino clover: Interactions. *Canadian Journal of Botany* 58:241–249.

Kopp, R. J., W. J. Vaughan, and M. Hazilla. 1984. Agricultural Sector Benefits Analysis for Ozone: Methods Evaluation and Demonstration. U.S. Environ. Protection Agency, Office of Air Quality Planning and Standards, Research Triangle Park, N.C. Report No. EPA-450/5-84-003.

Koziol, M. J., and C. F. Jordon. 1978. Changes in carbohydrate levels in red kidney bean (*Phaseolus vulgaris* L.) exposed to sulfur dioxide. *Journal of Experimental Botany* 29:1037–1043.

Koziol, M. J., and F. R. Whatley, eds. 1984. *Gaseous Air Pollutants and Plant Metabolism.* London: Butterworth. 446 pp.

Krizek, D. T., R. M. Mirecki, and P. Semenuik. 1986. Influence of soil moisture stress and abscisic acid pre-treatment in modifying SO_2 sensitivity in poinsettia. *Journal of American Society of Horticultural Science* 111:446–450.

Kuja, A., R. Jones, and A. Enyedi. 1986. A mobile rain exclusion canopy system to determine dose-response relationships for crops and forest species. *Water, Air, Soil Pollution* 31:307–315.

Larsen, R. I., and W. W. Heck. 1984. An air quality data analysis system for interrelating effects, standards, and needed source reductions—Part 8. An effective mean O_3 crop reduction mathematical model. *Journal of Air Pollution Control Association* 34:1023–1034.

Larsen, R. I., A. S. Heagle, and W. W. Heck. 1983. An air quality data analysis system for interrelating effects, standards, and needed source reductions—Part 7. An O_3-SO_2 leaf injury mathematical model. *Journal of Air Pollution Control Association* 33:198–207.

Lauenroth, W. K., and E. M. Preston, eds. 1984. *The Effects of SO_2 on a Grassland: A Case Study in the Northern Great Plains of the United States*. Ecological Studies Series, vol. 45. Berlin and New York: Springer-Verlag. 207 pp.

Laurence, J. A. 1981. Effects of air pollutants on plant pathogen interactions. *Zeitschrift für Pflanzenkrankheiten und Pflanzenschutz* 88:156–172.

Laurence, J. A., and K. L. Reynolds. 1984. Growth and lesion development of *Xanthomonas campestris* pv. Phaseoli on leaves of red kidney bean plants exposed to hydrogen fluoride. *Phytopathology* 74:578–580.

Laurence, J. A., A. L. Alusio, L. H. Weinstein, and D. C. McCune. 1981. Effects of sulfur dioxide on southern bean mosiac and maize dwarf mosiac. *Environ. Pollut. (Series A)* 24:185–191.

Laurence, J. A., K. L. Reynolds, and C. S. Greitner. 1985. Bioindicator of SO_2: Response of three plant species to variation in dosage-kinetics of SO_2. *Environmental Pollution (Series A)* 37:43–52.

Laurence, J. A., L. H. Weinstein, D. C. McCune, and A. L. Alusio. 1979. Effects of sulfur dioxide on southern corn leaf blight of maize and stem rust of wheat. *Plant Dis. Rept.* 63:975–978.

Lechowicz, M. J. 1987. Resource allocation by plants under air pollution stress: Implications for plant-pest-pathogen interactions. *Bot. Review* 53:281–300.

Lee, E. H., J. K. Byun, and S. J. Wilding. 1985. A new gibberellin biosynthesis inhibitor, paclobutrazol (pp $_{333}$), confers increased SO_2 tolerance on snap bean plants. *Environmental and Experimental Botany* 25:265–275.

Lee, J. J., E. M. Preston, and R. A. Lewis. 1978. A system for the experimental evaluation of the ecological effects of sulfur dioxide. Proceedings of 4th Joint Conference on Sensing of Environmental Pollutants, pp. 49–53. Washington, D.C. American Chemical Society.

Lefohn, A. S., and D. P. Ormrod, eds. 1984. *A Review and Assessment of the Effects of Pollutant Mixtures on Vegetation—Research Recommendations*. EPA-600/3-84-037. Corvallis Environmental Research Laboratory, Office of Research and Development, Corvallis, Oreg.: U.S. EPA. 104 pp.

Lefohn, A. S., H. P. Knudsen, J. Logan, J. Simpson, and C. Bhumralkar. 1987. An evaluation of the Kriging method, as applied by NCLAN, to predict 7-h seasonal ozone concentrations. *Journal of Air Pollution Control Association* 37:595–602.

Letchworth, M. B., and U. Blum. 1977. Effects of acute ozone exposure on growth, nodulation, and nitrogen content of *Ladino clover*. *Environmental Pollution* 14:303–312.

Leung, S. K., W. Reed, and S. Geng. 1982. Estimations of ozone damage to selected crops grown in southern California. *Journal of Air Pollution Control Association* 32:160–164.

Leuning, R., M. H. Unsworth, H. N. Newmann, and K. M. King. 1979. Ozone fluxes to tobacco and soil under field conditions. *Atmos. Environ.* 13:1155–1163.

Linzon, S. N. 1978. Effects of airborne sulfur pollutants on plants. In *Sulfur in the Environment: Part III. Ecological Impacts* ed. J. R. Nriagu, pp. 109–162. New York: John Wiley.

Lorenzini, G., A. Mimack, and M. R. Ashmore. 1985. Differential response of alfalfa strains to chronic and acute fumigations with O_3, SO_2, NO_2 and a mixture of SO_2 and NO_2. *Rivista di Patologia Vegetale* 21:13–27.

McCune, D. C., and T. C. Weidensaul. 1978. Effects of atmospheric sulfur oxides and related compounds on vegetation. In *Sulfur Oxides*, pp. 80–129. Washington, D.C. National Academy of Sciences.

McLaughlin, S. B., and G. E. Taylor. 1981. Relative humidity: Important modifier of pollutant uptake by plants. *Science* 211:167–169.

MacLean, D. C., R. E. Schneider, and L. H. Weinstein. 1982. Fluoride induced foliar activity in *Solanum pseudo-capsicum*: Its induction in the dark and activation in the light. *Environmental Pollution (Series A)* 29:27–33.

McLeod, A. R., and J. Fackrell. 1983. A prototype system for open-air fumigation of agricultural crops: 1. Theoretical design. CERL Note No. TPRD/L/2474/N83. Central Electrical Research Laboratory, Leatherhead, England.

McLeod, A. R., K. Alexander, and D. M. Cribb. 1986. Effects of open-air fumigation with sulfur dioxide on the growth of cereals. I. Grain yield of winter barley (*Hordeum vulgare*) cv. Sonja 1982–83. Report No. TPRD/L/3071/R86. Central Electricity Research Laboratories, Leatherhead, Surry, England.

Mandl, R. H., L. H. Weinstein, D. C. McCune, and M. Keveny. 1973. A cyclindrical open-top chamber for the exposure of plants to air pollutants in the field. *Journal of Environmental Quality* 2:371–376.

Mansfield, T. A., ed. 1976. *Effects of Air Pollutants on Plants*. Society for Experimental Biology-Seminar Series I, Cambridge University Press, Cambridge, Great Britain.

Mansfield, T. A., M. E. Whitmore, and R. M. Law. 1982. Effects of nitrogen oxides on plants: Two case studies. In *Air Pollution by Nitrogen Oxides* ed. T. Schneider and L. Grant, pp. 511–520. Amsterdam: Elsevier Scientific.

Markowski, A., S. Grezesiak, and M. Schramelo. 1975. Indexes of the susceptibility of various species of cultivated plants to sulphur dioxide action. *Bull. Pol. Acad. Sci., Ser. Sci. Biol. Sci.* 23:637–646.

Menchin, P. E. H., and R. Gould. 1986. Effects of SO_2 on phloem loading. *Plant Science* 43:179–183.

Meredith, F. I., C. A. Thomas, and H. E. Heggestad. 1986. Effect of the pollutant ozone in ambient air on lima bean. *J. Agric. Food Chem.* 34:179–185.

Miller, C. A., and D. D. Davis. 1981. Effect of temperature on stomatal conductance and ozone injury of pinto bean leaves. *Plant Disease* 65:750–751.

Miller, J. E., D. G. Sprugel, R. N. Muller, H. J. Smith, and P. B. Xerikos. 1980. Open-air fumigation system for investigating sulphur dioxide effects on crops. *Phytopathology* 70:1124–1128.

Miller, P. R. 1977. Ecosystems. In *Ozone and Other Photochemical Oxidants*, pp. 586–642. Washington, D.C.: National Academy of Sciences.

Mjelde, J. W., R. M. Adams, B. L. Dixon, and P. Garcia. 1984. Using farmers' actions to measure crop loss due to air pollution. *Journal of Air Pollution Control Association* 34:360–364.

Montes, R. A., U. Blum, and A. S. Heagle. 1982. The effects of ozone and nitrogen fertilizer on tall fescue, ladino clover, and a fescue-clover mixture. I. Growth, regrowth, and forage production. *Canadian Journal of Botany* 60:2745–2752.

Montes, R. A., U. Blum, A. S. Heagle, and R. J. Volk. 1983. The effects of ozone and nitrogen fertilizer on tall fescue, ladino clover, and a fescue-clover mixture. II. Nitrogen content and nitrogen fixation. *Canadian Journal of Botany* 61:2159–2168.

Mudd, J. B., S. K. Banerjee, M. M. Pooley, and K. L. Knight. 1984. Pollutants and plant cells: Effects on membranes. In *Gaseous Air Pollutants and Plant Metabolism.* ed. M. J. Koziol and F. R. Whatley, pp. 105–116. London: Butterworth.

Mulchi, C. L., D. J. Sammons, and P. S. Baenziger. 1986. Yield and grain quality responses of soft red winter wheat exposed to ozone during anthesis. *Agron. J.* 78:593–600.

Murray, F. 1983. Response of grapevines to fluoride under field conditions. *Journal of American Society of Horticultural Science* 108:526–529.

National Academy of Sciences. 1977. *Nitrogen Oxides,* pp. 197–214. Washington, D.C.: National Academy of Sciences.

Olszyk, D. M., and D. T. Tingey. 1984. Phytotoxicity of air pollutants. *Plant Physiology* 74:999–1005.

Olszyk, D. M., and D. T. Tingey. 1986. Joint action of O_3 and SO_2 in modifying plant gas exchange. *Plant Physiology* 82:401–405.

Olszyk, D. M., A. Bytnerowicz, G. Kats, P. J. Dawson, J. Wolf, and C. R. Thompson. 1986a. Crop effects from air pollutants in air exclusion systems vs field chambers. *Journal of Environmental Quality* 15:417–422.

Olszyk, D. M., G. Kats, P. J. Dawson, A. Bytnerowicz, J. Wolf, and C. R. Thompson. 1986b. Characteristics of air exclusion systems vs. chambers for field air pollution studies. *Journal of Environmental Quality* 15:326–334.

Omasa, K., Y. Hashimoto, P. J. Kramer, B. R. Strain, I. Aiga, and J. Kondo. 1985. Direct observation of reversible and irreversible stomatal responses of attached sunflower leaves to SO_2. *Plant Physiology* 79:153–158.

Ormrod, D. P. 1978. *Pollution in Horticulture.* Elsevier Amsterdam and New York: Scientific Publishing. 260 pp.

Ormrod, D. P., D. T. Tingey, M. L. Gumpertz, and D. M. Olszyk. 1984. Utilization of a response-surface technique in the study of plant responses to ozone and sulfur dioxide mixtures. *Plant Physiology* 75:43–49.

Oshima, R. J., M. P. Poe, P. K. Braeglemann, D. W. Baldwin, and V. Van Way. 1976. Ozone dosage-crop loss function for alfalfá: A standardized method for assessing crop losses for air pollutants. *Journal of Air Pollution Control Association* 26:861–865.

Oshima, R. J., O. C. Taylor, P. K. Braeglemann, and D. W. Baldwin. 1975. Effect of ozone on the yield and plant biomass of a commercial variety of tomato. *Journal of Environmental Quality* 4:463–464.

Page, W. P., G. Arbogast, R. G. Fabian, and J. Ciecka. 1982. Estimation of economic losses to the agricultural sector from airborne residuals in the Ohio River Basin region. *Journal Air Pollution Control Association* 32:151–154.

Pande, P. C. 1985. An examination of the sensitivity of five barley cultivars to SO_2 pollution. *Environmental Pollution (Series A)* 37:27–41.

Papple, D. J., and D. P. Ormrod. 1977. Comparative efficacy of ozone-injury suppression by benomyl and carboxin on turfgrasses. *Journal of American Horticultural Science* 102:792–796.

Pauls, K. P., and J. E. Thompson. 1981. Effects of *in vitro* treatment with ozone on the physical and chemical properties of membranes. *Physiol. Plant.* 53:255–262.

Pell, E. J., C. J. Arny, and N. S. Pearson. 1987. Impact of simulated acidic precipitation on quantity and quality of a field grown potato crop. *Environmental and Experimental Botany* 27:7–14.

Petolino, J. F., C. L. Mulchi, and M. K. Aycock, Jr. 1983. Leaf injury and peroxidase activity in ozone-stressed tobacco cultivars and hybrids. *Crop. Science* 23:1102–1106.

Plesnicar, M. 1983. Study of sulfur dioxide effects on phosphorus metabolism in plants using phosphorus-32 as indicator. *Int. J. Appl. Radiat. Isot.* 34:833–835.

Podleckis, E. V., C. R. Curtis, and H. E. Heggestad. 1984. Peroxidase enzyme markers for ozone sensitivity in sweet corn. *Phytopathology* 74:572–577.

Prasad, B. J., and D. N. Rao. 1980. Alterations in metabolic pools of nitrogen dioxide exposed wheat plants. *Indian Journal of Experimental Biology* 18:879–882.

Purvis. A. C. 1978. Differential effects of ozone on *in vivo* nitrate reduction in soybean cultivars: I. Response to exogenous sugars. *Canadian Journal of Botany* 56:1540–1544.

Rawlings, J. O., and W. W. Cure. 1985. The Weibull function as a dose-response model for studying air pollution effects on crop yields. *Crop Science* 25:807–814.

Reich, P. B., and R. G. Amundson. 1984. Low level O_3 and/or SO_2 exposure causes a linear decline in soybean yield. *Environmental Pollution (Series A)* 34:345–355.

Reich, P. B., A. W. Schoettle, and R. G. Amundson. 1985. Effects of low concentrations of O_3, leaf age and water stress on leaf diffusive conductance and water use efficiency in soybean. *Physiol. Plant.* 63:58–64.

Reich, P. B., A. W. Schoettle, R. M. Raba, and R. G. Amundson. 1986. Response of soybean to low concentrations of ozone: I. Reductions in leaf and whole plant net photosynthesis and leaf chlorophyll content. *Journal of Environmental Quality* 15:31–36.

Reilly, J. J., and L. D. Moore. 1982. Influence of selected herbicides on ozone injury in tobacco (*Nicotiana tabacum*). *Weed Science* 30:260–263.

Reinert, R. A. 1984. Plant response to air pollutant mixtures. *Annual Review of Phytopathology* 22:421–442.

Reinert, R. A., H. E. Heggestad, and W. W. Heck. 1982a. Response and genetic modification of plants for tolerances to air pollutants. In *Breeding Plants for Less Favorable Environments*. ed. M. N. Christiansen, pp. 259–292. New York: John Wiley.

Reinert, R. A., D. S. Shriner, and J. O. Rawlings. 1982b. Responses of radish to all combinations of three concentrations of nitrogen dioxide, sulfur dioxide, and ozone. *Journal of Environmental Quality* 11:52–57.

Rhoads, A., and E. Brennan. 1978. The effect of ozone on chloroplast lamellae and isolated mesophyll cells of sensitive and resistant tobacco selections. *Phytopathology* 68:883–886.

Rist, D. L., and J. W. Lorbeer. 1984. Moderate dosages of ozone enhanced infection of onion leaves by *Botrytis cinerea* but not by *B. squamosa*. *Phytopathology* 74:761–767.

Roberts, T. M. 1984. Effects of air pollution on agriculture and forestry. *Atmos. Environ.* 18:629–652.

Rogers, H. H., J. C. Campbell, and R. J. Volk. 1979. Nitrogen-15 dioxide uptake and incorporation by *Phaseolus vulgaris* L. *Science* 206:333–335.

Rowe, R. P., L. G. Chestnut, C. Miller, R. M. Adams, M. Thresher, H. O. Mason, R. E. Howitt, and J. Trijonis. 1984. Economic assessment of the effect of air pollution in the San Joaquin Valley. Draft report to the Research Division, California Air Resources Board. Boulder, Colo.: Energy and Resource Consultants.

Runeckles, V. C., and K. Palmer. 1987. Pretreatment with nitrogen dioxide modifies plant response to ozone. *Atmos. Environ.* 21:717–719.

Runeckles, V. C., and P. M. Rosen. 1977. Effects of ambient ozone pretreatment on transpiration and susceptibility to ozone injury. *Canadian Journal of Botany* 55:193–197.

Sakaki, T., N. Kondo, and K. Sugahara. 1983. Breakdown of photosynthetic pigments and lipids in spinach leaves with ozone fumigation—role of active oxygens. *Physiol. Plant.* 59:28–34.

Sakuri, S., K. Stai, and H. Tsunoda. 1983. Effects of airborne fluoride on the fluorine content of rice and vegetables. *Fluoride* 16:175–180.

Sekiya, J., L. G. Wilson, and P. Filner. 1982. Resistance to injury by sulfur dioxide. *Plant Physiology* 70:437–441.

Shinn, J. H., B. R. Clegg, and M. L. Stuart. 1977. A Linear-Gradient Chamber for Exposing Field Plants to Controlled Levels of Air Pollutants. UCRL Rep. No. 80411. Lawrence Livermore Laboratory, Livermore, Calif.

Shriner, D. S., W. W. Cure, A. S. Heagle, W. W. Heck, D. W. Johnson, R. J. Olson, and J. M. Skelly. 1984. An Analysis of Potential Agriculture and Forest Impacts of Long Range Transport Air Pollutants. ORNL-5910. Oak Ridge, Tenn.: Oak Ridge National Laboratory.

Shriner, D. S., C. R. Richmond, and S. E. Lindberg, eds. 1980. *Atmospheric Sulfur Deposition, Environmental Impact and Health Effects.* Ann Arbor, Mich.: Ann Arbor Science. 568 pp.

Shupe, J. L., H. B. Peterson, and N. C. Leone, eds. 1983. Fluorides: Effects on vegetation, animals, and humans. Proceedings of International Fluoride Symposium. Salt Lake City: Paragon Press. 370 pp.

Sigal, L. L., and O. C. Taylor. 1979. Preliminary studies of the gross photosynthetic response of lichens to peroxyacetyl nitrate fumigations. *Bryologist* 82:564–575.

Sinn, J. P., E. J. Pell, and R. L. Kabel. 1984. Uptake rate of nitrogen dioxide by potato plants. *Journal of Air Pollution Control Association* 34:668–669.

Smith, G., B. Greenhalgh, E. Brennan, and J. Justin. 1987. Soybean yield in New Jersey relative to ozone pollution and antioxidant application. *Plant Disease* 71:121–125.

Sprugel, D. G., J. E. Miller, R. N. Muller, H. J. Smith, and P. B. Xerikos. 1980. Sulfur dioxide effects on yield and seed quality in field-grown soybeans. *Phytopathology* 70:1129–1133.

Stanford Research Institute. 1981. An Estimate of the Nonhealth Benefits of Meeting the Secondary National Ambient Air Quality Standards. Prepared for the National Commission on Air Quality. Washington, D.C.

Starkey, T. E., D. D. Davis, E. J. Pell, and W. Merrill. 1981. Influence of peroxyacetyl nitrate (PAN) on water stress in bean plants. *HortScience* 16:547–548.

Steinberger, E. H., and S. Naveh. 1982. Effects of recurring exposures to small ozone concentrations on Bel W_3 tobacco plants. *Agric. Environ.* 7:255–264.

Strain, B. R., and J. D. Cure, eds. 1985. Direct Effects of Increasing Carbon Dioxide on Vegetation. Report No. DOE/ER-0238. Washington, D.C.: Office of Energy Research, U.S. Department of Energy.

Sutton, R., and I. P. Ting. 1977. Evidence for the repair of ozone-induced membrane injury. *American Journal of Botany* 64:404–411.

Taylor, G. E. 1978. Genetic analysis of ecotypic differentiation within an annual plant species, *Geranium carolinianum* L., in response to sulfur dioxide. *Bot. Gaz.* 139:362–368.

Taylor, G. E., Jr., and W. J. Selvidge. 1984. Phytotoxicity in bush bean of five sulfur-containing gases released from advanced fossil energy technologies. *Journal of Environmental Quality* 13:224–230.

Taylor, G. E., and D. T. Tingey. 1981. Physiology of ecotypic plant response to sulfur dioxide in *Geranium carolinianum. Oecologia* 49:76–82.

Taylor, G. E., D. T. Tingey, and H. C. Ratsch. 1982. Ozone flux in *Glycine max* (L). Merr-sites of regulation and relationship to leaf injury. *Oecologia* 53:179–186.

Temple, P. J., and O. C. Taylor. 1983. World-wide ambient measurements of peroxyacetyl nitrate (PAN) and implications for plant injury. *Atmos. Environ.* 17:1583–1587.

Temple, P. J., C. H. Fa, and O. C. Taylor. 1985a. Effects of SO_2 on stomatal conductance and growth of *Phaseolus vulgaris. Environmental Pollution (Series A)* 37:267–279.

Temple, P. J., O. C. Taylor, and L. F. Benoit. 1985b. Cotton yield responses to ozone as mediated by soil moisture and evapotranspiration. *Journal of Environmental Quality* 14:55–60.

Teramura, A. H. 1983. Effects of ultraviolet-B radiation on the growth and yield of crop plants. *Physiol. Plant.* 58:415–427.

Thomas, M. D. 1961. Effects of Air Pollution on Plants. World Health Organization, Monograph Ser. 46, 223–278.

Thompson, C. R., and G. Kats. 1978. Effects of continuous H₂S fumigation on crop and forest plants. *Environmental Science Technology* 12:550–553.

Tingey, D. T. 1979. Nitric oxide (NO) and nitrogen dioxide (NO₂) emissions from herbicide-treated soybean plants. *Atmos. Environ.* 13:1475.

Tingey, D. T., and W. E. Hogsett. 1985. Water stress reduces ozone injury via a stomatal mechanism. *Plant Physiology* 77:944–947.

Tingey, D. T., K. D. Rodecap, E. H. Lee, T. J. Moser, and W. E. Hogsett. 1986. Ozone alters the concentrations of nutrients in bean tissue. *Angewandte Botanik* 60:481–493.

Tingey, D. T., C. Standly, and R. W. Fields. 1976. Stress ethyene evolution: A measure of ozone effects on plants. *Atmos. Environ.* 10:969–974.

Tingey, D. T., G. L. Thutt, M. L. Gumpertz, and W. E. Hogsett. 1982. Plant water status influences ozone sensitivity of bean plants. *Agric. Environ.* 7:243–254.

Treshow, M., ed. 1984. *Air Pollution and Plant Life*. Chichester, England: John Wiley. 486 pp.

Trumble, J. T., J. D. Hare, R. C. Musselman, and P. M. McCool. 1987. Ozone-induced changes in host-plant suitability. *Journal of Chemical Ecology* 13:203–218.

Unsworth, M. H., and V. J. Black. 1981. Stomatal responses to pollutants. In *Stomatal Physiology* ed. P. G. Jarvis and T. A. Mansfield, pp. 187–204. New York: Cambridge University Press.

Unsworth, M. H., and D. P. Ormrod, eds. 1982. *Effects of Gaseous Air Pollution in Agriculture and Horticulture*. London: Butterworth. 522 pp.

Unsworth, M. H., A. S. Heagle, and W. W. Heck. 1984a. Gas exchange in open-top field chambers: I. Measurement and analysis of atmospheric resistance to gas exchange. *Atmos. Environ.* 18:373–380.

Unsworth, M. H., A. S. Heagle, and W. W. Heck. 1984b. Gas exchange in open-top field chambers: II. Resistance to ozone uptake by soybeans. *Atmos. Environ.* 18:381–385.

U.S. Environmental Protection Agency. 1976. *Diagnosing Vegetation Injury Caused by Air Pollution*. Research Triangle Park, N.C.: Applied Science Associates, Air Pollution Training Institute.

U.S. Environmental Protection Agency. 1978. Air quality criteria for ozone and other photochemical oxidants, chapter 10/11, EPA-600/8-79-004. Washington, D.C.: Office of Research and Development.

U.S. Environmental Protection Agency. 1982a. Air quality criteria for oxides of nitrogen, chapter 12, EPA-600/8-82-026. Research Triangle Park, N.C.: Office of Research and Development, Environmental Criteria and Assessment Office.

U.S. Environmental Protection Agency. 1982b. Air quality criteria for particulate matter and sulfur oxides. Vol. 3, chapter 8, EPA-600/8-82-029C. Research Triangle Park, N.C.: Office of Research and Development, Environmental Criteria and Assessment Office.

U.S. Environmental Protection Agency. 1986. Air quality criteria for ozone and other photochemical oxidants. Vol. 3, chapter 6, EPA-600/8-84-020CF. Research Triangle Park, N.C.: Office of Research and Development, Environmental Criteria and Assessment Office.

van Haut, H., and H. Stratmann. 1970. *Farbtafelatlas über Schwefeldioxid-Wirkungen an Pflanzen*. Essen, West Germany: Verlag W. Girardet, 206 pp.

Varshney, S. R. K., and C. K. Varshney. 1981. Effect of sulphur dioxide on pollen germination and pollen tube growth. *Environmental Pollution (Series A)* 24:87–92.

Weidensaul, T. C. 1980. N-(2-2-Oxo-1-Imidazolidinyl)-N-Phenylurea as a protectant against ozone injury to laboratory fumigated pinto bean plants. *Phytopathology* 70:42–45.

Wellburn, A. R., C. Higginson, D. Robinson, and C. Walmsley. 1981. Biochemical explanations of more than additive inhibitory effects of low atmospheric levels of sulfur dioxide plus nitrogen dioxide upon plants. *New Phytology* 88:223–237.

Wesley, M. L., and B. B. Hicks. 1977. Some factors that affect the deposition rates of sulfur dioxide and similar gases on vegetation. *Journal of Air Pollution Control Association* 27:1110–1116.

Wilson, G. B., and J. N. B. Bell. 1985. Studies on the tolerance to SO_2 of grass populations in polluted areas. III. Investigations on the rate of development of tolerance. *New Phytology* 100:63–77.

Winner, W. E., H. A. Mooney, and R. A. Goldstein, eds. 1986. *Sulfur Dioxide and Vegetation: Ecology, Physiology and Policy Issues.* Pala Alto: Stanford University Press.

Yoneyama, T., H. Sasakawa, S. Ishizuka, and T. Tatsuka. 1979. Absorption of atmospheric NO_2 by plants and soils. *Soil Science and Plant Nutrition* 25:267–275.

7

Economic Measures of the Impacts of Air Pollution on Health and Visibility

LAURAINE G. CHESTNUT AND
ROBERT D. ROWE

Air pollutants have long been known to affect human health and well-being. Primary and secondary national ambient air quality standards have been set by the U.S. Environmental Protection Agency (EPA) to protect human health and to minimize welfare impacts related to visibility, materials damage, crops, forests, vegetation, and other impacts. Although progress has been made, these air quality standards are still regularly exceeded in many parts of the United States, affecting millions of residents. This chapter summarizes the best available economic measures of health and visibility benefits that might be obtained with reductions in ambient concentrations of ozone and compounds of sulfur and nitrogen. It focuses upon the benefits of air pollution control. For convenience in economic analysis, damages due to the lack of air pollution control are sometimes calculated as "negative benefits." Since comparing the social importance of different physical impacts of changes in air pollution is difficult, economic measures of impacts in dollars provide one basis of comparison across different effects and allow comparison with costs of control.

Limited research has been conducted on the health impacts of acid aerosols. This work has not been translated to economic benefit measures of potential reductions in acid aerosols. For more, see a forthcoming report by U.S. EPA, Office of Air Quality Planning Standards, on an acid aerosols conference held in October 1987. Materials damage from air pollutants, especially particulate matter and possibly from acid deposition, also represents a potentially important welfare impact of the selected air pollutants. For more on these topics, see Rowe et al. (1986), U.S. EPA (1984, 1986a, 1987a, 1987b), Chestnut and Rowe (1988), and Horst et al. (1986).

Because of the broad scope of this chapter, a complete review of the literature on the effects of the selected pollutants upon human health and visibility cannot be provided. The reader is referred to the EPA criteria documents for extensive reviews of the health effects literature as well as other literature cited here.

Economic Concepts and Measures of Value

This section is intended to help the non-economist understand the economic concepts and measures of value that underlie economic benefit analysis in order to appropriately interpret the estimates. More detail on the economic theory behind benefit analysis for changes in environmental quality can be found in Freeman (1979).

Concepts of Value

Economic analysis of changes in environmental quality focuses upon determining monetary measures of value that society places upon such impacts as changes in health status or visibility. Any change that affects someone's well-being has value to that individual. The fact that health and visibility have value, even though there is little in the way of market prices for these values, is evident from residential location and employment choices, recreation behavior, costs incurred to combat adverse health impacts both before and after the fact, and changes in mood. Because they are not traded in markets, human health and visibility impacts due to changes in air pollution are often referred to as non-market goods.

This discussion looks at the attitudes of individuals because the impacts on human health and visibility of changes in air pollution are predominately experienced by individuals rather than firms. For other effects of air pollution, such as agriculture and materials damage, the impacts on companies can be dramatic. The total change in well-being due to changes in non-market goods like visibility and human health are generally discussed in terms of:

- Use values. These are the direct impacts to the affected individuals. They might include reduced enjoyment of recreation due to reduced visual aesthetics, or reduced well-being due to increased adverse health effects both now and in the future.
- Bequest and other indirect values. These include values by people indirectly affected by the reduced well-being of the directly affected individuals. Examples include bequesting to future generations clean air and good health or improved social well-being due to improved health and education of everyone in society.
- Nonuse and other preservation values. Individuals may value simply knowing that resources, such as visibility aesthetics or wildlife, are being preserved in their natural state even if no human use of these resources ever occurs.

Most economic valuations of nonmarket goods and services focus upon use values because these are more readily estimated. However, bequest and nonuse values are likely to be substantial for both human health and visibility aesthetics in certain circumstances, although estimation methods for these values are subject to more uncertainty and controversy, and estimates are not widely available. The components of value for human health effects and visibility aesthetics are discussed further in the relevant sections of this chapter.

Economic Measures of Value

Economists attempt to quantify changes in well-being using monetary measures. In general, the monetary measure of the change in an individual's well-being due to a change in a good such as health or visibility is the change in income or wealth that would yield the same or an offsetting change in the individual's well-being as the change in the good. This is usually referred to as the individual's maximum willingness to pay to obtain an improvement in a good, or willingness to pay to prevent a deterioration in a good. Alternatively, one might also measure the minimum willingness to accept compensation for a reduction in a good, or to accept compensation to forgo an improvement in a good.

When discussing measures of economic value, the technical literature most often uses the term *consumer's surplus*, which is directly related to willingness to pay or to accept compensation for impacts on human health and visibility resulting from changes in air pollution. A similar measure, called producer's surplus, applies to firms. The sum of consumer's and producer's surpluses is referred to as economic surplus. For more detailed discussion, see Freeman (1979) or almost any detailed benefit-cost analysis. Specific methods of quantifying economic measures of value for health and visibility impacts of changes in air pollution are briefly described later in the chapter.

Total benefits from a change in air pollution are traditionally calculated as the sum of all benefits to all individuals now and in the future, with future benefits usually discounted to a present value measure. This aggregation procedure is based upon what is known as the potential pareto improvement criteria used to determine socially efficient resource allocation within a benefit-cost analysis framework. Under these criteria, society's well-being is improved if the gains to those who benefit from a change exceed the losses to those who are worse off such that there is a net monetary gain. This net gain could result in a potential improvement to all parties if the gains to those who benefit are redistributed to compensate losers so that no one is worse off, and some are better off, under the resource reallocation.

Defining aggregate damages on an issue-by-issue basis in this fashion has important limitations. It implicitly makes interpersonal comparisons of changes in well-being using a monetary metric that relies on the existing distribution of wealth. As benefits (or damages) of an action are often valued differently, depending upon a person's wealth, this may bias the resource allocation decision. Further, even if wealth were equalized, the decision criterion weights each person equally; under a social rights perspective, individuals may have rights not to be injured by specific actions of others regardless of the total net social benefit. Finally, future generations are precluded from active input on the resource allocation decision. If the current estimation of future preferences is inaccurate, incorrect decisions may be made that are often more expensive to reverse than they would have been to prevent originally. When net current benefits are in fact not invested for future returns, the practice of using financial discount rates to obtain present values of future damages may be incorrect.

Health Effects

This section summarizes the current evidence on the health effects associated with ozone and compounds of sulfur and nitrogen, discusses the economic welfare implications of these health effects and their evaluation in the economics literature, and describes the results of several recent regional and national benefits studies.

Air Pollutants and Physical Health Effects

Short-term and long-term exposures to air pollutants are associated with increased acute and chronic health effects. These effects include increases in mortality as well as morbidity. Some people are more sensitive than others: those who already have chronic respiratory or cardiovascular conditions, the elderly, young children, and fetuses. The literature on all these potential health effects is vast; only the most significant adverse health effects that would be expected to be reduced due to lowered ambient levels of ozone and compounds of sulfur and nitrogen are highlighted here (see table 7.1). Comprehensive reviews can be found in the EPA criteria documents for each of these pollutants (U.S. EPA 1982a, 1982b, 1986b, 1986c).

Evidence concerning the health risks associated with air pollutants comes primarily from three types of studies: epidemiological, clinical, and animal toxicology. Each type has advantages and disadvantages. Epidemiological studies provide estimates that are more readily adapted to analysis of economic benefits than the results of the other two types of studies. Confidence in epidemiological results is bolstered when the evidence is consistent with other available clinical and toxicological evidence. All three approaches are described here briefly because it is important to understand the nature of the available evidence when evaluating the economic benefits of changes in exposure to air pollutants in terms of human health.

Epidemiological studies examine the incidence or prevalence of health problems in particular populations and attempt to determine whether exposures to certain pollutants are correlated with changes in health measures (or endpoints). These studies can use cross-sectional data for differences across a population at one point in time, or longitudinal data for changes in a specific population over time. The primary advantages are that epidemiological studies can

Table 7.1. *Summary of identified health effects.*

Pollutant	Health effect
Ozone	Aggravation of asthma and other chronic respiratory diseases Lung function reductions and respiratory irritation Acute respiratory illness
Total particulate matter (includes sulfate and nitrate particles)	Elevated risk of mortality Lung cancer Higher prevalence of chronic respiratory diseases Acute illness, including work loss and emergency room visits Lung function reductions and respiratory symptoms
Sulfur dioxide	Bronchoconstriction Also associated with acute morbidity and elevated mortality in epidemiology studies, although collinearity with particulates is suspected
Nitrogen dioxide	Aggravation of asthma and allergy symptoms Acute respiratory illness in children Increased airway resistance and reduced lung function

potentially capture actual effects on human populations at ambient pollution levels and can examine long-term as well as short-term exposures. These strengths result from the use of data on the health status of population groups in their normal environments. Yet this kind of data accounts for the major limitations of epidemiological studies as well. The complexity of the many influences on human health makes it difficult to isolate with confidence the effects of air pollutants both on a group and on individuals. Further, finding a statistically significant relationship between a change in health status and a change in pollution exposure does not prove causality.

Clinical studies examine the response of human subjects to air pollutant exposures in a controlled setting. The response of the individual can be monitored and the environment controlled so that the effects of one pollutant can be isolated with considerable confidence. These studies are limited, however, in what they can examine. Only reversible health effects of short-term exposures can be purposely induced in human subjects.

Animal toxicological studies use animal subjects to test the effects of exposures to pollutants in a laboratory setting. These can provide a great deal of information about potential human responses to pollution exposures, but quantitative results are often difficult to transfer from animal subjects to humans. Such studies can provide useful information on biological responses to pollutants and are often used to identify potential carcinogens, but the results are not always useful in determining the number of cases of a given illness to expect in a given human population exposed to a given amount of one or more pollutants.

Human Health Effects of Ozone. Clinical studies in which people have been exposed to a known amount of ozone in a controlled setting have demonstrated fairly conclusively that ozone is irritating to the nose and throat and causes temporary reductions in lung function. These effects have been found in healthy children and adults, and in adults with chronic respiratory conditions. There is a limited ability, however, to quantify these health effects in terms that can be clearly interpreted as changes in human welfare. Decreases in lung function are obviously undesirable physical changes, but the effects that have been observed appear to be reversible after ozone exposures are lowered. With temporary reductions in lung function, it is not clear at what ozone concentration an individual's ability to conduct his or her normal daily activities is impaired. It is also not clear what the effects of chronic or frequent exposures to elevated concentrations of ozone might be.

Epidemiological studies have found evidence that higher concentrations of ozone are associated with higher rates of acute respiratory illness and respiratory-related symptoms in healthy adults and in individuals with chronic respiratory illnesses. For example, Portney and Mullahy (1986) found a higher number of days with activities restricted due to respiratory illness associated with higher concentrations of ambient ozone in a cross-sectional study. Hammer et al. (1974) found higher frequencies of eye irritation, cough, and chest discomfort in a sample of student nurses in Los Angeles on days with elevated ozone concentrations. (Eye irritation that is associated with photochemical smog is apparently caused not by ozone but by other components of photochemical smog that form under the same conditions as ozone.) Whittemore and Korn (1980) found an association between daily ambient ozone levels and daily asthma attacks in the Los Angeles area. Those who suffer from other kinds of chronic respiratory diseases and allergies may also be more susceptible to ozone, but the quantitative evidence is limited. These kinds of epidemiological findings are consistent with the clinical evidence that ozone is an irritant to the respiratory system.

It is suspected, given the observed effects of ozone on lung function, that long-term exposures to elevated ozone might be associated with increased prevalence of chronic respiratory diseases. The evidence on this is as yet inconclusive. Animal toxicological studies suggest that prolonged exposure to elevated concentrations of ozone may be associated with accelerated aging of lung tissue and with cell damage. It is uncertain at what ozone concentrations such effects might be expected in humans.

No convincing association has been demonstrated between ozone concentrations and mortality rates. Ferris (1978) concluded that any association between ozone and mortality is more likely to be a result of the higher temperatures that are associated with elevated ozone levels than a result of ozone itself.

Human Health Effects of Compounds of Sulfur and Nitrogen. Sulfates and nitrates are two types of particles that are dispersed in the atmosphere. A large share of these particles are not emitted directly; they form in the atmosphere in the presence of gaseous sulfur dioxide and nitrogen oxides, which result primarily from the combustion of fuels. Sulfates and nitrates are typically in the size range of fine particles (less than 2.5 μm in diameter) that can be inhaled deeply into the lungs and therefore are most likely to be associated with adverse health effects. Considerable evidence exists in the literature on the health effects associated with all particulates, primarily total suspended particulates, but there is not as yet sufficient evidence to attribute with much confidence specific portions of these health effects to specific particle components such as sulfates and nitrates.

Higher death rates have been associated with short-term exposures to elevated concentrations of particles in studies conducted in a few cities in the United States and Europe (for example, Mazumdar et al. 1982). The EPA criteria document for particulate matter summarizes these studies as follows: concentrations of total suspended particulates (TSP) above 1000 μg/m^3 for 24 hours or more are clearly dangerous; concentrations above 500 μg/m^3 are probably dangerous; and concentrations above 200 might be dangerous. Before the primary standard was changed in 1987 to a measure of particulate matter less than or equal to 10 μm in diameter (PM$_{10}$), the ambient air quality standard for TSP was 260 μg/m^3 for 24 hours. Recent reanalysis of a sizable data set for London comparing daily mortality rates and daily particulate concentrations suggests that the association between higher rates and particulate levels extends to particulate levels as low as typically occur in many U.S. cities (Schwartz and Marcus 1986).

Cross-sectional studies of American cities also have found an association between mortality rates and particulate levels (for example, Lave and Seskin 1977). Typically 100 to 200 cities have been included in these studies. Questions continue to be raised about whether potential confounding factors have been sufficiently accounted for in these studies (Evans et al. 1984). Although some instability in the estimated magnitude of the relationship has been demonstrated when different estimation techniques are used, the association has yet to be convincingly demonstrated to be insignificant.

Epidemiological and clinical studies have found an association between particulate exposures and decrements in lung function as well as increases in respiratory symptoms in children, adults, and individuals with chronic respiratory diseases. Epidemiological studies also have found an association between particulate levels and emergency room visits and between particulate levels and restricted-activity days due to illness (for example, Samet et al. 1981, and Ostro 1987). Thus, there appears to be an association between acute illness rates in the general population and particulate levels to which people are exposed. These studies provide sufficient quantitative information to allow estimates of specific changes in illness rates to be made for specific changes in particulate levels.

The prevalence of chronic respiratory illnesses may be higher in locations where particulate levels are higher (for example, Ferris et al. 1976), but questions about potential confounding factors in these studies and about deriving a quantitative dose-response relationship from available results have yet to be fully addressed. Toxicological studies suggest that particulate matter exposures may be associated with increased risk of cancer, particularly lung cancer. These results remain tentative at present, and work continues to determine if certain constituents of ambient particulate matter are more likely to be responsible for the potential cancer risk.

Sulfates are a significant component of total particulates in many urban areas and are suspected of being a significant factor in the observed associations between particulate levels and the health effects just discussed. Efforts to determine the independent effects of sulfates on human health have not provided conclusive results. Some epidemiological results point more to sulfates and others point more to total particulates as the primary culprit. Because sulfates are a significant component of total particulates their presence is often highly correlated, making it difficult to separate them statistically.

Nitrates are also a component of total particulates, but less research has been done to try to isolate any independent effect of nitrates. When included separately in a few epidemiological studies having the necessary data, the nitrate variable frequently has not shown a statistically significant association with health effects that is independent of the effect associated with total particulates. However, to the extent that a reduction in nitrates would mean a reduction in total particulates, a health benefit would be expected.

Any control strategy to reduce compounds of sulfur and nitrogen would likely involve reductions in emissions of their gaseous precursors, sulfur dioxide and nitrogen dioxide, both of which are controlled under current national ambient air quality standards (unlike sulfates and nitrates, for which there are no separate federal ambient standards). Some epidemiological evidence can be used to develop quantitative estimates of the health benefits that would be expected if sulfur dioxide or nitrogen dioxide were reduced, and additional clinical evidence indicates the kinds of health effects involved with exposures to these pollutants.

The primary health effect associated with short-term exposures to sulfur dioxide is bronchoconstriction, a narrowing of the airways that is often associated with wheezing and shortness of breath. This effect has been observed in several studies with healthy individuals exercising at 1 ppm of sulfur dioxide. It has been confirmed for subjects with asthma or other chronic respiratory problems at 0.5 ppm. (See U.S. EPA 1982a for a review of these studies.) The current federal secondary standard for sulfur dioxide is a 3-hour average of 0.5 ppm and the primary standard is a 0.14 ppm average for 24 hours.

Some epidemiological studies also have examined the relationship between sulfur dioxide levels and morbidity or mortality (for example, Cropper 1981). Statistically significant relationships have been observed in some of the studies, but it is not clear whether this can be interpreted as an independent effect or whether it is a result of the correlation among levels of sulfur dioxide, sulfates, and total particulates.

Clinical studies show that short-term exposures to nitrogen dioxide can affect the functioning of the respiratory system. Healthy adults have shown increased airway resistance when exposed to 2.5 ppm nitrogen dioxide for 2 hours in clinical studies. No effects on healthy adults have been observed at 1.5 ppm or less. Subjects with asthma have shown effects at 0.5 ppm and possibly as low as at 0.1 ppm (see U.S. EPA 1982b for a review of the clinical evidence). Children also appear to have greater sensitivity than healthy adults to nitrogen dioxide. Higher incidences of respiratory illness and decreased pulmonary function have been measured in epidemiological studies of children living in homes where gas stoves are used (for example, Berkey et al. 1986). Lebowitz et al. (1985) also found a correlation between outdoor nitrogen dioxide levels and symptoms in asthmatic and allergic subjects.

Economic Welfare Implications of Physical Health Effects

The approach taken in most economic benefit studies concerning the effects of air pollution on human health is to estimate the change in the risk to each affected individual of each type of health effect that would be prevented under a given control strategy, and then to estimate a dollar value for this reduction in risk for each affected individual. Multiplying the dollar value by the number of individuals affected gives the total benefit of the prevention in health effects to the directly affected individuals (the use value). Limited work suggests that indirect values for improved health of others may also exist, but most benefit analyses have not yet begun to include these kinds of estimates.

Economic measures of value and the methods used to obtain these estimates for changes in morbidity and mortality have been reviewed for EPA by Violette and Chestnut (1983), Chestnut and Violette (1985), and Violette et al. (1986). Other reviews of this literature have been prepared by Mishan (1982), Blomquist (1982), and others. This section summarizes the primary estimation methods and their key advantages and disadvantages; for details on the results of specific studies, see the fuller reviews.

The conceptually desirable economic measure of value for a change in the risk of a given health effect is how much an individual would be willing to give up to avoid or prevent this risk and all associated effects on well-being. It is important to point out that changes in air pollution levels do not mean absolute changes in illness or premature death for any particular individual, but rather changes in the probability of illness or premature death for everyone in a given population group. It cannot be predicted exactly who would be affected, only on average how many people in a particular group are likely to be affected.

Ideally, the measure of value is for small changes in the probability of illness or premature death because it is a change in risk—not a certain death or illness—that the individual faces as a result of a change in environmental hazards. It would not be necessary to distinguish between values for changes in illness probabilities and values for changes in illness levels that are certain if it could be assumed that the values for a change in the probability of an illness equaled the expected value of the change in illness. In other words, if an individual were willing to pay $100 to prevent getting a respiratory infection, the expected value of lowering the probability of getting a respiratory infection from 15 percent to 10 percent would be $5. Owing to limited available information about values for changes in risks of morbidity, this assumption is often used but it has not been empirically demonstrated to be accurate. Estimates of value for changes in risks of mortality are available in the literature.

In general, illness and premature mortality are associated with the following effects on an individual's welfare: medical costs, productivity and income losses, reduction in enjoyment and quality of life, pain and discomfort, and worry and inconvenience to family and friends. All these effects would be reflected in a conceptually complete measure of value for changes in risks of morbidity or mortality. Quantification of these welfare effects of illness and premature mortality has typically followed one of two approaches: cost of illness or willingness to pay.

Cost-of-illness estimates include medical costs and lost income due to illness. These estimates represent a direct financial impact on society due to illness, but they are incomplete in terms of the total welfare impact. The estimates are frequently used, however, because they can be derived from available data and because they can be interpreted as a likely lower bound measure of the total welfare impact of illness.

Willingness-to-pay estimates for changes in risks of morbidity and mortality are conceptually complete measures of the welfare effect on an individual because the amount a person would be willing to pay to prevent, for example, an increase in the risk of respiratory illness can be expected to reflect all the factors listed above. Willingness-to-pay estimates for changes in risks of morbidity and mortality are more difficult to obtain than cost-of-illness estimates, however. Two types of studies are used to obtain willingness-to-pay estimates: market and survey studies.

Market studies involve analysis of actual behavior that involves trade-offs between income (or expenditures) and health. The primary difficulty in this approach is that the motives behind such decisions are complex and often involve more factors than simply health. It is difficult for an analyst to isolate values that can be attributed to health concerns alone. There often are also significant differences between observable situations and situations involving environmental hazards, such as the control the individual may feel in the situation.

One type of market study is the wage-risk study, in which values for preventing small changes in risks of death are estimated. These have analyzed differences between various jobs and have statistically estimated the portion of an individual's wage that can be considered a "risk premium"—the amount of additional income an individual requires before he or she is willing to accept additional risk of fatal injury at work.

Survey studies involve presenting hypothetical income/health trade-offs to respondents and asking them to determine as best they can how they would respond to that situation were it actually presented to them in their lives. These can be designed to address many kinds of environmental hazard issues, but they are limited in two important ways. First, it is difficult for respondents to know how they might react to a situation they may have never experienced. Even if the situation is familiar, the responses are still hypothetical and therefore subject to some amount of lessened credibility. Second, respondents sometimes find the premises to hypothetical dollar/health trade-offs objectionable. For example, it is very difficult to ask respondents the amount of additional income they would require to be willing to accept an increase in the risk of death. Even though people do in fact make these kinds of decisions in their lives, when put to them so bluntly they often object that such trade-offs are unethical.

Several willingness-to-pay studies have been conducted with regard to changes in morbidity, although they cover only a few of the types of symptoms that might be related to reductions in ozone and compounds of sulfur and nitrogen, including asthma attacks and minor respiratory symptoms in adults. Some unresolved questions remain regarding this estimation method and appropriate ways to transfer study results to specific pollution control policy questions. Table 7.2 indicates the morbidity values used in two benefits studies discussed in the next section. Krupnick (1986) drew upon available willingness-to-pay estimates to develop the values per case. Rowe et al. (1986) used cost-of-illness estimates reflecting medical costs and lost productivity. The types of morbidity quantified in these two studies overlap in a few cases, and the values used per case are roughly comparable. One important unresolved question regarding the willingness-to-pay estimates for morbidity is how individuals factor in lost productivity when many people receive some paid sick leave from their jobs. This may cause some of these estimates to be smaller than the cost-of-illness estimates for the same illness.

Table 7.2. *Typical estimates of the dollar values of pollutant-induced morbidity.*

Type of effect	Krupnick (1986)	Rowe et al. (1986)
Asthma attack	$10–43/attack	$59/attack
Allergy attack	$5–22/attack	—
Bronchitis/emphysema attack	$46–94/attack	—
Minor restricted activity	$12–32/day	$20/day
Symptom:		
Cough	$4–9/day	
Short breath	$8–19/day	
Chest congestion	$6–9/day	
Eye irritation	$5–12/day	
Headache	$5–22/day	
Work loss	—	$81/day
Emergency room visit	—	$267/visit
Emergency hospital admission	—	$4280/admission

Note: Values used in each study are adjusted here to 1986 dollars.

Recent willingness-to-pay survey studies that have examined changes in risks of death have focused on on-the-job injuries and traffic accidents. The results are within the range of results obtained in the wage-risk market studies. Violette et al. (1986) concluded that available willingness-to-pay results suggest a likely range of $1–$8 million per statistical life when small changes in risk of death are being evaluated. Questions do remain, however, about the applicability of these results for risks related to environmental hazards.

Results of Selected National and Regional Health Benefits Studies

Selected national and regional benefits studies that have developed dollar estimates for air pollution control alternatives that would be expected to result in changes in human health are shown in table 7.3. The results are not directly comparable due to differences in pollution changes considered, the health-effects evidence selected, the value per case estimates used, and geographic coverage. As a whole, however, these results suggest that health benefits of reductions in ozone and compounds of sulfur and nitrogen can be expected to be substantial, especially where the federal primary standards for these and related pollutants are being exceeded. In general, the results suggest that mortality effects dominate the benefits due to the relatively high value per case. In most cases, effects related to compounds of sulfur and nitrogen outweigh effects due to ozone, and effects related to sulfate and nitrate particulates can be expected to outweigh effects due to sulfur dioxide and nitrogen dioxide.

Our own recent study (1988) obtained the smallest estimates, primarily because it covered the smallest geographical region—metropolitan Denver. The scenario considered was current (1984–86) particulate matter and ozone levels versus federal primary standards for these pollutants. No health benefits of reductions below the standards were presumed. The ozone effects were a very small component of the total because ozone levels only slightly exceed federal standards in the Denver area, while particulate matter standards are substantially exceeded. Changes in compounds of sulfur and nitrogen were not addressed separately; the scenario presumed a proportional change in all particulates. The dollar estimates were dominated by those for the reduction in premature mortality if the federal particulate matter standards were met, which accounted for about 90 percent of the total dollar estimate. The best estimate of the total health benefit of meeting federal primary standards for particulate matter and ozone in the Denver area was about $105 million per year. This amounts to an average of $66 per resident of the Denver area.

The next smallest estimates were obtained in the most recent EPA study (1987a), which was part of the regulatory impact analysis for alternative primary ambient air quality standards for sulfur dioxide. The scenarios were based on estimates of changes in sulfur dioxide levels in 1990 with alternative ambient air quality standards versus what would be expected with the other types of controls, such as new source performance standards, already in place. The first scenario compares the effects of meeting current sulfur dioxide standards to

Table 7.3. Summary of selected national and regional health benefit studies.

Study	Area/population covered	Pollutant change considered	Valuation approach	Results in $1986/year	Additional comments

U.S. EPA (1987a)	Eastern United States: 31 states, approximate population 167 million (about 70% of total)	Three levels of SO_2 standards: 1. Current SO_2 standard 2. 1-hour 0.5 ppm 3. 1-hour 0.25 ppm
		
		Average estimated percentage change in sulfate particles for each alternative:
		1. 8–9%
		2. 14–16%
		3. 30–32%
		
		Meeting the SO_2 standards in 1990 relative to pollution levels expected with NSPS and other controls already in place
			COI for morbidity; WTP for mortality, $7 million/life	$ billions
				
				       SO_2 SO_4
				1. upper $.06 $.8
				   best .001 .7
				   lower 0 .6
				2. upper .1 1.4
				   best .001 1.2
				   lower 0 1.1
				3. upper .2 2.9
				   best .002 2.5
				   lower 0 2.2
					Only upper SO_2 estimates include any mortality. SO_4 estimates are for morbidity and soiling combined and are based on the change in total particulate matter that would be expected to result from the change in SO_4. Rowe et al. found materials damage to be one to three times morbidity in their California study

U.S. EPA (1984, 1986a)	Nationwide although predicted nonattainment areas are only a portion; 1985 U.S. population was approximately 239 million	Meeting the new federal PM_{10} standard in 1989 relative to pollution levels that would be obtained with NSPS and other controls already in place; five other possible standards also considered	COI for morbidity; WTP for mortality; $.4 to $7 million/life	lower estimate: $.6 to $5 billion per year upper estimate: $18.7 to $81.8 billion per year	The difference between the lower and upper estimates is in the different health effects studies used. The lower estimate implies a reduction of about 530 deaths each year. The range in the upper and lower estimates is due to the range in mortality risk values used

Study	Population/Area	Scenario	Valuation Method	Estimates	Notes
Krupnick (1986)	Nationwide, population approximately 236 million, assuming no ozone effects where there are no ozone monitors	Actual 1983 ozone levels versus meeting three alternative ambient ozone standards: 1. 0.14 ppm 2. 0.12 ppm (current) 3. 0.10 ppm	WTP for morbidity	1. best = $1.9 billion range = $.5–$5.9 2. best = $2.6 billion range = $.6–$7.9 3. best = $3.3 billion range = $.7–$10.3	No mortality effects presumed for ozone. These estimates presume there are health benefits when ozone is decreased even from levels that are already below the standards. If benefits are truncated at each standard, the totals are close to an order of magnitude smaller
Freeman (1982)	Urban areas, approximately 73% of 1978 population (U.S. population in 1978 was approximately 222 million); air quality change in reminder of U.S. presumed insignificant	TSP and sulfur compounds reduced by approximately 20% over the period 1970 to 1978 (based on actual pollution levels)	COI for morbidity; WTP for mortality, $1 million/life	$5 to $67 billion per year; best point estimate of $29 billion	Estimates are 80% for mortality. Deaths/year range from 2780 to 27,800, with a best point estimate of 13,900
Rowe et al. (1986)	Four major air basins in California; about 19 million people, 80% of state population	TSP, SO_2, and ozone *Scenario 1:* actual 1979 levels versus estimated levels in 1979 with only prevailing practice controls (no government-imposed controls); approximately a 20% reduction in TSP and SO_2, much smaller change in ozone	COI for morbidity; WTP for mortality, $2 million/life	Scenario 1: $.3 to $16 billion/year, with a best point estimate of $7 billion for mortality and morbidity Scenario 2: $.04 to $2.9 billion/year, with a best point estimate of $1.3 billion for mortality and morbidity	Best estimates are about 85% mortality. Lower estimates assume zero mortality effects. Ozone effects are only about 1% of morbidity due to the small change in ozone estimated (and the less serious nature of ozone health effects). Estimates for the first scenario reflect 0 to 5000 deaths/year, best estimate of 2500 deaths

(continued)

Table 7.3 (Continued)

Study	Area/population covered	Pollutant change considered	Valuation approach	Results in $1986/year	Additional comments
		Scenario 2: predicted levels in 1987 with planned controls versus 1982 controls only; planned controls were expected to bring levels close to current federal standards			
		Three other scenarios also considered			
Chestnut and Rowe (1988)	Metro-Denver area, population approximately 1.6 million	Actual 1984–86 TSP and ozone levels versus meeting the federal standards (the new PM$_{10}$ was used)	COI for morbidity, WTP for mortality $4.5 million/life	$8 to $160 million, best estimate of $105 million	Lower estimate assumes no mortality. Best estimate is approximately 90% mortality. Ozone has no mortality and reflects .2% to 4% of morbidity estimates due to relatively low ozone levels and less serious health effects. Estimates assume no health benefits of reductions in TSP and ozone and below standards

Notes: COI, cost of illness; NSPS, new source performance standards; PM$_{10}$, particulate matter $<$ or $=$ 10 μm in diameter; WTP, willingness to pay; TSP, total suspended particulates; ppm, parts per million; SO$_2$, sulfur dioxide; SO$_4$, sulfate.

what would be expected with existing controls. Even with the controls currently in place, it is expected that the current federal primary sulfur dioxide standard (0.14 ppm for a 24-hour average and 0.03 ppm for an annual average) will not be met in some locations in 1990. The second and third scenarios hypothesized meeting more stringent levels of 0.5 and 0.25 ppm for a 1-hour average. The effects of changes in sulfate particles as well as sulfur dioxide were considered for the eastern United States (about 70 percent of the national population). Even if we account for the inclusion of materials effects in the reported estimates for sulfate particles, the health effects attributable to sulfate particles exceed those for sulfur dioxide. The estimates are small relative to some other benefit studies in table 7.3, primarily because elevated mortality risks are included only in the upper-bound sulfur dioxide estimates. In most of the other studies considering changes in particulates, mortality effects were included in the best estimates. The best estimate of the annual health and materials benefits of meeting the current federal sulfur dioxide standard in the eastern United States is about $700 million. This amounts to an average of about $4 per person for the entire 31-state area.

Rowe et al. (1986) covered the major air basins in California (80 percent of the state population), and the results varied significantly across the different pollution change scenarios. Changes in total particulates, sulfur dioxide, and ozone were considered, although separate health effects estimates for sulfur dioxide were not made. Sulfur dioxide was included only through its effect on total particulates. The second scenario reported is most comparable to the scenarios considered in the EPA regulatory impact analyses (U.S. EPA 1984, 1986a, 1987a) in that it compares controls that were in place in 1982 with controls that were being considered for implementation in 1987, which were expected to bring most of the area into compliance with the federal standards. The best annual estimate for the second scenario is about $1.3 billion, or $68 per capita. The first pollution change scenario reported in Rowe et al. (1986) is more comparable to the scenario used in Freeman (1982), which reflected pollution changes due to all pollution controls through the 1970s. The best annual estimate for the first scenario is about $7 billion, or $368 per capita. Ozone reflects a very small portion of total health effects due largely to a relatively small predicted change in ozone levels. The total best estimates are about 85 percent for mortality related to changes in particulate matter.

Krupnick (1986) provides estimates of total health benefits of meeting alternative ambient ozone standards relative to actual 1983 ozone levels throughout the United States. The best estimates are a few billion dollars per year even though no mortality effects are expected. The best annual estimate for meeting the current federal primary ozone standard is about $2.6 billion, or $11 per capita for the whole country. An important assumption behind these estimates is that benefits are presumed to exist for reductions in ozone levels below the standards, which would be expected to occur in many areas in order to bring the highest spots to the standards. When health effects are cut off at the standard, the benefits estimates are an order of magnitude smaller. For example, the best estimate for meeting the current ozone standard, assuming no benefits below the standard, is about one-sixth the estimate allowing benefits below the standard. Whether and where a health-effects threshold may lie is an unresolved issue in the literature.

The Environmental Protection Agency (1984, 1986a) conducted a regulatory impact analysis related to the new federal particulate matter standard. The federal ambient air quality standard for particulate matter was changed in 1987 from total suspended particulates to particulate matter less than or equal to 10 μm in diameter. The scenario reported in table 7.3 is for meeting what is now the new standard versus the levels expected in 1989 with other controls that are already in place, such as new source performance standards. The range in the results for this scenario illustrates quite dramatically how much the estimates are influenced by the interpretation of the health effects evidence available in the literature. When $7 million per life is used, the results go from $5 billion to $82 billion depending on the health effects studies used. The studies used for the higher estimate are more comparable to those used by Rowe et al. (1986) and Chestnut and Rowe (1988). On a per capita basis, the estimates using a $7 million value per life imply annual benefits of meeting the new particulate matter standard of between $21 and $342 per person for the United States as a whole.

Freeman (1982) conducted one of the earliest nationwide benefits assessments for air pollution controls. He considered a fairly substantial change in total particulate matter based on the actual reduction from 1970 to 1978. The estimate of number of deaths prevented was somewhat higher than in many subsequent benefits studies because more recent health effects studies have found a smaller mortality effect associated with total particulates. Freeman's best estimate implies an annual health benefit of the actual pollution reduction of about $29 billion, or some $179 per capita for the U.S. urban population.

Visibility Effects

This section summarizes the effects of air pollutants on visibility, describes the methods used to estimate economic measures of value for changes in visibility, and presents results of selected national or regional benefits studies with regard to changes in visibility due to pollution controls.

Effects of Air Pollutants on Visibility

Sulfur and nitrogen compounds are two of the most common air pollutants associated with visibility degradation. High concentrations of ozone are sometimes associated with a white haze, but at typical ambient levels in the United States, ozone is not considered to have a major visibility impact. This section focuses on compounds of sulfur and nitrogen.

Gaseous emissions of sulfur dioxide do not affect visibility directly but affect it through the secondary formation of sulfate particles. Together with nitrate particles, these form a major share of the fine particles (less than 2.5 μm in diameter) in the atmosphere that are most efficient at scattering light and thereby reducing visual range. Gaseous nitrogen dioxide also absorbs light, reducing visual range, and discolors the atmosphere to a reddish brown.

The most common and widespread visual effect of these pollutants occurs in the form of a regional haze that covers a large area and can be a significant distance from the original source of the pollution emissions. When less mixing of the atmosphere occurs due to more stable meteorological conditions, these pollutants can sometimes be seen as a layered haze, with a noticeable differentiation between more and less polluted air layers. Gaseous nitrogen dioxide and particulates can also be seen sometimes as a plume that extends immediately from a stationary source. The focus here is on regional haze, the most widespread problem. The impacts of plumes and layered hazes tend to be very site specific and are difficult to generalize about, although some economic benefits have been estimated with regard to these impacts in a few specific circumstances.

Immediately upon emission, sulfur oxides and nitrogen oxides begin to convert to aerosols of various compositions and sizes. Sulfur oxides convert to sulfuric acid and ammonium sulfate, and nitrogen oxides convert to nitric acid and ammonium nitrate. Near a source region (within 100 kilometers) such as an urban center, haze is a mixture of gases and aerosols. After the pollutants are transported hundreds of kilometers, regional haze is primarily made up of fine primary and secondary aerosols. Therefore, large-scale regional haze, as experienced in much of the eastern United States and perhaps in the Los Angeles area, is likely to have a high share of sulfates and nitrates; smaller hazes, such as in the Denver area, may also contain significant shares of other pollutants.

Two measures of visibility are commonly used. The most familiar is visual range, which is the distance at which a black object just disappears against the background sky. An object is usually referred to as at threshold contrast when the difference between the brightness of the sky and the brightness of the object is reduced to such a degree that an observer can just see the object. Historically, a 2 percent difference in contrast was believed to be the minimum difference perceptible to the human observer. Recent work has suggested that a 5–10 percent contrast threshold may be more accurate.

The other common measure of visibility is light extinction, which is a measure of the amount of light lost between the target and the observer due to the light absorption and scattering of particles and gases in the air. There is an inverse relationship between visual range and light extinction. This relationship is typically approximated with the Koshmeider equation. Assuming a 5 percent contrast threshold, the equation is

$$VR \text{ (miles)} = 23.6/B,$$

where B equals total light extinction in units of $10^{-4}m^{-1}$.

The issue of the contrast threshold is important from a policy perspective because a 2–5 percent change in contrast translates into about a 5–15 percent change in visual range. Therefore, about a 5–15 percent change in visual range may be required before it is perceptible to the human viewer. Although a change in visual range that is not perceptible cannot be expected to have any value from the point of view of the human observer, how this threshold should be interpreted with respect to policies that would cause changes in visual range is not entirely clear. For example, a given change in emissions might cause only a 4 percent change in the annual average visual range, but it might change the distribution of visual range such that there are 10 more very hazy days and 10 fewer clear days each year. If the difference between the two distributions is perceptible, we cannot say for sure that the change is insignificant just because the change in the average visual range is not perceptible. Another consideration is that visibility typically changes gradually over time. For example, if visual range declines 4 percent each year for 3 years, it could be argued that the change is insignificant because the decline each year is not perceptible, but this misses the possibility that the change over the 3-year period may be perceptible and significant.

Research has been conducted to try to quantify the relationship between fine particles in the atmosphere and light extinction. This has been found to vary according to the composition of the particles and atmospheric conditions such as humidity. For details on the estimated relationships see U.S. EPA (1979, 1985) and Trijonis (1982, 1987).

Economic Measures of Welfare Effects of Visibility Impairment

From the perspective of human welfare, visibility impairment due to air pollutants from human activities is largely an aesthetic issue. The average person may not automatically separate the aesthetic effects from concern about health effects, but conceptually these are two different categories. There may at times be travel safety concerns about visibility impairment, especially with regard to aircraft, but current pollution levels have not been shown to be associated with degradation to this degree on any sort of regular basis. This section therefore focuses on aesthetics.

Visibility is not something that can be directly bought or sold, but evidence suggests that people value good visibility. In urban and residential areas, a nice view can add value to a property. In recreation areas people drive and hike to reach certain overlooks and view the scenery. There is considerable complexity involved in people's perceptions of visibility conditions. From an aesthetic point of view, people may be as concerned with what they can see through the air as with what is in the air. The content of the view and the interests of the individual seeing it can affect the perception of visibility conditions, as can many other circumstances such as weather and time of day. Given the need for an objective scale with which to measure visibility conditions, most benefits studies simply measure changes in visibility in terms of visual range or light extinction.

Two types of approaches have been used for estimating the value of changes in visibility in a residential or recreational setting: property value studies and contingent valuation studies. Rowe and Chestnut (1982), Mitchell and Carson (1986), and Freeman (1979) review these methods in depth.

Property value studies have been conducted in residential areas to obtain estimates of value for differences in air pollution levels. This approach uses relationships between property values and air quality conditions to infer values for differences in air quality. The approach is used to determine an implicit price for air quality in the residential housing market based on the theoretical expectation that differences in property values associated with differences in air quality reveal how much households are willing to pay for different levels of air quality. A neighborhood's air quality is one characteristic of a property, and preferences concerning different air quality levels will be reflected in the property market just as preferences concerning school quality or crime rates are.

Probably the most important limitation with the property value approach for estimating the value of changes in visibility is that it is difficult to separate concerns about health effects from displeasure with the aesthetic effects of air pollution. Even when a measure of visibility such as visual range, light extinction, or coefficient of haze, is used in the estimation process, it is likely that air pollution impacts upon property values will reflect concern about health as well as about visibility because the perceived existence of health and aesthetic effects can be expected to be statistically correlated.

Contingent valuation approaches use survey techniques to elicit values that respondents place on changes in visibility. The most common of these is the bidding method, in which individuals are directly asked to estimate how much they would be willing to pay for different changes in visibility levels. The change in visibility is usually presented to the respondents with photographs and verbal descriptions. Respondents are then asked to bid their maximum willingness to pay to prevent a specific deterioration in visibility or to obtain a specific improvement. It is important that a realistic payment scenario be hypothesized in the survey. A variation on this is called the referendum approach and involves asking whether the respondent would be willing to pay some specified amount. If we vary the amount in this question asked of many respondents, the yes/no answers can be used to determine marginal willingness to pay for the hypothesized change in visibility.

A major problem in applying contingent valuation approaches is ensuring that the respondent is separating visual from other effects of air pollution. For example, if an individual sees a large amount of pollution and dislikes it in part due to concern about breathing it, the person may have trouble separating concern about visibility from concern about potential health effects. This may be more of a problem with evaluating changes in visibility in urban and residential areas than in parks and recreation areas, where visitors experience only short-term exposures.

An important advantage of contingent valuation approaches is that they are not dependent on the availability of market data and can be designed to address visibility-related policy issues for which relevant market data may not be available. For example, these approaches have been applied to potential visibility changes in parks and recreation areas as well as in residential areas. This strength, however, parallels the major weakness of the method: its hypothetical nature. No matter how carefully designed and executed to resemble a realistic trade-off that the individual might face, the survey elicits responses that are often viewed with skepticism because they are based on what people say they would do, not what they actually have done.

Most of the available literature on values for changes in visibility focuses on use values. A limited amount of research has been conducted regarding bequest and nonuse values for visibility in recreation areas. Results suggest that such values may be substantial relative to use values in scenic recreation areas. For example, Schulze et al. (1981) obtained in a contingent valuation study an estimate of about $6 billion per year for all residents of the United States to protect visibility from significant deterioration in the southwest region of the country, where the Grand Canyon and many other scenic national parks are located. Use values represented less than 10 percent of the total value. Since people spend most of their time in residential areas, we could expect use values to represent a much greater share of total values for protecting visibility in residential areas, but there is essentially no research evidence on this issue.

Results of Selected National and Regional Benefits Studies

Selected national and regional benefits studies that have developed dollar estimates for pollution control alternatives that would be expected to result in changes in visibility conditions are shown in table 7.4. (All four of these studies also estimated health benefits and were included in table 7.3.) Again, differences in the pollution control scenarios and geographic coverage make these results difficult to compare.

In general, the estimates of visibility benefits are smaller than the health benefits for the same change in pollution. Rowe et al. (1986), Chestnut and Rowe (1988), and Freeman (1982) found visibility to be about 15 percent of total health benefits when comparing the "best" estimates for each study. In all three of these studies, the health benefits reflect reductions in total particles and include estimates of reductions in mortality. EPA (1987a) obtained visibility benefits about equal to morbidity and materials (combined) benefits for a reduction in sulfate particles.

The size of the visibility benefit estimates shown in table 7.4 roughly follows the size of the geographic area covered in the study. The smallest estimates are for metropolitan Denver and the largest estimates are for the country as a whole. The results from EPA (1987a) and Rowe et al. (1986) show, however, that the size of the change in pollution also has a significant impact on the visibility benefits.

The values for a unit change in visual range used in each of these studies vary somewhat, but not as much as the variation in values for changes in mortality risks shown in table 7.3. Chestnut and Rowe (1988) and EPA (1987a) used essentially the same values for changes in visual range. These were based on a review of the results of contingent valuation studies by Chestnut et al. (1986). Rowe et al. (1986) relied upon several contingent valuation and residential property value studies conducted in California. Freeman (1982) derived value estimates entirely from residential property value studies conducted in several U.S. cities. The estimates from residential property value studies can be expected to overstate values for visual aesthetics because they reflect concerns about health effects of air pollution as well.

As with the scientific uncertainty surrounding the physical health effects of air pollutants, there is also uncertainty regarding the change in visibility conditions that would accompany a specific change in air pollution levels. The estimates used in these studies are based on available evidence regarding the relationship between particulates and visual range, but considerable uncertainty remains regarding the accuracy of the predicted changes.

The value estimates in each of these studies are based on values held by private individuals for changes in visibility in residential areas. Rowe et al. (1986) is the only study that included values for changes in visibility in recreation areas. The latter were found to be quite small compared with the former, but it is important to note that the recreation estimates reflect use values only. None of these studies includes benefits that may be related to business and industry, because there is little empirical information available on the magnitude of such possible benefits.

The Environmental Protection Agency (1987a) obtained a best estimate of about $700 million per year as the visibility benefit of meeting current federal primary sulfur dioxide standards. An average change in visual range of about 4 percent was estimated for the 8–9 percent reduction in sulfate particles in the eastern United States that was predicted to be associated with the change in sulfur dioxide. The results for meeting current standards imply an annual per capita benefit of about $4.

Freeman (1982) used a different approach for estimating visibility benefits. He did not estimate a change in visual range, but used an estimated change in sulfate particles and property value studies that have found a relationship between sulfate levels and property values. The change in sulfate levels used was the actual change observed on average in urban areas from 1970 to 1978 (about a 20 percent decline). The resulting best estimate was about $3.9 billion per year, or about $24 per capita per year. These estimates are overstated due to the use of property value estimates, which as indicated can reflect concerns about health as well as visual asthetics.

Rowe et al. (1986) obtained a best estimate for roughly meeting current federal standards for particulate matter and sulfur dioxide of about $500 million per year for the four major air basins in California. The estimated improvement in visual range for this scenario was an average of about 30 percent. The results imply an average annual per capita value of about $26. For the pollution change scenario more comparable to that used by Freeman (the actual reduction in these

Table 7.4. Summary of selected national and regional visibility benefit studies.

Study	Area/population covered	Pollutant change considered	Visibility change	Valuation approach	Results in $1986/year	Additional comments

U.S. EPA (1987a)	Eastern United States: 31 states, approximate population 167 million	Meeting 3 alternative SO_2 standards versus expected levels with NSPS and other controls already in place with current SO_2 standards
		1. 1-hour 0.5 ppm
		2. 1-hour 0.25 ppm
		
		Average estimated change in sulfate particles:
		1. 8–9%
		2. 14–16%
		3. 30–32%
			Change in annual median visual range for each state was estimated given the change in sulfate particles. Average change in visual range:
			
			1. 4%
			2. 7%
			3. 16%
				WTP surveys conducted in several U.S. cities, specifically addressing aesthetics	$ billions
					1. upper 0.9
					   best 0.7
					   lower 0.6
					2. upper 1.6
					   best 1.3
					   lower 1.1
					3. upper 3.4
					   best 2.8
					   lower 2.4	

Freeman (1982)	Urban areas, approximately 73% of 1978 population (U.S. population in 1978 was about 222 million); air quality change in remainder of U.S. presumed insignificant	TSP and sulfur compounds reduced by approximately 20% over the period 1970–78 (based on actual pollution levels)	Not explicitly estimated; presumed associated with 20% change in sulfate particles	Property value studies in several U.S. cities that provide estimates of marginal WTP for reductions in sulfation levels	$1.5 to $11.6 billion each year, best point estimate $3.9 billion	Property value studies can be expected to reflect perceived health risks as well as aesthetic effects. Survey results suggest that 30–50% may be attributable to aesthetics alone. Reported values are therefore overstated.

| Rowe et al. (1986) | Four major air basins in California; about 19 million people, 80% of state population | TSP, SO$_2$, *Scenario 1:* actual 1979 levels versus estimated levels in 1979 with no government-imposed controls (prevailing practice only). Approximately a 20% reduction in TSP and SO$_2$. *Scenario 2:* predicted levels in 1987 with planned controls versus 1982 controls only. The planned controls were expected to bring levels close to current federal standards. These are two of five scenarios considered | Changes in annual average visual range estimated based on predicted change in TSP, sulfates, and nitrates; change in visual range varied across sites but averaged the prevention of roughly a 40% decrease for scenario 1 and the obtainment of roughly a 30% improvement for scenario 2 | WTP surveys and property value studies conducted in California cities, and WTP surveys conducted in U.S. recreation areas | *Scenario 1:* $0.4 to $2.1 billion each year, with a best point estimate of $0.9 billion. *Scenario 2:* $0.2 to $1.6 billion each year, with a best point estimate of $0.5 billion | Values for residential areas were 90% or more of the total in each air basin, but recreation area values covered only use values |
| Chestnut and Rowe (1988) | Metro-Denver area, population approximately 1.6 million | Actual 1986 TSP levels versus meeting the federal standards (the new PM$_{10}$ was used) | Changes in annual average visual range predicted based on change in TSP, presuming a proportional change in fine particles; estimates roughly a 25% improvement in visual range (miles) for the metropolitan area | WTP surveys conducted in several U.S. cities, specifically addressing aesthetics | $20 to $110 million per year with a best point estimate of $27 million | Relationships both between particulate matter and visual range and between visual range and values were based on national, not Denver-specific, data |

Notes: See table 7.3.

pollutants in the 1970s), the best estimate was about $900 million, or about $47 per capita. This is larger than Freeman's estimate primarily as a result of the larger change in pollution estimated for California for this scenario.

Chestnut and Rowe (1988) obtained a best estimate of about $27 million per year as the visibility benefit of meeting new federal primary particulate matter standards in the metropolitan area of Denver. The estimated average improvement in visual range for this scenario was about 25 percent. The results imply an average annual per capita value of about $17.

Summary

The results of this review suggest that the health and visibility impacts of ozone and compounds of sulfur and nitrogen are substantial at current ambient levels of these pollutants. Of this group of pollutants, sulfate and nitrate particles appear to be having the most serious economic impacts, with the effects on human health outweighing those on visibility. In terms of total dollars, the health effects are dominated by potentially elevated risks of mortality associated with these air pollutants. The uncertainty with regard to the magnitude of this risk is substantial. For changes in levels of particulates, human morbidity and visibility benefits tend to be of a similar order of magnitude. A few studies have estimated the benefits for human health, agriculture, materials, and forests of reducing ozone and have found that economic measures of these benefits fall in the order listed, with human health being the greatest.

References

Berkey, C. S., J. H. Ware, D. W. Dockery, B. G. Ferris, and F. G. Speizer. 1986. A longitudinal study of indoor air pollution and pulmonary function growth of pre-adolescent children. Working paper, Harvard School of Public Health.

Blomquist, G. 1982. Estimating the value of life and safety: Recent developments. In The Value of Life and Safety, ed. M. W. Jones-Lee, pp. 27–40. New York: North-Holland.

Chestnut, L. G., and R. D. Rowe. 1988. Ambient particulate matter and ozone benefit analysis for Denver. Draft report prepared for the U.S. Environmental Protection Agency, Denver, Colo. January.

Chestnut, L. G., R. D. Rowe, J. Murdoch, D. Ross, and J. Trijonis. 1986. Review of Establishing and Valuing the Effects of Improved Visibility in Eastern United States. Energy and Resource Consultants. Report to the U.S. Environmental Protection Agency, Office of Policy, Planning, and Evaluation, Washington, D.C., October.

Chestnut, L. G., and D. Violette. 1985. Estimates of Willingness to Pay for Changes in Pollution Induced Morbidity. EPA-230-07-85-008. Report prepared by Energy and Resource Consultants, for the U.S. Environmental Protection Agency, Washington, D.C.

Cropper, M. L. 1981. Measuring the benefits from reduced morbidity. AEA Papers and Proceedings 71 (May):235–240.

Evans, J. S., T. Tosteson, and P. L. Kinney. 1984. "Cross-Sectional Mortality Studies and Air Pollution Risk Assessment." Environment International 10:55–83.

Ferris, B. G., Jr., H. Chen, S. Puelo, and R. L. H. Murphy. 1976. Chronic nonspecific respiratory disease in Berlin, New Hampshire. *American Review of Respiratory Disease* 113:475.

Ferris, B. G., Jr. 1978. Health effects of exposure to low levels of regulated air pollutants: A critical review. *Journal of Air Pollution Control Association* 28 (May):482–497.

Freeman, A. M., III. 1979. *The Benefits of Environmental Improvement: Theory and Practice.* Baltimore: Johns Hopkins Press for Resources for the Future.

Freeman, A. M., III. 1982. *Air and Water Pollution Control: A Benefit-Cost Assessment.* New York: John Wiley.

Hammer, D. I., V. Hasselblad, B. Portnoy, and P. F. Wehrle. 1974. Los Angeles student nurse study: Daily symptom reporting and photochemical oxidants. *Archives of Environmental Health* 28 (May):255–260.

Horst, R. L., Jr., R. M. Black, K. M. Brennan, E. H. Manuel, J. K. Tapiero, and H. C. Duff. 1986. A damage assessment of building materials: The impact of acid deposition. Prepared by Mathtech, for the U.S. Environmental Protection Agency, Office of Policy, Planning, and Evaluation, Washington, D.C., May.

Krupnick, A. J. 1986. Benefit estimation and environmental policy: Setting the NAAQS for photochemical oxidants. Paper presented at the American Economic Association Meetings, New Orleans, La., December.

Lave, L. B., and E. P. Seskin. 1977. *Air Pollution and Human Health.* Baltimore: Johns Hopkins University Press for Resources for the Future.

Lebowitz, M. D., C. J. Holberg, B. Boyer, and C. Hayes. 1985. Respiratory symptoms and peak flow associated with the indoor and outdoor air pollution in the Southwest. *Journal of the Air Pollution Control Association* 35:1154–1158.

Mazumdar, S., H. Schimmel, and I. T. T. Higgins. 1982. Relationship of daily mortality to air pollution: An analysis of 14 London winters, 1958/59–1971/72. *Archives of Environmental Health* 37 (July/August):213–220.

Mishan, E. J. 1982. Recent contributions to the literature of life valuation: A critical assessment. In *The Value of Life and Safety,* ed. M. W. Jones-Lee, pp. 81–94. New York: North-Holland.

Mitchell, R. C., and R. T. Carson. In press. Using surveys to value public goods: The contingent valuation method. Resources for the Future, Washington, D.C.

Ostro, B. D. 1987. Air pollution and morbidity revisited: a specification test. *Journal of Environmental Economics and Management* 14 (March):87–98.

Portney, P. R., and J. Mullahy. 1986. Urban air quality and acute respiratory illness. *Journal of Urban Economics* 20 (July):21–38.

Rowe, R. D., and L. G. Chestnut. 1982. *The Value of Visibility: Economic Theory and Practice for Air Pollution Control.* Cambridge, Mass.: Abt Books.

Rowe, R. D., L. G. Chestnut, D. C. Peterson, C. Miller, R. M. Adams, W. R. Oliver, and H. Hogo. 1986. The benefits of air pollution control in California. Prepared for the California Air Resources Board, Sacramento, Calif., December.

Samet, J. M., Y. Bishop, F. E. Speizer, J. D. Spengler, and B. G. Ferris, Jr. 1981. The relationship between air pollution and emergency room visits in an industrial community. *Journal of Air Pollution Control Association* 31 (March):236–240.

Schulze, W. D., D. S. Brookshire, E. G. Walther, and K. Kelley. 1981. *Methods Development for Environmental Control Benefits Assessment.* Vol. 8, *The Benefits of Preserving Visibility in National Parklands of the Southwest.* Washington, D.C.: U.S. Environmental Protection Agency, Office of Research and Development.

Schwartz, J., and A. H. Marcus. 1986. "Statistical Reanalysis of Data Relating Mortality to Air Pollution During London Winters 1958–1972." Draft paper presented in December to the U.S. Environmental Protection Agency, Clean Air Science Advisory Committee, Research Triangle Park, NC.

Trijonis, J. 1987. National relationship between visibility and NO$_2$ emissions. Santa Fe

Research Corporation. Draft research report for the U.S. Environmental Protection Agency, Office of Policy, Planning, and Evaluation, Washington, D.C.

Trijonis, J. 1982. Visibility in California. *Journal of Air Pollution Control Association* 32 (February):165–169.

U.S. Environmental Protection Agency. 1979. *Protecting Visibility: An EPA Report to Congress.* EPA-450-5-79-008. Prepared by the Office of Air Quality Planning and Standards, Research Triangle Park, N.C., October.

U.S. Environmental Protection Agency. 1982a. *Air Quality Criteria for Particulate Matter and Sulfur Oxides.* EPA-600-82-029. Research Triangle Park, N.C.: Environmental Criteria and Assessment Office.

U.S. Environmental Protection Agency. 1982b. *Air Quality Criteria for Oxides of Nitrogen.* EPA-600-882-026. Research Triangle Park, N.C.: Environmental Criteria and Assessment Office. September.

U.S. Environmental Protection Agency. 1984. Regulatory impact analysis on the national ambient air quality standards for particulate matter. Prepared by the Strategies and Air Standards Division, Office of Air, Noise, and Radiation, Research Triangle Park, N.C., February 21.

U.S. Environmental Protection Agency. 1985. *Developing Long-Term Strategies for Regional Haze: Findings and Recommendations of the Visibility Task Force.* Research Triangle Park, N.C.

U.S. Environmental Protection Agency. 1986a. Regulatory Impact Analysis on the National Ambient Air Quality Standards for Particulate Matter: Second Addendum. Prepared by the Strategies and Air Standards Division, Office of Air, Noise, and Radiation, Research Triangle Park, N.C., December.

U.S. Environmental Protection Agency. 1986b. *Air Quality Criteria for Ozone and Other Photochemical Oxidants.* EPA-600-8-84-0206F. Research Triangle Park, N.C.: Environmental Criteria and Assessment Office.

U.S. Environmental Protection Agency. 1986c. Review of the national ambient air quality standards for particulate matter: Updated assessment of scientific and technical information. Addendum to the 1982 OAQPS Staff Paper. EPA 450-05-86-012. Office of Air Quality Planning and Standards, Research Triangle Park, N.C., December.

U.S. Environmental Protection Agency. 1987a. Regulatory Impact Analysis on the National Ambient Air Quality Standards for Sulfur Oxides (Sulfur Dioxide). Draft prepared by the Strategies and Air Standards Division, Office of Air, Noise, and Radiation, Research Triangle Park, N.C., May.

U.S. Environmental Protection Agency. 1987b. Report of the Clean Air Scientific Advisory Committee (CASAC): Review of the Office of Policy, Planning and Evaluation's Material Damage Assessment. Washington, D.C., June.

Violette, D. M., and L. G. Chestnut. 1983. Valuing Reductions in Risks: A Review of the Empirical Estimates. EPA-230-05-83-003. Prepared by Energy and Resource Consultants, for the Office of Policy Analysis, U.S. Environmental Protection Agency, Washington, D.C., June.

Violette, D. M., L. G. Chestnut, and A. Fisher. 1986. Valuing risks to human health. *Toxics Law Reporter.* September.

Whittemore, A. S., and E. L. Korn. 1980. Asthma and air pollution in the Los Angeles area. *American Journal of Public Health* 70 (July):687–696.

8

Policies to Reduce Damages to Forests and Crops

JAMES J. MACKENZIE AND
MOHAMED T. EL-ASHRY

Summary of Pollution Damages to Forests and Crops

Across the United States, air pollution is contributing to the decline and ultimate death of forest trees and to widespread losses in crop yields. Extensive mortality induced by a combination of ozone air pollution and subsequent attack by bark beetles and root-rotting fungi is occurring among ponderosa and Jeffrey pines in southern California. Growth reduction and visible damage are taking a toll among certain sensitive genotypes of eastern white pines—also the result of ozone pollution. And there is growing evidence that acid deposition and ozone are likely contributors to the declines of several other tree species in the East, especially the high-elevation red spruce in the Appalachian Mountains from Vermont to North Carolina. Air pollution is also suspected as a factor in the decline of Fraser fir in the Southeast, sugar maple and beech in Vermont's mountains, and in the reduced growth (a common sign of ozone injury) of commercial yellow pines in much of the Southeast.

Trees are not the only plants being injured. For sensitive crops such as kidney beans, peanuts, soybeans, cotton, and winter wheat, current ozone-induced losses vary from between 5 and 20 percent. Total agricultural benefits from decreasing ozone concentrations by 50 percent have been estimated as high as $5 billion a year.

As documented in the preceding chapters, the concentrations of ozone and other air pollutants are high where tree and crop injuries are occurring and various direct and indirect damage mechanisms have been identified by which air pollution injures vegetation. On the mountains where the trees are dying, the acidity of the cloud moisture is about 10 times greater than that at low elevations and about 100 times greater than that of unpolluted precipitation. Similarly, total wet acidic deposition at high elevations is about 10 times greater than at nearby lower elevations. Ozone levels are also high both on these mountains and in agricultural areas.

Scientific evidence clearly indicates that ozone injures foliage directly. In addition, acids can leach nutrients from both trees and soils. Trees weakened by air pollution are more susceptible to insects (for example, the balsam wooly adelgid), disease, drought, or frost than they would otherwise be. Thus, although natural stress may deliver the final blow, air pollution readies the ground (or the tree itself) for forest decline.

Air pollution can travel tens and even hundreds of miles before it is removed from the atmosphere through wet or dry deposition. Consequently, long-distance transport of ozone and acid pollutants has become a national and even international problem as trees and crops are threatened far from the sources of emissions.

In view of the inherent complexity of forests and the lack of long-term monitoring and study in the United States, it is understandable that some of the more recent forest declines are still not fully understood. Still, a consensus is emerging that current injuries arise from multiple stresses acting together, some of which are natural (for example, weather extremes and insects) and some man-made (air pollution). As a consequence, the processes of decline vary by region with local climate conditions, rates of pollutant deposition, soil composition, and biological factors.

The significance for these forest systems goes far beyond the decreased growth or death of the damaged trees. If air pollution's effects on an ecosystem become progressively severer, the whole ecosystem can be expected to deteriorate. Unlike a healthy ecosystem, which can normally rebound from damage by fire and insects, a system exposed to chronic air pollution gradually loses diversity and, hence, the capacity to cope with changing natural stresses. Some scientists believe that ecosystems chronically exposed to air pollution can eventually collapse, as is happening on Camels Hump in Vermont and Mount Mitchell, North Carolina.

Of great concern is the possibility that the forest damages now occurring primarily at high elevations could spread to lower elevations and to other tree species. This pattern has been observed in Europe and could occur in the United States if, for example, long-term acidification robs the soils at lower elevations of vital nutrients. According to researchers at Oak Ridge National Laboratory, about 40 percent of the soils in the eastern United States are susceptible to substantial nutrient leaching.

Implicated Pollutants: Sources, Deposition, and Future Trends

The pollutants most implicated in forest damage are ozone, sulfuric acid, and nitric acid. Ozone also poses the most serious threat to crops. The precursors of these pollutants—sulfur dioxide (SO_2), nitrogen oxides (NO_x), and volatile organic compounds (VOCs)—are emitted primarily from power plants, vehicles, industrial boilers, and other fossil fuel burning facilities.

Total emissions of sulfur dioxide, nitrogen oxides, and volatile organic compounds have grown considerably during the twentieth century, reflecting long-term trends in both population growth and the substitution of energy sources for animal and human labor.

The principal anthropogenic sources of these three pollutants are shown in table 8.1. Power plants are the largest source (almost two-thirds) of sulfur emissions while both transportation (44 percent) and power plants (28 percent) contribute substantially to nitrogen oxide releases. Unlike oxides of sulfur and nitrogen, VOCs are emitted by all manner of activities. The largest single source of man-made VOCs is transportation (38 percent), mostly from gasoline-powered vehicles, followed by industry and a variety of other, smaller sources.

Sulfur Dioxide

Sulfur compounds are released to the atmosphere from both natural and man-made sources. Natural sources—soils, vegetation, oceans, and volcanoes—together contribute less than 4 percent to total U.S. sulfur emissions and less than 6 percent to combined U.S. and Canadian emissions (NAPAP II 1987).

Anthropogenic emissions of sulfur dioxide (SO_2) result principally from fossil fuel combustion (NAPAP II 1987). Sulfur dioxide emissions peaked in 1973 at about 30 million metric tons per year (MMT/Y) and by 1986 had declined to about 21 MMT/Y. The electric utilities' growing reliance on coal along with the use of tall stacks to disperse emissions have helped spread the acid deposition problem. Indeed, more than two-thirds of all sulfur dioxide emissions come from stacks over 73 meters (240 feet) high. Although the use of tall stacks may lower SO_2 concentrations locally, it can also lead to greater dry and wet deposition of sulfur compounds, including sulfuric acid, far from the power plants.

Most sulfur dioxide is emitted in the eastern half of the country. Indeed, nine of the ten states that are largest sources of sulfur dioxide are eastern. These nine states accounted for 59 percent of national emissions in 1985 (see table 8.2).

Table 8.1. *Sources of important air pollutants, 1980 (millions of metric tons).*

	Sulfur dioxide		Nitrogen oxides		VOCs	
Power plants	15.9	(65%)	5.8	(28%)	—	—
Transportation	1.1	(4)	9.2	(44)	8.2	(38%)
Other combustion	4.3	(18)	4.0	(19)	2.5	(12)
Smelters	1.1	(4)	—	—	—	—
Industry processes	2.1	(9)	—	—	5.8	(27)
Other	—	—	1.7	(8)	4.9	(23)
Total	24.5	(100%)	20.7	(100%)	21.4	(100%)

Source: NAPAP, *Interim Assessment, the Causes and Effects of Acidic Deposition,* Vol. 2, *Emissions and Control,* pp. 1–25, 1–44, 1–58.

Table 8.2. Ten largest emitters of SO_2
in 1985 (millions of metric tons,
and percent of total emissions).

Ohio	2.43	11.3%
Indiana	1.67	7.8
Pennsylvania	1.50	7.0
Illinois	1.27	5.9
Texas	1.01	4.7
Missouri	1.00	4.7
West Virginia	0.99	4.6
Georgia	0.96	4.5
Kentucky	0.90	4.2
Tennessee	0.89	4.1

Total: 12.62 million metric tons (59% of
total emissions)

Source: NAPAP, Interim Assessment, vol.
2, p. 1–35.

Nitrogen Oxides

Nitric oxide (NO) is the principal man-made nitrogen pollutant emitted to the atmosphere. The Environmental Protection Agency estimates total NO_x emissions in 1985 at 20 MMT/Y (USEPA 1987b). Because of its important roles in both ozone formation and acid deposition, NO emissions are of great concern in air pollution abatement. This gas, comprising about 93 percent of anthropogenic nitrogen oxide emissions, is produced when nitrogen in the air combines with atmospheric oxygen during fuel combustion.

The principal natural sources of nitrogen oxides (which include both nitric oxide and nitrogen dioxide) are lightning and biological processes in soils. Hard to measure, natural processes are believed to contribute about 10 percent of total NO_x releases.

Unlike SO_2 releases, which come primarily from coal combustion, NO_x pollution can result when any fuel is burned. Nitrogen oxides also tend to be generated much closer to ground—from highway vehicles—than SO_2 is. Almost 70 percent of NO_x releases occur at elevations less than 120 feet (NAPAP II 1987). For this reason, nitrogen oxides and associated ozone levels also tend to trouble cities more than sulfur oxides and their related conversion products do. Ultimately, though, nitrogen oxides are converted to nitric acid and, like sulfuric acid, can be deposited great distances from where they are emitted.

Table 8.3 lists the ten states with the highest NO_x emission levels. High emission rates reflect high levels of fuel burning, especially in transportation and power generation. Nonetheless, compared to sulfur dioxide emissions, releases of nitrogen oxides are more dispersed. Also of note is that nitrogen oxide emissions tend to be constant year round since transportation and electric power-production activities vary little seasonally.

Table 8.3. *Ten leading emitters of* NO_x
in 1985 (millions of metric tons
of NO_2, *and percent of total).*

Texas	2.47	12.9%
California	1.11	5.8
Ohio	1.02	5.3
Pennsylvania	0.86	4.5
Illinois	0.81	4.2
Indiana	0.71	3.7
Florida	0.66	3.5
New York	0.65	3.4
Michigan	0.62	3.2
Louisiana	0.60	3.1

Total: 9.51 million metric tons (50% of
total emissions)

Source: NAPAP, *Interim Assessment,* vol.
2, p. 1–52.

Volatile Organic Compounds

The term *volatile organic compounds* covers a bewildering variety of chemicals that collectively play a critical role in ozone formation. Total anthropogenic VOC emissions rose between 1940 and 1970 and then gradually declined, thanks to automotive controls. Today's levels compare with those of 1950.

Natural sources in the United States, principally vegetation, also emit large amounts of organic compounds, perhaps more than man-made sources do. Approximately 90 percent of the natural VOCs are emitted from forests—60 percent from conifers and 30 percent from deciduous trees (NAPAP II 1987). Summertime emissions from natural sources are greatest, estimated at 18 million tons. Total VOC emissions from natural sources in the United States could total 35 million tons annually, compared with 21 million tons from man-made sources.

The 11 states bordering Canada and the Great Lakes accounted for about 30 percent of total anthropogenic VOC emissions in 1980 (see table 8.4). Significantly, the south central states and the Southeast are major sources of both anthropogenic and natural VOCs. Of the 11 states that produce the most anthropogenic VOCs, 9 also number among the largest emitters of nitrogen oxides.

Formation of Secondary Pollutants

The pollutants of most concern in forest decline and crop damage are formed in the atmosphere following the release of NO_x, VOCs, and SO_2. These secondary pollutants—ozone (O_3), nitric acid (HNO_3), and sulfuric acid (H_2SO_4)—are the important contributors to crop and tree damage.

Table 8.4.   *Eleven leading emitters*
of man-made VOCs in 1980 (millions
of metric tons, and percent of total).

California	2.01	9.6%
Texas	1.97	9.4
New York	1.05	5.0
Pennsylvania	0.95	4.5
Ohio	0.91	4.3
Illinois	0.90	4.3
Michigan	0.83	4.0
Florida	0.75	3.6
New Jersey	0.63	3.0
Louisiana	0.57	2.7
Missouri	0.57	2.7

Total: 11.14 million metric tons (53% of
total emissions)

Source: NAPAP, *Interim Assessment,* vol.
2, p. 1–62.

Ozone and Nitric Acid Formation

Ozone occurs naturally in the atmosphere at relatively low concentrations
and arises from several processes. A major source is the stratosphere, where
oxygen atoms, formed when ultraviolet radiation breaks up oxygen molecules,
combine with molecular oxygen to form ozone. This stratospheric ozone is
"good" ozone because it protects life on earth from the sun's intense ultraviolet
radiation. But some finds its way to lower altitudes, where it contributes to
background ozone levels. (Primarily as a result of burning fuels, global background
ozone levels appear to have doubled over the past 100 years from 10 to
about 20 parts per billion (Volz and Kley 1988).

Ozone is also formed in the lower atmosphere through chemical reactions
involving sunlight, oxygen, and hydrocarbons and NO_x from both natural and
anthropogenic sources. The complex reactions creating ozone continue through
the day, leading to higher levels of NO_2 (the reddish-brown gas commonly seen
over urban areas) and ozone. Gradually, the NO_2 is removed from the atmosphere,
primarily through the formation of nitric acid, an important element of
acid deposition.

The reactions involving NO_x occur rather quickly. In summer, NO_x has a
lifetime of less than a day; in winter, about a week. Nitric acid is removed from
the atmosphere through precipitation or surface deposition over 1–10 days (Singh
1987). Ozone is removed from the lower atmosphere (the troposphere) as it is
deposited on various surfaces and then broken up by sunlight. During summer,
ozone lasts about 2 days; in winter, about a month.

Sulfuric Acid Formation

Sulfuric acid (H_2SO_4) forms as sulfur dioxide (SO_2) oxidizes. This conversion can occur either in gas reactions or in liquids (clouds, fog, or rain). The resulting sulfuric acid is removed either in precipitation or by dry deposition. An estimated 20 percent of the SO_2 emitted in the eastern United States and Canada is deposited in precipitation over this area; the rest is either deposited dry over land or else carried out over the Atlantic Ocean (Chamberlain et al. 1985). Since dry-deposited sulfur compounds eventually end up as sulfuric acid, estimating total acid deposition rates involves great uncertainties. Undoubtedly, estimates based solely on wet acid deposition (rain and snow) significantly understate the total deposition rate.

Patterns of Pollution and Patterns of Damages

How do the patterns of pollutant concentrations and wet deposition rates at both low and high elevations correlate with observed forest and crop damages?

Acid Deposition

The acidity of precipitation at low elevations in the eastern United States is about 10 times greater than would be expected for rain falling through unpolluted air. As figure 8.1 shows, even at low elevations the lowest pH values (about pH 4.2) occur in the Northeast. At high elevations in the eastern half of the

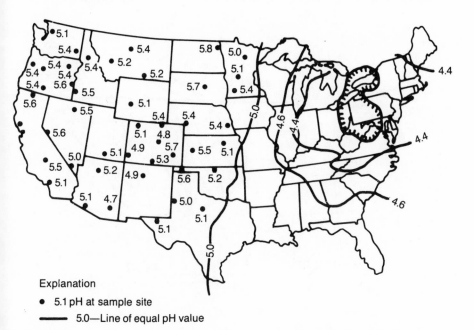

Explanation

- 5.1 pH at sample site
- 5.0—Line of equal pH value

Figure 8.1. Acidity (pH) of wet deposition for 1986. Source: 1986 Annual Report, National Acid Precipitation Assessment Program, p. 78.

country, much of the moisture is deposited by clouds and the acidity is much greater. The average pH value for cloudwater at Mount Mitchell in 1986 was only 3.3, more than 10 times as acidic as the precipitation at low elevations. Some samples were as low as 2.3 to 2.6 (Bruck chapter 4), 100 times more acidic than at lower elevations and more than 2000 times more acidic than unpolluted precipitation. High acidity also characterizes the other eastern mountains where spruce and fir are declining. Clearly, in the high-altitude environments where trees are dying, moisture is much more acidic than at lower elevations.

Similar observations hold true for total wet acid deposition. As figure 8.2 shows, annual wet acidic deposition values at low elevations in the eastern United States range from 0.3 to 0.6 kilograms of H^+ per hectare. The pattern resembles that for pH values. At high elevations, where red spruce are dying, hydrogen deposition is much greater. On Mount Mitchell, annual H^+ deposition ranges between 2.0 and 4.6 kilograms per hectare, about 10 times the deposition rate at low elevations (Saxena et al. 1989). By either measure—pH of moisture or total acid deposition—the mountains where forests are declining receive exceedingly high doses of acidic moisture.

Ozone Concentrations

Although ozone is not formed and transported like acid substances, the patterns of high exposures where trees are dying are similar. Urban areas have

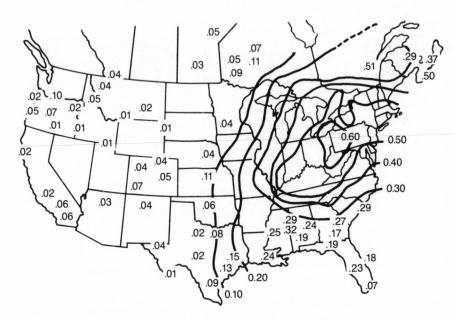

Figure 8.2. Annual wet hydrogen (H^+) deposition (in kilograms per hectare) for 1985. Source: NAPAP III, p. 5-42.

numerous sources of ozone precursors—NO_x, CO, and VOCs—so in the presence of sunlight, ozone forms readily. These precursors and the ozone they help create can travel great distances, threatening crops and materials hundreds of miles from the pollution source (USEPA 1986). Pollution from the San Francisco Bay area, for example, has been traced 300 km (160 miles) to Yosemite National Park (Miller chapter 3). The Los Angeles plume has been followed for 350 km (190 miles) eastward across the Mohave Desert to the Colorado River.

In rural areas that are not downwind from population centers, substantial ozone levels have also been measured. Recent research suggests that naturally occurring hydrocarbons, primarily from trees, in the presence of almost trace amounts of NO_x (about 1 ppb) can produce fairly large amounts of ozone (Trainer et al. 1987). As figure 8.3 shows, average *rural* ozone levels in the United States are relatively uniform, which strongly suggests that unless NO_x emissions are significantly reduced, rural ozone and nitric acid deposition levels cannot be expected to decline.

An examination of the pattern of injury shows that agricultural losses are occurring in or near states that are large sources of ozone precursors (NO_x and VOCs) and that states where forest declines or anomalous growth reductions are significant lie downwind (generally east or northeast) of states with large emis-

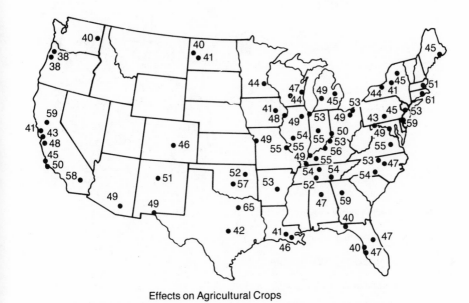

Effects on Agricultural Crops

Figure 8.3. Average daily 7-hour maximum ozone concentration (parts per billion) in rural areas during the growing season. Source: Interim Assessment, vol. IV, National Acid Precipitation Assessment Program, September 1987, p. 6-5.

sions of sulfur dioxide and/or ozone precursors. (See figure 8.4.) This correlation does not establish causality but, when coupled with the extensive scientific evidence reviewed here, it supports the conclusion that ozone and acid deposition are seriously contributing to U.S. tree and crop losses.

Future Pollution Trends

Those who oppose further emission controls argue that SO_2 releases in the United States have dropped by about 25 percent since the Clean Air Act was enacted and will continue to decline as new and cleaner power plants and vehicles replace older ones. They conclude that acid deposition and ozone problems will eventually disappear without further measures to reduce emissions. What do future trends for the implicated pollutants suggest?

Many factors will influence future sulfur dioxide emissions. These include future demand for electricity, the sulfur content of the fuels burned, the kinds of

■ Areas with Crop Damages

▲ Areas with Forest Damage

☐ State with Large SO2 Emissions

◯ State with Large Voc Emissions

△ State with Large Nox Emissions

Figure 8.4. Comparison of states where forests and crops are being damaged with those where air pollution emissions are high. Source: Authors.

technologies that power plants use, and, especially, the retirement age of coal plants built before the New Source Performance Standard (NSPS) took hold. Although the National Coal Association (NCA) estimates that sulfur emissions will continue to decline, at least through 1990 (National Coal Association 1987), the Electric Power Research Institute (EPRI) recently concluded that total sulfur emissions nationwide are likely to remain constant or increase slightly through about 2010 (Shepard 1988). A similar conclusion was reached by the recent Interim Assessment of the National Acid Precipitation Assessment Program (NAPAP) (NAPAP II 1987). One of the most important factors in forecasting future emissions is the assumption about the retirement age of older plants: Since the Clean Air Act imposes stricter SO_2 emission standards on new sources than on older ones, refurbishing aging facilities to extend their lifetimes to 60 years (at costs of less than 50 percent of a new generating facility) keeps plants exempt from the stricter NSPS regulations. Retirement age, therefore, will be a critical factor in assessing when total sulfur emissions will begin to decline.

The trend is up for NO_x emissions too. Without further regulatory controls, NO_x pollution from power plants is projected to increase through 2030, roughly doubling emissions over present values (NAPAP II 1987). (See figure 8.5.) Non-utility emissions—including those from transportation, industry, residential

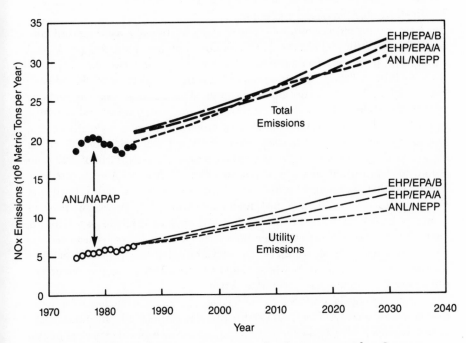

Figure 8.5. Various projected emission trends of nitrogen oxides. Source: NAPAP II, p. 3-12.

and commercial heating, and waste incineration—could increase by 30 percent between 1990 and 2030 unless technologies, regulations, or both are improved (NAPAP II 1987).

Like NO_x emissions, VOCs released mainly from transportation, industrial processes, and other dispersed activities have been declining, primarily because vehicle emissions have been cut. However, by 1990 such emissions are expected to begin rising again until by 2030 they are 25 percent above current levels (NAPAP II 1987). Because vehicle releases are strictly regulated, future VOC emissions will increasingly be from sources outside transportation.

These long-range projections, coupled with current evidence on the multiple adverse impacts of air pollution, make it clear that the nation's air pollution problems will worsen without new efforts to lower emissions. Unless pollutant levels are significantly reduced, we can only look forward to further forest damage, crop losses, ill health, water and soil acidification, damage to materials, and visibility degradation.

Economic Benefits from Air Pollution Reduction

Although the economic damage to forests from air pollution cannot yet be quantified, financial losses from reduced crop yields have been estimated carefully. And other studies have been completed on the economic impacts of adverse health effects and decreased visibility.

Determining the agricultural economic benefits from decreasing ozone concentrations is difficult because of the complex interplay of markets and federal agricultural programs. The most recent and complete analyses, based on data from the National Crop Loss Assessment Network, reveal that ozone is causing extensive crop losses. According to one study of agricultural losses, reductions in ozone levels to 0.04 ppm (7-hour average) would result in an annual benefit of about $2.1 billion ($1987) from improved productivity in corn, soybeans, cotton, wheat, and peanuts (USEPA 1986). Conversely, a further deterioration in air quality leading to an increase in ozone levels to an assumed level of 0.08 ppm would lead to further losses of about $5.3 billion ($1987). Another analysis concluded that a 25 percent reduction in 1980 ozone levels would result in a benefit of about $2.4 billion ($1987) for 6 major crops. A 25 percent increase in ozone levels was calculated to lead to $3.3 billion in additional losses (USEPA 1986). The most recent and complete analysis to date concludes that a 40 percent reduction in ozone levels would lead to a $3.5 billion ($1987) annual benefit in increased yields for 8 crops (Heck chapter 6). If all crops were considered, estimates of economic losses would be larger.

Air pollution has long been recognized as a threat to public health. Indeed, the national primary Ambient Air Quality Standards are specifically intended to protect human health while the secondary standards are meant to protect public welfare (materials, vegetation, visibility, and so forth). The pollutants of concern (sulfur dioxide, nitrogen oxides, and hydrocarbons) pose threats both directly in their emitted form, and indirectly after they have been converted into other air pollutants including acids, ozone, and a variety of organic compounds such as PAN (peroxyacetyl nitrate). These pollutants have been associated with various adverse health effects including respiratory illnesses, aggravation of asthma and allergy symptoms, lung cancer, acute illnesses, and elevated risk of mortality (Chestnut and Rowe chapter 7).

In terms of health benefits, deciding how much money pollution reduction is worth is a difficult task. Most researchers estimate how much reducing pollution by X amount would reduce various health risks to an individual. They then extrapolate the estimated dollar values of these risk reductions to the total affected population. A number of studies have been completed using this approach. In one such analysis, the health benefits were estimated of reducing Denver's existing ozone and particulate pollution levels to meet federal standards. The researchers calculated an annual benefit of $105 million per year, about $66 per person in the Denver area (Chestnut and Rowe chapter 7). In a study of the major air basins in California, the annual benefits of further pollution controls to meet federal air quality standards were estimated at $1.3 billion per year, about $68 per affected person. According to one estimate of ozone's effects, controls strong enough to simply meet the current primary ozone standard would yield benefits of $2.6 billion a year nationwide (Krupnick 1986). A fourth study examined the benefits in the eastern United States from meeting current and stricter sulfur dioxide standards. Reductions in sulfate pollution were also taken into account. Achieving the current standard was estimated to yield about $700 million in health benefits annually.

Visibility degradation from air pollution also has high costs. The emission of sulfur and nitrogen compounds into the air leads to the formation of nitrogen dioxide, sulfates, and nitrates. These compounds are major contributors to the regional hazes that affect the urban areas of the country.

Degradation in visibility from air pollution is essentially an aesthetic issue. Although the economic worth of clear air is not easily measured, the fact that people stop at scenic vistas, climb mountains to enjoy distant panoramas, and pay more for homes that offer attractive views is evidence that visibility is of value. According to EPA estimates, the benefits of improved visibility in the East from meeting current standards for SO_2 would be $700 million per year (USEPA 1987a), while other researchers estimate benefits from meeting current particulate and sulfur dioxide standards at $500 million per year just for California's four major air basins (Rowe et al. 1986).

Structural Shortcomings of the Clean Air Act

The Clean Air Act and its amendments have failed to conquer air pollution partly because when the legislation was enacted in 1970 and 1977, understanding of how pollutants form, travel, and interact was limited. Ambient concentrations of such primary pollutants as sulfur dioxide and nitrogen oxides were then deemed the most relevant indices of the overall risks that these gases would pose. Too little consideration was given to the ultimate form of the compounds, air pollution's regional nature, the increased toxicity of the chemicals that primary pollutants eventually become, the long distances pollutants can travel before being deposited, and the cumulative impacts of acid deposition on soils and aquatic systems. The then-approved use of tall stacks to achieve local ambient air quality standards ultimately dispersed pollutants high into the atmosphere, where they were transformed into acidic compounds and deposited many miles downwind.

The Clean Air Act also failed to anticipate how much the *number* of pollutant sources would grow. The focus on sulfur fuel limits for individual boilers and emission limits for individual vehicles initially reduced pollution levels. But this approach proved limited as the number of cars, trucks, factories, and power plants increased with population and economic growth. Continuing this tack—focusing on technology-prescriptive regulatory limits on individual sources—will prove increasingly ineffective and costly.

Now is an opportune time to reexamine the basic approaches to reducing air pollution, bearing in mind several important facts. First, the costs of emission-control technologies in conventional vehicles and boilers can be expected to continue rising as ever more stringent limits are prescribed to meet ambient air quality standards. Second, long-term economic and population growth will increase the number of dispersed pollution sources—vehicles, small industrial facilities, homes, and commercial buildings—so the progress made by reducing emissions from individual sources will continue to erode. And, third, other pressing, energy-related national concerns discussed briefly below will have to be taken into account in developing pollution control strategies.

Air Pollution and Other National Issues

Air pollution problems (both the failure to meet urban air-quality goals and the long-range transport of airborne acids and ozone) are intimately linked to two other national issues, national oil security and greenhouse-induced climate change.* All of these national concerns relate directly to patterns and rates of growth of fossil fuel combustion. As a result, long-term planning to control air pollution must take a broader perspective than simply reducing air pollution emissions. The connections among these issues are particularly significant in transportation (which depends almost totally on oil, ever more of which is

---

*The greenhouse problem derives from the buildup in the lower atmosphere of largely man-made gases (carbon dioxide, methane, chlorofluorocarbons, ozone ["bad ozone"], and nitrous oxide). These gases are present in trace amounts in the atmosphere and are trapping an increasing fraction of the earth's escaping heat, threatening a global rise in the earth's temperature and long-term climate change. The depletion of the ozone layer ("good ozone") in the upper atmosphere (stratosphere) is primarily the result of chlorofluorocarbons and nitrous oxide and could lead to serious ecological and health effects as well as enhanced greenhouse impacts.

imported) and in electric power generation (which is becoming more dependent on coal, a major source of air pollution and carbon dioxide).

A simple example makes these links clear. The automobile accounts for about 43 percent of total U.S. oil use. As such it contributes heavily to growing U.S. reliance on foreign oil that is imported increasingly from the Middle East. Cars are also a major source of carbon dioxide (CO_2) and ground-level ozone. Since these gases are both "greenhouse gases"—threatening long-term climate change—cars figure centrally in our climate problem. Finally, ground-level ozone in combination with other pollutants (including acid rain) poses risks to public health, crops, materials, and forests. Hence, cars also play an important part in health and environmental problems.

Clearly, public policies affecting air pollution must take these other considerations into account. If not, actions to mitigate one problem could easily neglect or even exacerbate others.

An Integrated Approach to Pollution Control

As with other environmental questions, the connections among air pollution, forest and crop damage, the erosion of health and materials, and the costs and benefits of corrective measures are not without uncertainty. But prudence and common sense dictate that, in a climate of scientific uncertainty, policies and regulatory actions should prevent harm while researchers strive for greater certainty. Doing nothing in the meantime places sensitive ecosystems at the risk of irreversible damage.

Over the next ten to twenty years, air pollution control strategies should focus largely on reducing sulfur and nitrogen emissions from power plants, transportation systems, and other large sources of air pollutants. At the same time, the groundwork should be laid—through an ambitious research and development program—for the inevitable long-term transition to new energy sources that are inherently cleaner than the fossil fuels on which we now depend so heavily. A national strategy to reach these goals should have four major thrusts.

Controlling Stationary Source Emissions

Sulfur dioxide emissions must be cut substantially below present levels. A reasonable national goal for SO_2 emissions reduction is 10 million tons per year to be achieved in two phases, 5 million tons per year each, over a period of 10 years. This cut represents, roughly, a 50 percent reduction from 1986 emissions—a goal consistent with the National Academy of Sciences's contention that a 50 percent reduction in acid (H^+) deposition would probably protect sensitive aquatic ecosystems (National Research Council 1983). Still, even a reduction of this magnitude may not protect all important resources. In 1984, the Office of Technology Assessment (OTA) concluded that wet sulfur deposition should be reduced by 50–80 percent in areas of high deposition and by 20–50 percent in other areas of eastern North America (U.S. Congress 1984).

Scientific uncertainty exists over the extent of NO_x reductions needed to protect health and other resources. NO_x emissions—three-fourths of which derive from power plants and vehicles—lead to both ozone formation and nitric acid deposition. Primarily as a result of fuel burning, global ozone levels in the lower atmosphere have approximately doubled over the past century. There can be little question that significant cuts in emissions of NO_x—the essential precursor of ozone—are needed. An initial reduction goal of 5 million tons per year to be achieved over the next 10 years should be considered. This goal represents a 25 percent reduction from 1986 levels but a much larger reduction from projected growth in NO_x emissions.

The cost of modifying power plants to reduce emissions will be considerable—up to $7.3 billion a year, according to one estimate (Baasel 1988). This estimate is based on the use of wet scrubbers, the most expensive, but proven means of SO_2 removal ($600–$700 per ton of sulfur dioxide removed). The costs could be far lower if other technologies are used instead. For example, employing limestone injection multistage burners (LIMB) could bring the cost down to $400–$500 per ton; with advanced slagging combustors, it would drop further to $50 to $100 per ton (NAPAP II 1987).

To reduce compliance costs, a regulatory program must be flexible enough to permit creative combinations of measures to reach desired clean air goals. Overly prescriptive regulations, such as rules that in effect require wet scrubbers or that stipulate specific limits on the sulfur content of fuels, are unlikely to reduce emissions as efficiently or cost effectively as possible.

One cost-effective strategy for reducing emissions from power plants—indeed, by far the cheapest means for reducing pollution—is to improve energy efficiency. In a study of the East Central Area Reliability (ECAR) power pool (including Ohio, Michigan, Indiana, Kentucky, West Virginia, and small sections of Maryland and Pennsylvania), the effects of various conservation options on emission levels from power plants were evaluated (Geller et al. 1987). Options included more efficient residential appliances, electric motors, lighting, and measures to reduce heating and cooling needs in buildings. More efficient end-use technologies like these, the study found, could reduce electricity consumption in the ECAR region by 26 percent. Greater efficiency could, in turn, reduce SO_2 emissions by 7–11 percent during the 1990s. The study found that midwestern consumers could realize a *net* savings of $4–$8 billion if both emission controls and conservation measures were pursued, compared to a do-nothing scenario.

To encourage efficiency and other cost-effective options for reducing pollution, new regulations must allow full credit for the resulting pollutant decreases. One effective way is to prescribe pollutant caps for SO_2 and NO_x for each state and to allow the states (through their state implementation plans) flexibility in meeting the caps. States should be encouraged to consider various options to meet their goals: improved efficiency, fuel switching, retrofitting with clean coal technologies, transportation planning, and so forth. Establishing absolute caps on statewide emissions would also force states to consider longer term issues related to growth, land use, and local economics. Instead of relying solely on traditional rules and engineering fixes for pollution control, new regulations should include the use of incentives to reach the many dispersed sources of pollution and should allow emissions trading among sources within a state as well as between bordering states. New regulations could also rely on disincentives, such as a stiff tax on each ton of excess emissions beyond established caps. These taxes could be used to help defray compliance costs.

States should be allowed ten years to meet the two-phase emission reduction program. A decade provides a cushion against high initial costs and allows states time to develop innovative strategies to reach designated targets.

Reducing Vehicular Emissions

Transportation activities remain the single largest source of nitrogen oxides, volatile organic compounds, and carbon monoxide emissions. This is true even though autopollution controls have reduced emissions-per-vehicle-mile of nitrogen oxide, carbon monoxide, and hydrocarbons by 76, 96, and 96 percent, respectively, since 1968 (Walsh 1988). Unfortunately, increases in total vehicle miles traveled (VMT) have almost offset the reductions from individual automobiles. Partially as a result, more than 80 million Americans live in areas where the ozone air standard is still exceeded, while 30 million live in areas where the carbon monoxide standard is routinely violated.

Losing the transportation battle for clean air is by no means inevitable. Transportation emissions can be cut significantly over the next 10 to 20 years through a combination of more stringent (but quite technologically feasible) emission limitations for cars, buses, and trucks; strengthened inspection and maintenance procedures; stronger measures to prevent tampering with pollution-control equipment; and the use of cleaner fuels (such as compressed natural gas) in commercial fleets and urban buses.

If the most stringent standards recently proposed in Congress† were adopted along with better inspection and maintenance procedures, actual VOC emissions from mobile sources would drop from 1985 values by almost half by the year 2010 (Walsh 1988). Without these measures, VOC releases would increase slightly. With the proposed control effort, nitrogen oxide emissions would be cut by 30 percent by 2010 relative to 1985 values rather than increasing by about 15 percent, an overall reduction of 45 percent. The tighter emission limits (for VOC and NO_x) are already being met by about half all new passenger cars and, if fully implemented, would cost roughly $50–$125 per car (Walsh pers. comm.).

Programs that reduce the total number of vehicle miles traveled and improve traffic flow would help reduce transportation emissions further. Promis-

---

†The new standards for hydrocarbons would cut emissions by 40 percent relative to the current standard. NO_x emissions for automobiles would be cut by 60 percent.

ing measures include greater use of public transit; preferred parking spaces for car- and van-pools; the removal of subsidies for parking spaces except for car- and van-pools; provisions to encourage high-occupancy and nonmotorized vehicles (bicycles) in all plans for new road construction; and the installation of dedicated lanes for high-occupancy vehicles on urban roadways.

Reducing the number of vehicle miles traveled would also slow the rate of climate change and enhance U.S. oil security. Today, transportation accounts for two-thirds of domestic oil consumption. As a result of increasing oil demand (gasoline consumption grew 5 percent between 1985 and 1987), U.S. oil imports are also growing and in 1988 represented more than 37 percent of the U.S. oil supply.

Improved vehicle fuel efficiency can be encouraged through various measures, including such regulatory requirements as the CAFE (Corporate Average Fuel Economy) standards, through financial tools (fuel taxes, graduated taxes on the purchase price of new vehicles according to efficiency, variable annual registration fees according to efficiency), and through federal and state government purchases of ultraefficient vehicles.

Planning for the Long Term

Because air pollution is so intimately linked with energy use, long-term pollution-control planning must take into account two other national, energy-related problems: oil security and climate change. United States oil production in the lower–48 states peaked in 1970 and has been declining ever since. Prudhoe Bay production in Alaska is also nearing its peak. The depletion of U.S. oil resources and the increasing threat of climate change from the greenhouse effect are two long-term, and essentially irreversible, trends that will profoundly influence U.S. energy use and, hence, air pollution emissions.

Unless the United States is willing to import increasing amounts of oil from OPEC, it must sharply curb its petroleum consumption. Although the nation has huge coal resources and could make synthetic crude oil or methanol (to replace gasoline) from coal, such a policy would further increase carbon dioxide emissions (DeLuchi et al. 1987) and accelerate climate change.

The policy implications for transportation planning are clear. Over the short term, national policies should emphasize more highly efficient conventionally powered vehicles with greatly reduced pollution emissions. At the same time, the development of electric and hydrogen-powered vehicles needs to be accelerated. Increased research and work on nonfossil fuel power sources for vehicles should be a top national priority. Introducing such inherently clean vehicles— with the hydrogen or electricity ultimately derived from nonfossil energy sources—would alleviate a whole range of national problems, including urban air pollution, acid deposition, climate change, and foreign oil dependency.

Electric power production is the other sector destined to see profound changes over the decades ahead. Driving these changes will be the need to reduce conventional pollutants and the risk of global climate change. To understand electricity's growing role in these issues, consider that in 1973 electricity generation accounted for 27 percent of U.S. primary fuel consumption and that, by 1986, U.S. power production accounted for 36 percent of primary energy consumption, about 70 percent of total SO_2 emissions, and one-third of all U.S. carbon dioxide emissions. By 1986, 70 percent of U.S. electricity was generated in fossil fuel plants, 55 percent in coal-fired plants alone. The Department of Energy expects a 50 percent increase in coal consumption between 1985 and the year 2000, with most of the coal being burned in electric power plants (U.S. Department of Energy 1987).

Almost certainly, the new generation of clean coal technologies (integrated gasification combined cycle, fluidized bed combustion, in-duct sorbent injection, and so on) would allow power production with very few emissions of sulfur oxides, nitrogen oxides, or hydrocarbons. The use of these technologies would materially reduce the problem of acid deposition across the country. However, the problem of carbon dioxide emissions—the greenhouse effect—remains. Nothing now available or foreseeable can remove and dispose of the enormous quantities of carbon dioxide that coal-burning power plants would produce in the coming years. As a result, clean coal technologies can and should play an important, but ultimately transitional, role in our energy future.

Besides the evolution of clean new transportation technologies, the climate problem will also force the introduction of nonfossil fuel power generation. The renewable-energy technologies—solar cells, wind turbines, hydropower, geothermal energy—are strong candidates to assume the burden of future power production. The prospects of so-called second generation nuclear technologies (smaller, passively safe, fuel-efficient, standardized fission reactors) offer a second, although less certain, option. In either case, air pollution as we know it today—high levels of ozone, acid deposition, carbon monoxide, and particulates—would be greatly reduced, along with the adverse effects now being visited on public health and welfare.

Improving Our Knowledge and Capabilities

The long-term ramifications of the forest declines occurring in the United States cannot be determined with great certainty until the mechanisms involved are better understood. The federal government's principal investigation (the National Acid Precipitation Assessment Program) is scheduled to terminate in 1990. (The related federal program on crops, the National Crop Loss Assessment Network, ended in 1987.) It is highly unlikely that the unresolved scientific issues related to forest decline can be settled by 1990. Given the complexity and diversity of forest ecosystems, the host of possible contributing factors (pollution, weather extremes, biotic factors, and other anthropogenic causes), the importance of information for improving decision making, and the prospect of long-term climate modification and elevated levels of ultraviolet radiation from stratospheric ozone depletion, a long-term research program must be maintained.

NAPAP provides an umbrella mechanism for coordinating federal research in these areas. It should be reauthorized as the National Atmospheric Pollution Assessment Program (NAPAP) and its scope expanded to support broad ecological research. Emphasis should be on gathering baseline data and understanding the ecological issues relevant to decision makers.

As for air pollution's impacts on forest ecosystems, research should be expanded on both declining and apparently healthy forests to examine nutrient and pollutant flows, including those through soils. A number of representative forest ecosystems should be studied and monitored over time so that subtle changes can be detected early.

Long-term solutions to air pollution, acid precipitation, and greenhouse problems will ultimately require major innovations in energy technology. Since the need is pressing to move toward a more energy-efficient society powered primarily by nonfossil energy, research and development of energy technologies—particularly the renewable sources—should be given a very high priority. Accelerated development of transportation systems that do not depend on fossil fuels, such as electric or hydrogen-powered vehicles, should be supported. As these technologies become technically feasible, incentives for their introduction should be provided. One promising possibility is a tax on conventional transportation fuels; such a tax would reflect the enormous ecological, health, security, and climate risks not reflected in current fuel prices.

Research in energy storage also needs to be expanded. The absence of an inexpensive, light-weight, high energy-density battery is the principal obstacle to the widespread introduction of electric vehicles, a technology that could dramatically reduce air pollution emissions. Improved methods for storing hydrogen should also be developed.

References

Baasel, W. D. 1988. Capital and operating costs of wet scrubbers installed on coal-fired utilities impacting the east coast. *Journal of Air Pollution Control Association* 38 (March):327–332.

Chamberlain, J., et al. 1985. *Acid Deposition.* JASON, MITRE Corporation, McLean, Va.

DeLuchi, M. A., R. A. Johnston, and D. Sperling. 1987. Transportation fuels and the greenhouse effect. UER-180. University of California, Davis, Division of Environmental Studies.

Geller, H. S., et al. 1987. *Acid Rain and Electricity Conservation.* Washington, D.C.: American Council for an Energy Efficiency Economy.

Krupnick, A. J. 1986. Benefit estimation and environmental policy: Setting the NAAQS for photochemical oxidants. Meeting of American Economic Association, New Orleans, December.

National Acid Precipitation Assessment Program (NAPAP II). 1987. *Interim Assessment, the Causes and Effects of Acid Deposition.* Vol. 2, *Emissions and Control.* Washington, D.C.: U.S. Government Printing Office.

National Acid Precipitation Assessment Program (NAPAP III). 1987. *Interim Assessment, the Causes and Effects of Acid Deposition.* Vol. 3, *Atmospheric Processes.* Washington, D.C.: U.S. Government Printing Office.

National Coal Association. 1987. *Reduction in Sulfur Dioxide Emissions at Coal-Fired Electric Utilities, the Trend Continues.* Washington, D.C.: National Coal Association.

National Research Council. 1983. *Acid Deposition, Atmospheric Processes in Eastern North America.* Washington, D.C.: National Academy Press.

Rowe, R. D., et al. 1986. The benefits of air pollution control in California. Air Resources Board, Sacramento.

Saxena, V. K. et al. 1989. Monitoring the chemical climate of the Mt. Mitchell State Park for evaluating its impact on forest decline. *Tellus* (February) 416:92–109.

Shepard, Michael. 1988. Coal technologies for a new age. *EPRI Journal* (January/February):4–17.

Singh, H. B. 1987. Reactive nitrogen in the troposphere. *Environmental Science and Technology* 21 (April):320–327.

Trainer, M., et al. 1987. Models and observations of the impact of natural hydrocarbons on rural ozone. *Nature* 329 (October 22):705–707.

U.S. Congress, Office of Technology Assessment. 1984. *Acid Rain and Transported Air Pollutants, Implications for Public Policy.* Washington, D.C.: U.S. Government Printing Office.

U.S. Department of Energy. 1987. *Energy Security: A Report to the President of the United States.*

U.S. Environmental Protection Agency, Office of Air, Noise, and Radiation (USEPA). 1987a. Regulatory impact analysis on the national ambient air quality standards for sulfur oxides (sulfur dioxide). Research Triangle Park, N.C. Draft.

U.S. Environmental Protection Agency (USEPA). 1987b. *National Air Quality and Emissions Trends Report, 1985.* Washington D.C.: U.S. Government Printing Office.

U.S. Environmental Protection Agency (USEPA). 1986. *Air Quality Criteria for Ozone and Other Photochemical Oxidants.* Washington, D.C.: U.S. Government Printing Office.

Volz, A., and D. Kley. 1988. Evaluation of the Montsouris series of ozone measurements made in the nineteenth century. *Nature* 332 (March):240–242.

Walsh, M. P. 1988. *Pollution on Wheels.* Washington, D.C.: American Lung Association.

Walsh, M. P. Telephone conversation with James MacKenzie. July 1988.

Contributors

Robert Ian Bruck
Department of Plant Pathology
North Carolina State University
Raleigh, North Carolina

Lauraine G. Chestnut
RCG/Hagler, Bailly, Inc.
Boulder, Colorado

Mohamed T. El-Ashry
World Resources Institute
Washington, D.C.

Walter W. Heck
Department of Botany, and
Chairman, Research Committee
National Crop Loss Assessment
Network
North Carolina State University
Raleigh, North Carolina

Reinhard F. Huettl
Institute of Soil Science and Forest
Nutrition
Albert-Ludwig University
Freiburg, Federal Republic of
Germany

Arthur H. Johnson
Department of Geology
University of Pennsylvania
Philadelphia, Pennsylvania

James J. MacKenzie
World Resources Institute
Washington, D.C.

Paul R. Miller
USDA Forest Service
Riverside, California

Robert D. Rowe
Vice President
RCG/Hagler, Bailly, Inc.
Boulder, Colorado

Thomas G. Siccama
Yale School of Forestry and
Environmental Studies
Yale University
New Haven, Connecticut

Advisory Panel

William Becker
Executive Director
State and Territorial Air Pollution
Program Administrators,
Washington, D.C.

F. Herbert Bormann
School of Forestry and Environmental
Studies
Yale University, New Haven,
Connecticut

Ellis B. Cowling
Associate Dean for Research
School of Forest Resources
North Carolina State University
Raleigh, North Carolina

Katherine Y. Cudlipp
Chief Counsel
Committee on Environment and Public
Works
U.S. Senate
Washington, D.C.

David Hawkins
Senior Attorney
Natural Resources Defense Council
Washington, D.C.

Walter W. Heck
Department of Botany, and
Chairman, Research Committee
National Crop Loss Assessment
Network
North Carolina State University
Raleigh, North Carolina

Thomas Jorling
Commissioner
Department of Environmental
Conservation
Albany, New York

Richard Klimisch
Executive Director
Environmental Activities Staff
General Motors Technical Center
Warren, Michigan

William H. Lawrence
Senior Research Advisor
Weyerhaeuser Corporation
Centralia, Washington

Orie Loucks
Director, Holcomb Research Institute
Butler University
Indianapolis, Indiana

James Lyons
Staff Assistant
Committee on Agriculture
U.S. House of Representatives
Washington, D.C.

Ralph Perhac
Director of the Environmental Science
Department
Electric Power Research Institute
Palo Alto, California

Benjamin B. Stout
National Council of the Paper Industry
for
Air and Stream Improvement
New York, New York

Index

Acid clouds, 241; acid deposition by means of in Appalachian Mountains, 9–10, 128–129, 135, 168–169, 175–177; and acid rain, 174–177; pH of moisture in, 129, 135, 146–147, 343, 349–350; and red spruce decline in Northeast, 191, 218; and red spruce and Fraser fir decline in Southeast, 145–149, 160–161, 177–182, 185; throughfall and stemflow as indicators of acid deposition by, 160–167. See also Acid deposition; Acid precipitation

Acid deposition: and crop productivity, 15, 241–242; via dry deposition, 9–10, 42, 44, 96, 100–101, 165, 166, 245, 349; effect of elevation on, 9–10, 343–344, 349–350; and forest decline, 2, 47–48, 113–114, 186, 349–350, 352; interaction of ozone with in crop losses and forest decline, 11–12, 15, 38; inventory of in West, 96, 100–103; necessity of investigating effects of on basic physiological processes, 108–109; and nutrient leaching, 11, 12–13, 14, 47, 131; and red spruce and Fraser fir decline in Appalachian Mountains, 9–10, 135, 145, 153–154, 218, 221, 343, 350; and soil acidification, 12–13, 37–38, 47–48, 52–53, 127–129, 219–221; soil solution, throughfall, and stemflow as indicators of, 155–167; welfare impact of, 316; via wet deposition, 11, 28, 42, 96, 100–102, 160, 169, 176; and yellow pine decline in Southeast, 7, 184. See also Acid clouds; Acid fog; Acid precipitation; Acid rain

Acid fog, 9–10, 11, 38, 44–45, 49–50, 52, 102, 126, 241. See also Acid deposition; Acid precipitation

Acidification. See Soil: acidification of

Acid precipitation: and crop productivity, 240–242; and foliar leaching of nutrients, 52, 221, 241; and forest decline, 11–12, 37, 38, 49–50, 64, 223–225, 349–350; and soil leaching of nutrients, 12–13, 37–38. See also Acid clouds; Acid deposition; Acid fog; Acid rain

Acid rain: and acid clouds, 174–177; and foliar leaching of nutrients, 11; and forest decline, 9–10, 23, 42, 55, 132; pH of, 38, 44, 100–102, 349–350; and soil acidification, 47–48; throughfall and stemflow as indicators of acid deposition by, 162–167. See also Acid deposition; Acid precipitation

Adirondacks, 13, 114, 194, 208, 210, 213, 215, 216, 221–222. See also Whiteface Mountain

Age: as predisposing factor in forest decline, 80, 105, 108, 121, 193; and red spruce decline in Northeast, 203, 206–207, 218; and yellow pine decline in Southeast, 7, 183, 184

Air pollution: acute vs. chronic effects of, 9, 37–38; developing an integrated national approach to control of, 356–362; economic measures of health impacts of, 316–332, 336, 340; economic measures of visibility impacts of, 316–319, 332–340; failure of Clean Air Act to conquer, 355–356; fertilization and liming as correctives to effects of on soils,

White ash, 7, 191–193
Whiteface Mountain, 12, 195–196, 197–
  207, 217, 222–223, 227–228
White Mountains, 114, 195, 208
White pine blister rust, 78
Whitetop Mountain, 10, 12
Wind damage, 1, 78, 80, 84, 89
Winter damage, 213–216, 217, 218, 225–
  227. *See also* Desiccation; Frost; Ice
  breakage

Winter hardening: interference of air pollution with, 13–14, 132, 225–226

"x-disease," 4, 85–86

Yosemite National Park, 94, 97, 351

Zinc, 5, 45, 52, 59, 129, 218